Human Error in Medicine

Human Error in Medicine

Edited by
Marilyn Sue Bogner
U.S. Food and Drug Administration

1994

LAWRENCE ERLBAUM ASSOCIATES, PUBLISHERS
Hillsdale, New Jersey   Hove, UK

Lawrence Erlbaum Associates, Inc., Publishers
365 Broadway
Hillsdale, New Jersey 07642

Cover design by Mairav Salomon-Dekel

Library of Congress Cataloging in Publication Data
Human error in medicine / edited by Marilyn Sue Bogner.
         p.   cm.
      Includes bibliographical references and index.
      ISBN 0-8058-1385-3 (acid-free paper)/ -- ISBN 0-8058-1386-1 (pbk.
   : acid-free paper)
      1. Medical errors.   I. Bogner, Marilyn Sue.
   R729.8.H86   1994
   610'.69—dc20   94-2007
                                                                       CIP

Books published by Lawrence Erlbaum Associates are printed
on acid-free paper, and their bindings are chosen for
strength and durability.

Printed in the United States of America
10 9 8 7 6 5 4 3 2 1

Contents

Foreword

James T. Reason
University of Manchester

This last decade or so has seen a growing openness on the part of the medical profession regarding the part played by human error in mishaps to patients. The pursuit of safety is a multidisciplinary enterprise, and this is as true for patient safety, hitherto an exclusively medical domain, as for any other kind. This new spirit of *glasnost* has led to an increasing number of fruitful research partnerships between doctors (particularly anesthetists) and human factors specialists. One of the most important results of these collaborations has been the awareness that medical accidents share many important causal similarities with the breakdown of other complex sociotechnical systems, such as the Chernobyl Unit 4 reactor. Some idea of what factors are involved in these accidents is provided by the sad story of Valeri Legasov, the chief Soviet investigator of the Chernobyl accident.

FRONTLINE ERRORS ARE NOT THE WHOLE TRUTH

In August 1986, 4 months after the world's worst nuclear accident at Chernobyl, a Soviet team of investigators, headed by academician Valeri Legasov, presented their findings to a meeting of nuclear experts, convened in Vienna by the International Atomic Energy Agency. In a verbal report lasting 5 hours, Legasov described both the sequence of events leading up to the accident and the heroic attempts to contain its aftermath. In his address, Legasov put the blame for the disaster squarely on the errors and especially the procedural violations committed by the plant operators. The report was acclaimed for its full and frank disclosures. "At last," said the headline in a Viennese newspaper, "a Soviet scientist who tells the truth" (Read, 1993).

On the morning of April 27, 1988, 2 years to the day after the Chernobyl accident, Valeri Legasov hanged himself from the balustrade of his apartment building. He had just learned that his proposal for establishing an autonomous institute of industrial safety had been turned down by the Academy of Sciences. Some time before his suicide, he confided to a friend, "I told the truth in Vienna, but not the whole truth."

What was the "whole truth?" Fortunately, Legasov had an opportunity to give us an inkling. Just prior to his death, he began dictating his innermost thoughts about the Chernobyl accident into a tape recorder. The excerpt that follows is taken verbatim from the ninth of 10 tape recordings.

> After being at Chernobyl, I drew the unequivocal conclusion that the Chernobyl accident was the apotheosis, the summit of all the incorrect running of the economy which had been going on in our country from many decades. There are not abstract but specific culprits for what happened at Chernobyl, of course. We now know that the reactor protection control system was defective, and proposals were made on how to eliminate this defect. Not wishing to get involved in quick additional work, the designer was in no hurry to change the protection control system.

> What happened at the Chernobyl power station itself had been going on for a number of years: Experiments were carried out, the program for which had been drawn up in an extremely negligent and untidy way. There were no rehearsals of possible situations before the experiments were conducted. . . . The disregard for the viewpoint of the designer and scientific leader was total, and the correct fulfillment of the technical procedures had to be fought for. No attention was paid to the state of the instruments or the state of equipment before it was time for planned preventive maintenance. One station director actually said: "What are you worried about? A nuclear reactor is only a samovar; it's much simpler than a heat station. We have experienced personnel and nothing will ever happen."

> When you look at the chain of events, at why one person acted in this way while another acted in that way, and so forth, it is impossible to name a single guilty person, the initiator of the events . . . because it is precisely a closed chain: The operators made a mistake because they wanted without fail to complete the experiment—they considered this a "matter of honour"; the plan for conducting the experiment was drawn up in a very substandard and undetailed manner and was not sanctioned by those specialists who should have sanctioned it.

> I keep in my safe a record of the operators' telephone conversations on the eve of the accident. It makes your flesh creep when you read such records. One operator telephoned another and asked: "The program here states what must be done, but then a great deal has been crossed out, so what am I to do?" His

interlocutor thought for a moment, then said: "Act according to what has been crossed out." This was the standard of preparation of serious documents at a project such as a nuclear station: Someone had crossed something out, and the operator might interpret correctly or incorrectly what has been crossed out and perform arbitrary actions. The whole weight of the blame cannot be shifted onto the operator because someone drew up the plan and crossed something out, someone signed it, and someone did not agree to it. The very fact that station personnel could independently carry out some actions not sanctioned by professionals is a flaw in the professionals' relations with this station. The fact that representatives of the state committee for the supervision of safe working practices in the atomic power industry were present at the station, but were not apprised of the experiment being conducted or of the program is not only a fact of this station's biography.

How is this story concerning nuclear power operations in a now-defunct state relevant to human errors in medicine? The answer is in many ways and at many levels. Some of these lessons for the study and prevention of medical mishaps are discussed in the following.

THE BLAME TRAP

The first lesson is that blaming the fallible individuals at the sharp end (i.e., those people in direct contact with vulnerable parts of the system—in medicine, this would be surgeons, physicians, anesthetists, etc.) is universal, natural, emotionally satisfying, and legally (also, in this instance, politically) convenient. Unfortunately, it has little or no remedial value. On the contrary, blame focuses our attention on the last and probably the least remediable link in the accident chain: the person at the sharp end.

Blaming also leads to ineffective countermeasures: disciplinary action, exhortations to "be more careful," retraining, and writing new procedures to proscribe those actions implicated in some recent accident or incident. These measures can have an impact at the outset of some necessary safety program, but they have little or no value when applied to a well-qualified and highly motivated work force. Indeed, they often make matters worse.

People do not mean to commit errors, and the best practitioners sometimes make the worst mistakes. Violations of procedures, on the other hand, do involve an intentional component, but the intention is usually to bend the rules for what seem like good or necessary reasons at the time. Except in the case of saboteurs (and these lie well outside the scope of this book), there is no wish to bring about the damaging outcomes that occasionally follow these deviations. Most people understand this very well, at least in regard to their own errors and violations, so why are we so quick to blame others?

There are various tangled issues here that are worth unravelling if we are to break out of this unproductive "blame trap." One very clear issue is that, from a legal perspective, it is much easier to pin the responsibility for an accident upon the perpetrators of those unsafe acts that had an immediate impact on the integrity of the system or the well-being of the patient. The connection between these proximal actions and the bad outcome is far more easily proved than are any possible links between prior management decisions and the accident, as was clearly shown in Britain by the failed prosecution of the shore-based managers of the capsized ferry, the *Herald of Free Enterprise.*

The convenience for lawyers in chasing individual errors rather than collective ones is further reinforced by the willingness of professionals, such as pilots, ships' masters, and doctors, to accept responsibility for their actions. As part of a professional package that contains (after a long and expensive period of training) considerable power, high status, and large financial rewards, they both expect and are expected to carry the can when things go wrong. This also suits the victims and their relatives who, in trying to come to terms with an injury or a bereavement, find identifiable people a more satisfactory target for their anger and grief than some faceless organization.

But why is the need to assign blame so strong in almost all of us, even when, as so often is the case, it is both misplaced and maladaptive? There are three powerful psychological forces working together to create this reaction.

The first is called the *fundamental attribution error.* Should we observe someone acting in an odd or unsatisfactory fashion, we are automatically inclined to attribute this behavior to some dispositional feature of the individual. The person is viewed as careless, incompetent, or reckless. But if we asked the people so observed why they were acting in that fashion, they would almost certainly emphasize the situational aspects that forced them to do what they did. The truth, of course, lies somewhere in between these dispositional and situational extremes.

The second influence is the *illusion of free will.* We all place a large value on personal autonomy, the feeling of being in control of one's actions. People denied this sense of free will can become mentally ill. We also attribute this freedom of choice to others, so that when we learn that someone has committed an error, we are inclined to think that the individual deliberately chose an error-prone course of action. Bad actions presumed to be deliberate attract exhortations, warnings, and sanctions. These have little or no effect on the organization's error rates, but their continued occurrence fuels greater anger and exasperation on the part of system managers, because the work force has now been warned yet still makes errors. And so the blame cycle goes round and round.

The third factor is our strong tendency to match like with like, termed *similarity bias.* In seeking explanations for puzzling events, the human mind, as Bacon noted in 1620, "is prone to suppose the existence of more order and regularity in the world than it finds" (Anderson, 1960, p. 50). One means of simplification is to presume a symmetry of magnitude between causes and consequences. In the face of horrific manmade catastrophes like Bhopal, Chernobyl, and the King's Cross Underground fire, it seems natural to suppose some equally monstrous act of incompetence or irresponsibility as the primary cause. But a detailed examination of the causes of these accidents reveals the insidious concatenation of often relatively banal factors, hardly significant in themselves, but devastating in their combination.

ACTIVE AND LATENT HUMAN FAILURES

Perhaps the most important lesson to be learned from the Legasov tapes is that Chernobyl, like other accidents, was the product of many different failures, distributed widely in both space and time over the whole Soviet nuclear power generation system. The errors and violations of the operators on the night of April 25–26, 1988 simply added the finishing touches to a disaster that had been in the making for many years. They supplied just those ingredients necessary to breach the various defenses-in-depth. Some of these barriers and safeguards were actively switched off by the operators, others proved to be inadequate for their purpose, and others were slowly eroded through time and custom.

It is convenient to divide the human contributions to the breakdown of well-defended complex systems into two categories: active failures and latent failures (Reason, 1990). The distinction hinges both on who initiated the failures and how long they take to have an adverse effect.

Active failures are errors and violations committed by those in direct contact with the human–system interface. Their consequences are apparent almost immediately, or at least within a few hours. These are what Legasov focused on in his Vienna speech, and—until quite recently—occupied center stage in most accident investigations in any country and in any sphere of operation.

Latent failures are the delayed-action consequences of decisions taken in the upper echelons of the organization or system. They relate to the design and construction of plant and equipment, the structure of the organization, planning and scheduling, training and selection, forecasting, budgeting, allocating resources, and the like. The adverse safety effects of these decisions may lie dormant for a very long time. For example, the wooden escalator installed in the King's Cross Underground Station in the late 1930s was recognized as a fire hazard over 50 years ago; many of the latent failures leading to the Chernobyl explosions originated with the inception of the Soviet Union's atomic power program in the late 1940s and early 1950s; the history of the Challenger disaster goes back at least 12 years prior to the actual disaster, and so on.

Latent failures often only become apparent when they combine with active failures and local triggering events to penetrate or bypass the system's defenses and safeguards. However, once they are recognized for what they are, latent failures can be diagnosed and remedied before they combine to cause bad outcomes. This is what makes them so important relative to active failures, whose precise modes and times of occurrence are extremely difficult, if not impossible, to predict in advance. Who can say exactly when a surgeon's scalpel will slip and cut a nerve, or when an anesthetist will perform an incorrect intubation, or when a nurse will misread a prescription and administer the wrong drug or an inappropriate dose?

It is important to appreciate that the decisions giving rise to latent failures in the system at large need not necessarily be mistaken or ill-judged ones, though they sometimes are. It is probably true to say that all strategic decisions, even those judged as good ones, will have a downside for someone, somewhere in the system, at some time. Top-level decisions are rarely, if ever, optimal. They invariably involve some degree of compromise. Like designers, decision makers are perpetually hedged around by local constraints. Resources, for example, are rarely allocated equitably; there are always winners and losers. Similarly, in forecasting uncertain futures, managers will sometimes call the shots wrongly.

They key point is this: Latent failures are inevitable. Like resident pathogens in the human body, they will always be present. The trick, then, is not to commit limited resources to the futile task of trying to prevent their creation; rather, we should strive to make their adverse consequences visible to those who currently manage and operate the system in question. Doing this is the essence of what Hollnagel (in press) termed *synchronous risk management*; that is, the continuous assessment of the parameters that determine a system's "safety health" and then remedying those most in need of attention.

Many organizations treat safety management like a negative production process. They assess their negative outcome data (accidents, incidents, near misses) and then set themselves reduced targets for the coming accounting period. The trouble with this is that errors and accidents, by their nature, are not directly manageable. Much of their determining variance lies outside the control of system managers and in the realms of chance, Sod, and Murphy's Law. A most effective model for safety management is that of a long-term fitness program, designed to bring about continuous, step-by-step improvement in the system's intrinsic resistance to chance combinations of latent failures, human fallibility, and hazards. This entails managing the manageable; that is, the organizational factors lying within the direct spheres of influence of the system operators and managers.

THE ORGANIZATIONAL ACCIDENT

Legasov's final tapes tell us that Chernobyl, like other well-documented disasters, was an *organizational accident*. That is, it had its origins in a wide variety of latent failures associated with generic organizational processes: designing, building, operating, maintaining, communicating, managing, and the like.

The etiology of an organizational accident divides into five phases: (a) organizational processes giving rise to latent failures, (b) the consequent creation of error- and violation-producing conditions within specific workplaces (operating rooms, intensive care units, pharmacies, etc.), (c) the commission of errors and violations by individuals carrying out particular tasks at the sharp end, (d) events in which one or more of the various defenses and safeguards are breached or bypassed, and (e) outcomes that can vary from a "free lesson" to a catastrophe.

Viewed from this perspective, the unsafe acts of those in direct contact with the patient are the end result of a long chain of causes that originates (for the purposes of systemic improvement, at least) in the upper echelons of the system. In theory, of course, one could trace these precursors back to the Big Bang, but that would have little practical value. One of the basic principles of error management is that the transitory mental states associated with error production—momentary inattention, distraction, preoccupation, forgetting, and so on—are the last and least manageable links in the error chain because they are both unintended and largely unpredictable. Such states can strike anyone at any time. We can moderate human fallibility, but it will never be eliminated entirely. The reason is that errors have their cognitive origins in highly adaptive mental processes.

Correct performance and errors are two sides of the same coin. Human fallibility is not the result of some divine curse or design defect, rather it is the debit side of a cognitive balance sheet that stands heavily in credit. Each entry on the asset side carries a corresponding debit. Absent-minded slips and lapses are the penalties we pay for the remarkable ability to automatize our recurrent perceptions, thoughts, words, and deeds. If we were perpetually "present minded," having to make conscious decisions about each small act, we would never get out of bed in the mornings. The resource limitations on the conscious "workspace" that allow us to carry through selected plans of action in the face of competing situational demands also lead to informational overload and the leakage of memory items. A long-term memory store that contains specialized theories of the world rather than isolated facts makes us liable to confirmation bias and cognitive lockup. An extraordinarily rapid memory retrieval system, unmatched by any known computer, leads our interpretations of the present and anticipations of the future to be overly influenced by the recurrences of the past.

Unsafe acts then are like mosquitoes. You can try to swat them one at a time, but there will always be others to take their place. The only effective remedy is to drain the swamps in which they breed. In the case of errors and violations, the "swamps" are equipment designs that promote operator error, bad communications, high workloads, budgetary and commercial pressures, procedures that necessitate their violation in order to get the job done, inadequate organization, missing barriers and safeguards . . . the list is potentially very long, but all of these latent factors are, in theory, detectable and correctable before a mishap occurs.

Whereas even the simplest accident, such as tripping on the stairs, has some organizational roots, it is probably the case that certain systems are far more prone to organizational accidents than others. Indeed, it could be argued that for certain complex, automated, well-defended systems, such as nuclear power plants, chemical process plants, modern commercial aircraft, and various medical activities, organizational accidents are really the only kind left to happen. Such systems are largely proof against single failures, either human or technical. But they can often be quite opaque to the people who work with them (see Woods, Johannesen, Cook, & Sarter, 1994). Perhaps the greatest risk to these high-technology systems is the insidious accumulation of latent failures, hidden behind computerized, "intelligent" interfaces, obscured by layers of management, or lost in the interstices between various specialized departments.

THERE IS NOTHING UNIQUE ABOUT MEDICAL ACCIDENTS

The main reason for beginning with the Legasov story was to emphasize what many contributors to this book have done in a far more detailed and relevant fashion, namely that nearly all accidents have organizational and systemic root causes, and that these latent factors are more amenable to diagnosis and remediation than are the ephemeral error tendencies of those at the sharp end. It would be tragic if this burgeoning field of medical error studies should have to repeat the mistakes made in the longer established areas of aviation or nuclear power generation and become fixated on the medical equivalent of pilot or operator error. The pressures to pursue this largely unprofitable path are very great, as discussed earlier, but the penalties are also considerable. Even though the human contribution to accidents has been studied over many decades in these more established areas, only over the past few years have they broken free from the trap of concentrating on the psychology of fallible individuals at the human–system interface rather than on the situational and organizational latent failures that they inherited. It is, of course, true that people at the sharp end commit errors and violations, but—as Legasov came to appreciate—it is not the whole truth, nor even the most important part of that truth.

REFERENCES

Anderson, F. H. (1960). *Bacon: The new organon*. Indianapolis: Bobbs-Merrill.
Hollnagel, E. (in press). Computer-supported risk management. In E. Beroggi (Ed.), *Computer-supported risk management*.
Read, P. P. (1993). *Ablaze: The story of Chernobyl*. London: Mandarin.
Reason, J. T. (1990). *Human error*. New York: Cambridge University Press.
Woods, D., Johannesen, L., Cook, R., & Sarter, N. (1994). *Behind human error: Cognitive systems, computers, and hindsight* (CSERIAC State-of-the-Art-Report). Wright Patterson Air Force Base, OH: Crew Systems Ergonomics Information Analysis Center.

1 Introduction[1]

Marilyn Sue Bogner
U.S. Food and Drug Administration

Human error is a fact of life. It occurs in all aspects of life and in every occupation with varying consequences. For those providing medical care, the consequences of an error may be serious injury or death for the very individuals they intend to help. The common reaction to an error in medical care is to blame the apparent perpetrator of the error. Blaming the person does not necessarily solve the problem; more likely, it merely changes the players in the error-conducive situation. The error will occur again, only to be associated with another provider. This will continue until the conditions that induce the error are identified and changed. The purpose of this book is to stimulate interest and action in reducing the incidence of error in medical care by identifying the conditions that incur error so they might be alleviated.

This chapter provides a context for the remainder of the book by first discussing why human error in medicine is the topic of this book. That is followed by a description of audiences for whom the book is useful. Next, working definitions of human error and of medical care are given, and then academic disciplines referred to in this book are briefly described. Applications of those disciplines to the analysis and reduction of human error in medical care are described in the discussion of opportunities for error.

---

[1]The opinions expressed are solely the author's and do not necessarily represent those of the Food and Drug Administration or the Federal Government.

1

WHY ERROR IN MEDICINE

The topic, human error in medicine, was selected for several reasons. First, it is important: Nearly everyone will at some point in life be the recipient of medical care and will want that care to be as error-free as possible. Second, human error provides a flag indicating that a problem exists so that efforts to reduce error in medical care can be instituted. Third, errors in medical care are a factor in the ever-increasing cost of medical care, albeit a rarely considered factor. Because of this, the reduction of medical error merits consideration as an avenue for cost containment. The cost of error in medicine is discussed in chapter 17 of the book by Bogner.

INTERDISCIPLINARY APPROACH

This book consists of 18 chapters, each written by an individual who is experienced in health care. This experience was gained through actual hands-on provision of medical care or research into factors contributing to error in such care. Because of the experience of the authors, their systematic consideration of the issues in this book affords the reader an insightful, applied approach to human error in medicine—an approach fortified by academic discipline.

The interdisciplinary approach of the book makes it useful to a variety of audiences. First, this book addresses issues that are relevant to a number of academic departments (e.g., medicine, nursing, public health, psychology, human factors). Thus, the book is useful as a text for students in those as well as related disciplines.

Second, the book is a tool to be used by practitioners and government agencies in addressing the incidence of human error in medicine. As efforts are instituted to reduce such error and the incidence of human error is impacted, the cost of medical care will be reduced. That reduction will reflect a decrease in hospital stays and out-patient medical treatments necessitated by problems resulting from error.

Third, the book is useful in stimulating research by raising issues related to human error in medicine. One stimulant for research has been public concern resulting from media coverage of catastrophic loss of life resulting from human error, such as in transportation and nuclear power accidents. Except for highly publicized malpractice litigation for flagrant negligent error, deaths and injuries resulting from error in medicine receive little, if any, media attention. In medical care, sick people are made sicker or die from error, one person at a time. These singular deaths are not newsworthy, hence they do not evoke public concern.

WORKING DEFINITIONS

Because this book focuses on the identification of conditions that contribute to the occurrence of human error, *human error* is considered within the context in which it is observed. It is to be noted that from the perspective of this book, the person associated with an error is not automatically considered as the precipitating factor, the cause of that error. Rather, human error is the manifestation of a problem, and the error is the starting point for the analysis of the problem. The analysis is of the context for the error, however deep that context may be, so that the factors actually inducing the error might be identified and remediated. The disquieting fact, which Leape discusses in his chapter, is that much of human error is preventable.

Continuing this perspective, human error in medicine is considered as the mismanagement of medical care induced by factors such as:

* Inadequacies or ambiguity in the design of a medical device or setting for the delivery of medical care.
* Inappropriate responses to antagonistic environmental conditions such as crowding and excessive clutter in institutional settings, and, in a home or field setting, extremes in weather and lack of power and water.
* Cognitive errors of omission (not doing something) and commission (doing something) precipitated by inadequate information, inappropriate mental processing of information, inappropriate actions stemming from adequate information, or situational factors, such as stress, fatigue, or excessive cognitive workload.
* Compliance with unproductive and inappropriate policy.

Because of the focus on identifying factors that contribute to human error rather than assigning blame to the person associated with the error, an extensive consideration of negligent error does not appear in this book. For the same reason, and to foster a nonlitigious hence more positive approach to adverse events in medical care, malpractice is discussed only briefly. The discussions of negligence and malpractice in the chapter by Perper point to the importance of conscientious, systematic investigation of such allegations.

The term *medical care* in this book refers to the diagnosis and treatment of a health problem as well as the maintenance of patients with chronic conditions. Such care may be provided by trained health-care professionals: physicians, nurses, allied health personnel, nurses aides; quasi-trained people such as home health aides; lay persons such as friends and relatives of the patient; and the patients' own self-care. In addition to considering the types of providers of care, a range of settings in which medical care is provided—

offices and clinics, hospitals, nursing homes, and the patients' homes—are considered.

ACADEMIC DISCIPLINES

The perspectives and research from several academic disciplines are particularly applicable to the consideration of human error in medicine. The disciplines are systems analysis, human factors, and cognitive science. Each of these disciplines is briefly described to provide the reader background for understanding the material in the chapters.

Systems analysis as used in this discussion is considered more as art than discipline. As a discipline, the methodology of systems analysis describes an entity such as an organization in terms of mathematical equations, which then can be used in developing a model. Systems analysis as an art, however, defines the entity under examination as a system with interrelated subsystems. The problem at hand is analyzed with respect to the contributions to the problem by the various system components.

The delivery of medical care in the United States is referred to as the health care delivery system. Is that actually the case? Several chapters in this book, including that by Van Cott, address that question.

Systems analysis is an outgrowth of World War II operations research (Hoag, 1956) and has been used in its various iterations by the military since that time. Systems analysis is also used successfully by some sectors of industry. It is reasonable that the health care industry, the largest single industry in the U.S. excluding the federal government, might benefit from the application of a systems approach.

The discipline of *human factors* (also referred to as ergonomics, applied experimental psychology, and human factors engineering) addresses ways in which peoples' performance might be improved. It was not until during World War II, according to Chapanis, Garner, and Morgan (1949), that a gap between the technology of complex machines and the resources available to operate them was spotlighted.

As was typical during WWII, people from diverse backgrounds and academic settings were brought together to solve real-world problems. This was the case in the effort to resolve human–machine performance issues. That effort, however, began too late to have much impact on the operation of battlefield weapons for the war effort. The work to address human–machine issues continued, however, in the development of weapon systems for defense. Findings from that work also have been applied to consumer goods such as automobiles and office equipment.

In recent years, the explosion of highly sophisticated technology and its integration into weapons systems, together with the all-volunteer military, reemphasized the need to address the people issues in the development and use of weapon systems. This need was so great that a special program was developed in the Army to ensure that those issues would be addressed. That program, referred to as MANPRINT, has human factors as one of its basic components and is applicable to medical care. (See Booher, 1990, for a discussion of that program.)

Human factors, more particularly ergonomics and human factors engineering, has become focused almost exclusively on the human–machine interaction. That interaction centers around the placement and direction of hardware such as knobs and dials to enhance operator performance. This focus has been to such a degree that human factors engineering is referred to by some as "knob-ology." Such was not the original orientation of human factors:

> We have tried to advance the general thesis that machines do not work alone. They work in environments. And so do the men who operate them. . . . Our human requirements and our expectations of what a man can do in a system depend to a great extent on the kinds of environment into which the system is thrown. . . . Too little attention has been given to this problem in the past, and it seems almost certain that a considerable amount of improvement can be made by altering many characteristics of the environment in which men work. (Chapanis et al., 1949, p. 420)

References to human factors in this book are in a sense going back to the future by being in the spirit of the definition by Chapanis et al.

There is a history of the application of human factors to health care. Rappaport (1970) pointed to several problem areas in medicine that could benefit from human factors applications, such as medical information system needs. Apparently the tenor of that time, the *zeitgeist,* was not accepting of a human factors approach. A search of the literature finds no evidence of actual applications.

The application of the human factors engineering aspect of human factors to the design of hospitals was advocated by Ronco (1972). Once again, a literature search found no indications of actual applications. Pickett and Triggs (1975) edited a book on human factors in health care. In that book, Rappaport's overview noted many of the issues that plague medical care today.

What differentiates this book from the one by Pickett and Triggs is the focus on a specific problem in medical care, that of human error, rather than the health-care system itself. Also, this book differs in its interpretation of a systems approach. This book, by using human error as a flag of a systems problem, addresses the impact of the system on that problem. The systems approach is discussed extensively in Moray's chapter in this book. In her chapter, Applegate demonstrates the value of that approach in considering the factors that impinge on medical care.

The discipline of human factors employs various methods to analyze the person–machine interaction. Among those methods is functional analysis, which explores the functions to be performed, and task analysis, which examines the components of the function, the tasks. The tasks are decomposed into subtasks for further analysis of performance. The chapter by Serig presents an abbreviated task analysis and demonstrates its utility in addressing nonhardware medical misadministration problems within their environmental contexts.

Cognitive science is an amalgamation of disciplines including artificial intelligence, neuroscience, philosophy, and psychology. Within cognitive science, cognitive psychology is an umbrella discipline for those interested in cognitive activities such as perception, learning, memory, language, concept formation, problem solving, and thinking.

OPPORTUNITIES FOR ERROR

Medical Decision Making

There are many domains of medical care that provide a real-world setting for the application of cognitive science and cognitive psychology. Providers' understanding of information that they present to patients and the potential for erroneous conclusions from such information is discussed in the chapter by Klatzky, Geiwitz, and Fischer. Factors that impact medical care providers and lead to wrong decisions for right reasons, or conversely, right decisions for wrong reasons, are addressed throughout this book.

The phenomenon of perceiving information one expects to perceive, referred to as Einstelling or "set," affects medical diagnosis. Altman (1992) reported a situation that demonstrates how physicians' specialty training may result in a mindset for diagnosis that approaches tunnel vision.

The saga related by Altman is of a physician who experienced symptoms, although not all the classical symptoms, which he diagnosed as appendicitis. Because he could not operate on himself, he consulted a gastroenterologist colleague who diagnosed his symptoms as viral gastroenteritis. He then consulted a cardiologist friend who admitted him to a coronary care unit. Ultimately, after a battery of high technology tests, surgeons prepared him for surgery to remove a large area of presumably damaged bowel. The physician-patient finally convinced the surgeons to explore his appendix, which they did and found it ruptured. The patient was quite ill, but survived.

Another example of "set" is medical care providers' forming an impression of patients based on appearances and interacting with them accordingly. When incorrect, this could lead to inappropriate diagnosis and treatment. Issues related to this for an ever-growing segment of the population, the elderly, are addressed in the chapter by Vroman, Cohen, and Volkman.

Failure to diagnose correctly occurs in adverse drug reactions and can have deadly consequences. When a patient presents to a physician with adverse drug reaction, the physician may fail to recognize the symptoms as caused by a drug and erroneously diagnose the problem (Lucas & Colley, 1992). Another aspect of diagnosis is that physicians tend to have difficulty recognizing the diagnostic value of information (Gruppen, Wolf, & Billi, 1991). This can result in decisions based on information that has no diagnostic value.

Laboratory Reports

Variations in observing and interpreting patterns can have significant consequences when they occur in interpreting laboratory tests. Variations were found in observers' repeated interpretations for the same microscopic slide. Furthermore, the reliability of the observers was not improved by comparing the laboratory test results with digitized captured images (Jagoe et al., 1991).

Not only is it important to avoid error in interpreting laboratory tests, but it is also necessary that the requested information on the correct patient be available to the appropriate physician at a time when it can be considered in making decisions (Aller, 1991). It is not uncommon for physicians to order laboratory tests and not use the results. This sets the stage for medical decision making without the full complement of information. Technology can be employed for electronic transfer of information, such as laboratory results, to reduce the time for hard copy information transfer. As ideal as this may seem, there is an issue of acceptance by medical care providers.

Technology: Friend or Foe

Technology has the potential to assist in the provision of medical care as discussed in the chapter by Sheridan and Thompson. Medical care providers have been reluctant to seek assistance from technology such as computerized diagnostic aiding. That reluctance is apparently on the wane (Lindberg, Siegel, Rapp, Wallingford, & Wilson, 1993).

High technology in medical diagnostic and treatment procedures is a double-edged sword in terms of human error; that is, it can reduce error and it can induce error. This issue, among others, is discussed in the chapter by Hyman.

Medication

There are aspects of medical care, such as medication, to which knowledge from both human factors and cognitive science is applicable. Issues related to errors associated with the means by which a drug is administered are discussed in the chapter by Senders. Drug delivery systems such as containers with difficult-to-open caps and tablets too small to handle afford opportunities for error in self-administration of medication by the elderly (Kottke, Stetsko, Rosenbaum, & Rhodes, 1991).

Drug Names

The indecipherability of physicians' handwriting has been a long-standing joke; however, the potential for error from such handwriting is not a laughing matter. When it comes to prescribing drugs, whether on charts of hospitalized patients or on the prescription form handed to patients in providers' offices, it is important that the name of the drug be intelligible. The importance of intelligibility is underscored by similarity of drug names.

It is easy to miss the difference in names between Norflox and Norflex in the printed word. It is much easier to confuse the names when written. Substitution of drugs due to confusion of names can have serious implications for patients (Pincus & Ike, 1992). Concern about problems resulting from the confusion between acetazolamide and acetohexamide prompted suggestions by Huseby and Anderson (1991) that drug companies change the size, name, and coloring of one of these medications to reduce the likelihood of error in dispensing and administering them.

The problem with very similar drug names is that similarity of names does not suggest similarity of purpose of the drugs. Indeed, drugs with very similar names often are intended for very different purposes. For example, Ansaid is an anti-inflammatory agent, and Axsain is a topical pain killer; Hiprex is an urinary anti-infective drug, whereas Herplex is ophthalmic solution. Confusion caused by the names of these medications can be offset by alert people dispensing them attending to the form of the drug. In the first example, the form is tablet for one and cream for the other, and in the second example, the form is tablets and solution, respectively.

The dose of both similarly named drugs, Trexan and Tranxene, are in tablet form, albeit tablets of differing shape and dose (Teplitsky, 1992), yet there is confusion. A clue of varying shades of the same color is not sufficient to differentiate between doses of a given drug when dispensing tablets of the same shape and size (Klaber, 1992). The potential for confusion is great in a pharmacy. In complex, dynamic, and stressful medical environments such as in a hospital, the potential for error is even greater.

Situated Environments

There are two units in a hospital where the factors discussed in this chapter come together. This makes those units, the operating room (OR) and the intensive care unit (ICU), hotbeds for human error. These environments are particularly appropriate applications of the current realization of the importance of the situation, the environment, on the individual's performance.

After a number of years in which cognitive processing has been presumed to occur within people's minds by means of symbols, there is considerable debate within the cognitive science community as to whether that is the appropriate location. The upstart position is that of situated action, or cognition related to the situation in the world (Greeno & Moore, 1993). Bringing cognition into the situation necessitates the consideration of factors within the situation impacting cognition. That there are situational factors in the OR that impinge on, and indeed, reduce the acuity of cognition, has been realized by some who work in that environment. Knowing that the environment can be conducive to error, Gaba, Howard, and Fish (1994) developed a catalog of critical events in anesthesiology. That catalog is the first systematized source of event management to reduce the likelihood of error.

Another situation-oriented approach is referred to as situation awareness, that is, being aware of all that is occurring in the situation and what might be causing it. Situation awareness has been found by the Air Force to reduce the likelihood of error in flying planes. Appreciation of the impact of the situation and the importance of being aware of what is going on in the operational environment underscores the potential of such an approach for the consideration of human error in medical settings, especially the OR and ICU.

Both the OR and ICU environments are complex, with many diverse activities; both are dynamic, with constant change and time stress. A wide variety of high-technology equipment is used in both environments. Such equipment, which may be viewed as aiding the operator, can create additional demands in the full context of the user's environment (van Charante, Cook, Woods, Yue, & Howie, 1992). The anesthesiology community is especially aware of the demands of the situational factors in their work environment. Those and other specific factors are discussed in the chapters by Gaba, Helmreich and Schaefer, Cook and Woods, and Kreuger.

The concern about error in anesthesiology is not unique to the United States and Switzerland (where Helmreich & Schaefer are conducting their research). The Australian Patient Safety Foundation has embarked on a study of anesthesia-related error (1993) by collecting and analyzing data for over 2,000 incidents.

The ICU is a near cousin to the OR and has many analogous situational demands, but in multiples, for there are multiple patients in an ICU. Although there has been some research on the topic of error in the ICU (Donchin et al., 1989), there has not been a medical specialty to champion such activity as the anesthesiologists did for the OR. This may reflect the lack of a specialty's being considered as perpetrator of excess error in the ICU. The ICU, however, is emerging as a fertile environment for error research. The emergency room, with its myriad activities and extreme time demands, also is emerging as a candidate for error research.

SUMMARY

This chapter has provided pieces to the picture of human error in medicine. Most of the pieces mentioned in this chapter are described more fully in other chapters in this book. Other pieces were described in this chapter to add dimension to the picture. As pieces of the picture come together, it will become obvious that the picture of human error in medicine is not a pretty one.

REFERENCES

Aller, R. D. (1991). Information for medical decision making. *Clinics in Laboratory Medicine, 11*(1), 171–186.

Altman, L. K. (1992, May 12). How tools of medicine can get in the way. *New York Times*, p. C.

Australian Incident Monitoring Study. (1993) Symposium issue. *Anaesthesia and Intensive Care, 21*(5), 501–695.

Booher, H. R. (Ed.). (1990). *MANPRINT: An approach to systems integration*. New York: Van Nostrand Reinhold.

Chapanis, A., Garner, W. R., & Morgan, C. T. (1949). *Applied experimental psychology: Human factors in engineering design*. New York: Wiley.

Donchin, Y., Biesky, M., Cotev, S., Gopher, D., Olin, M., Badihi, Y., & Cohen, G. (1989). The nature and causes of human errors in a medical intensive care unit. *Proceedings of the 33rd Annual Meeting of the Human Factors Society* (pp. 111–118). Santa Monica, CA: Human Factors Society.

Gaba, D., Howard, S., & Fish, K. (1994). *Crisis management in anesthesia*. New York: Churchill-Livingstone.

Greeno, J. G., & Moore, J. L. (1993). Situativity and symbols: Response to Vera and Simon. *Cognitive Science, 17,* 49–59.

Gruppen, L. D., Wolf, F. M., & Billi, J. E. (1991). Information gathering and integration as sources of error in diagnostic decision making. *Medical Decision Making, 11*(4), 233–239.

Hoag, M. W. (1956). An introduction to systems analysis. In S. T. Optner (Ed.), *Systems analysis* (pp. 37–52). Middlesex, England: Penguin.

Huseby, J. S., & Anderson, P. (1991). Confusion about drug names. *The New England Journal of Medicine, 325*(8), 588–589.

Jagoe, R., Steel, J. H., Vuicevic, V., Alexander, N., Van Noorden, S., Wootton, R., & Polak, J. M. (1991). Observer variation in quantification of immunocytochemistry by image analysis. *Histochemical Journal, 23,* 541–547.

Klaber, M. (1992, March 14). Methotrexate tablet confusion. *The Lancet, 339,* 683.

Kottke, M. K., Stetsko, G., Rosenbaum, S. E., & Rhodes, C. T. (1991). Problems encountered by the elderly in the use of conventional dosage forms. *Journal of Geriatric Drug Therapy, 5*(2), 77–91.

Lindberg, D. A. B., Siegel, E. R., Rapp, B. A., Wallingford, K. T., & Wilson, S. R. (1993). Use of MEDLINE by physicians for clinical problem solving. *Journal of the American Medical Society, 269*(24), 3124–3129.

Lucas, L. M., & Colley, C. A. (1991). Recognizing and reporting adverse drug reactions. *Western Journal of Medicine, 156,* 172–175.

Pickett, R. M., & Triggs, T. J. (1975). *Human factors in health care.* Lexington, MA: Lexington.

Pincus, J. M., & Ike, R. W. (1992). Norflox or Norflex? *The New England Journal of Medicine, 326*(15), 1030.

Rappaport, M. (1970). Human factors applications in medicine. *Human Factors, 12*(1), 25–35.

Ronco, P. G. (1972). Human factors applied to hospital patient care. *Human Factors, 14*(5), 461–470.

Teplitsky, B. (1992, January). Avoiding the hazards of look-alike drug names. *Nursing92,* pp. 60–61.

van Charante, E. M., Cook, R. I., Woods, D. D., Yue, L., & Howie, M. B. (1992). Human–computer interaction in context: Physician interaction with automated intravenous controllers in the heart room. In H. G. Stassen (Ed.), *Evaluation of man-machine systems 1992* (pp. 263–274). New York: Pergamon.

2 The Preventability of Medical Injury

Lucian L. Leape
Harvard School of Public Health

BACKGROUND

When patients enter a hospital, they reasonably assume that their treatments will make them better, or, at the least, not make them worse. But modern hospital medical care is complex, involving many human interactions between patients and nurses, doctors, pharmacists, technicians, and others. Each encounter, indeed, each treatment or diagnostic maneuver, presents an opportunity for error. And, errors inevitably occur. Even doctors, nurses, and pharmacists who are trained to be careful sometimes make mistakes, mistakes that sometimes cause injuries. These injuries are technically referred to as *iatrogenic*, that is, caused by treatment, but a more appropriate term is probably *accidental*, for they are truly unintended.

Accidental injury is being increasingly recognized as a significant cause of morbidity and mortality in hospitalized patients (Barr, 1956; Bedell, Deitz, Leeman, & Delbanco, 1991; Brennan et al., 1991; Couch, Tilney, Rayner, & Moore, 1981; Dubois & Brook, 1988; Friedman, 1982; Leape et al., 1991; McLamb & Huntley, 1967; Mills, Boyden, Rubsamen, & Engle, 1977; Shimmel, 1964; Steel, Gertman, Crescenzi, & Anderson, 1981). Recently, the Harvard Medical Practice Study reported the results of a population-based study in New York state that found that nearly 4% of patients hospitalized in New York in 1984 suffered an adverse event, defined as an unintended injury caused by treatment that resulted in prolongation of hospital stay or measurable disability at the time of discharge (Brennan et al., 1991; Leape et al., 1991). For the state of New York, this amounted to

13

98,600 injuries. If rates in other states are similar, more than 1.3 million people are injured annually in the United States by treatments intended to help them.

An important question is, which of these injuries are preventable? Unpredictable allergic reactions, for example, are not preventable at the current state of medical knowledge. However, other injuries result from errors, which, in theory, should be preventable. These may be errors in diagnosis, in decisions about treatment, or in the implementation of treatment decisions. Prevention of medical injury requires finding ways to prevent errors as well as developing mechanisms to protect patients from injuries when errors inevitably occur. Interestingly, there have been relatively few studies of the preventability of medical injury. Bedell et al. (1991) showed that a substantial fraction of in-hospital cardiac arrests were preventable. Eichhorn (1989) observed that anesthetic accidents declined with increased use of intraoperative monitoring.

To provide a more comprehensive appraisal, we reviewed the records of patients who suffered adverse events in the Medical Practice Study to determine which injuries might be preventable. First, I briefly review the methods and findings of adverse events. Most of this material has been previously reported (Brennan et al., 1991; Leape, 1991; Leape, Lawthers, Brennan, & Johnson, 1993; Medical Practice Study, 1990; Weiler et al., 1993).

MEASURING IATROGENIC INJURY

The Medical Practice Study was a population-based study of patients hospitalized in the state of New York in 1984. We randomly sampled all types of acute care hospitals and randomly selected patients within those study hospitals. To insure adequate numbers of adverse events, we oversampled patients cared for by high-risk specialties, such as neurosurgery and vascular surgery. Final results were then weighted to compensate for this sampling strategy.

Hospital medical records were reviewed by trained nurse-abstractors who searched for one of 18 screening criteria that indicated an injury was possible. Each medical record that met a screening criterion was then independently reviewed by two physicians who were certified in either internal medicine or general surgery. The physician-reviewers made judgments as to whether an adverse event (AE) had occurred and if the AE could have been caused by a reasonably avoidable error, defined as a mistake in performance or thought. They then made a judgment as to whether the AE was caused by negligence. Physician-reviewers also estimated the extent of disability caused by the AE and identified the location in the hospital or elsewhere where the treatment occurred that resulted in the injury.

Findings

A total of 30,195 medical records were reviewed by nurse-abstractors. Of these, 7,743 records that met one or more screening criteria were reviewed by physician-reviewers. Adverse events were identified in 1,133 patients in the sample. From this finding, we estimate that of the 2.7 million patients discharged from hospitals in New York in 1984, 98,609 suffered an AE, an adverse event rate of 3.7% of all hospital discharges. Twenty-eight percent of these AEs were judged to be caused by negligence.

Although over half of the AEs resulted in disability that lasted less than a month, in 7% disability was prolonged or permanent, and 14% of patients died as a result of their injuries. Adverse events were twice as common in patients over the age of 65 as in young adults and children, and patients undergoing vascular, cardiac, or neurosurgical operations had significantly higher rates of injury than those undergoing other types of specialty care. Individual hospital rates of adverse events varied more than ten-fold, even after adjustment for disease severity and age.

However, hospital AE rates were not correlated with hospital size or region of the state, although AE rates were higher in teaching hospitals.

Nearly half of AEs were related to a surgical operation. Wound infections were the most common type of surgical complication. Complications of the use of medications were the most common nonsurgical adverse event, accounting for 19.4% of all AEs in the sample.

PREVENTABILITY OF MEDICAL INJURY[1]

The study records of all adverse events were reviewed, and each AE was classified as preventable, unpreventable, or potentially preventable. We classified an AE as *preventable* if it resulted from an error. This included all AEs due to errors that were identified by reviewers, as well as AEs that are commonly recognized to be caused by failure to follow accepted practices. Examples of the last category include wound separation, recurrent spinal disc herniation, and aspiration of gastric contents during induction of anesthesia.

---

[1]Most of this section is reprinted from QRB (Leape et al., 1993). © Quality Review Bulletin, May 1993. Joint Commission, Oakbrook Terrace, IL. Reprinted with permission.

We classified an AE as *unpreventable* if it resulted from a complication that cannot be prevented at the current state of knowledge. These may be unpredictable (such as an allergic reaction in a patient not known to be allergic to a drug) or predictable (such as marrow depression following chemotherapy). An AE was classified as *potentially preventable* if no error was identified but it is widely recognized that a high incidence of this type of complication reflects low standards of care or technical expertise. Unexplained wound infections and postoperative bleeding were the most common potentially preventable AEs.

We further subclassified AEs due to error according to the type of error. Table 2.1 lists these errors grouped in several categories.

Diagnostic errors included failure to make a diagnosis as well as errors in the ordering or use of tests. An example of an error in diagnosis (Item 1) was the failure to diagnose septicemia in a febrile patient. *Treatment* errors included failures in all aspects of the actual provision of care, both in decision and execution. In this category were technical errors (Item 5) that embraced both surgical AEs, such as laceration of the ureter during hysterectomy, and procedural AEs, such as creation of a pneumothorax while

TABLE 2.1
Types of Errors

Diagnostic

1. Error in **diagnosis** or delay in diagnosis.
2. Failure to **employ indicated tests.**
3. Use of **outmoded tests** or therapy.
4. **Failure to act** on the results of monitoring or testing.

Treatment

5. **Technical** error in the performance of an operation, procedure, or test.
6. Error in **administering** the treatment (including preparation for treatment or operation).
7. Error in the **dose of a drug** or in the method of use of a drug.
8. **Avoidable delay** in treatment or in responding to an abnormal test.
9. **Inappropriate (not indicated) care.** Considering the patient's disease, its severity, and comorbidity, the anticipated **benefit** from the treatment did not significantly exceed the known risk, or a superior alternative treatment was available.

Preventive

10. Failure to provide indicated **prophylactic** treatment.
11. Inadequate **monitoring or follow-up of treatment.**

Other

12. **Failure in communication.**
13. **Equipment failure.**
14. Other **systems failure.**

---

inserting a central venous catheter. Also in this category were errors in the use of a drug and provision of care that was not indicated. *Preventive* errors included both failure of prophylaxis and failure to monitor or follow up treatment. An example of a preventive error (Item 10) was failure to prescribe anticoagulants to prevent pulmonary embolus following hip replacement. *"Other"* errors included failures in communication, equipment, or systems.

Each preventable AE was assigned to a single error. When more than one type of error was present, we chose the more serious error. For this purpose, we established a hierarchy of severity for the errors listed in Table 2.1 so that the four major categories, 1—Error in diagnosis, 5—Technical error, 7—Drug error, and 9—Inappropriate care, took precedence over the other specific types of error, as well as the general category, 6—Error in administering treatment. For example, failure to obtain a pregnancy test in a woman of childbearing age who presents to the emergency room with lower abdominal pain, missed menses, and a questionable pelvic mass would be classified as 1—Error in diagnosis, rather than 2—Failure to employ indicated tests. If the multiple errors were of similar severity, we randomly selected one. Population estimates and percentages were calculated by reweighting the sample numbers with the inverse of the sampling weights. Extrapolations to national figures were calculated by multiplying the New York results by the United States/New York population ratio in 1984.

Findings

More than two thirds (70%) of AEs were found to be preventable, 24% were judged unpreventable, and the remaining 6% were classified as potentially preventable. There were no significant differences in the percentages of AEs that were preventable according to age.

Types of Errors.   The estimated number of preventable errors and the percentage accounted for by each type of error are listed in Table 2.2. The most common types of preventable errors were errors in diagnosis (17%), technical errors (44%), errors in the use of a drug (10%), and failures to prevent injury (12%). It is important to note that most errors were not considered to represent negligence. Although technical errors were by far the most frequent type of error, relatively few (20%) technical errors were judged to be due to negligence. For example, injury of the ureter during hysterectomy was not considered to be due to negligence by physician-reviewers unless the injury was unrecognized and the patient developed symptoms that were ignored. By contrast, 71% of diagnostic errors were deemed negligent, such as failure to make a diagnosis in a woman with a possible ectopic

TABLE 2.2
Frequency of Preventable Errors

	No.	Frequency (%)	% of Errors Due to Negligence
Diagnostic			
1 Error in diagnosis	11,731	17.1	71.1
2 Failure to use indicated tests	782	1.1	91.4
3 Use of outmoded tests	944	1.4	56.4
4 Failure to act on results of tests	1,579	2.3	55.2
Treatment			
5 Technical error	30,373	44.4	19.8
6 Error in administering treatment	776	1.1	9.1
7 Error in use of a drug	6,988	10.2	37.1
8 Delay in treatment	3,154	4.6	69.4
9 Inappropriate management	141	0.2	0.0
Preventive			
10 Failure to prevent injury	7,943	11.6	50.3
11 Inadequate monitoring	3,172	4.6	36.9
Other			
12 Failure in communication	244	0.4	52.6
13 Equipment failure	422	0.6	77.2
14 Other systems failure	136	0.2	0.0
15 Unclassified	260		
Totals	68,645	100.0	39.7

pregnancy. Half of preventive errors were negligent. A major fraction of AEs in the "failure to prevent" category were falls. Most (63%) errors in the use of a drug were not judged to be due to negligence.

Preventability by Type of AE.   Preventability varied considerably by type of AE (Table 2.3). Surgical AEs were more likely to be preventable (74%) than those resulting from nonsurgical management (65%; $p < .05$). Over 90% of late surgical failures, diagnostic mishaps, and nonprocedural therapeutic mishaps were preventable, as were falls and systems failures. Technical complications of surgery were also usually (87%) considered to be preventable. Even nontechnical postoperative complications (such as phlebitis or pulmonary embolus) were deemed preventable in more than half of the cases.

Diagnostic mishaps (98.8%), falls (92.2%), therapeutic mishaps (91.3%), and surgical failures (94.0%) had the highest levels of preventability. Drug-related adverse events were the only group in which more than half were not

TABLE 2.3
Preventability According to Type of Adverse Event

	Adverse Events		Preventable		S.E. (%)
Type	No.	(%)	No.	(%)	
Operative					
Wound infections	13,411	13.6	9,659	71.9	4.9
Tech complication[1]	12,721	12.9	10,981	86.7	3.3
Late complication[2]	10,453	10.6	6,998	67.2	5.7
Nontechnical complication[3]	6,903	7.0	3,767	54.5	7.6
Late failures[4]	3,550	3.6	3,356	94.0	3.0
All operative	47,038	47.7	34,761	74.0*	2.7
Nonoperative					
Drug-related	19,130	19.4	8,644	45.2	5.2
Diagnostic mishap	7,987	8.1	7,866	98.8	1.2
Therapeutic mishap	7,396	7.5	6,785	91.3	4.2
Procedure-related	6,903	7.0	3,637	52.8	5.8
System failures	1,362	1.4	1,164	85.5	0.0
Fall	2,662	2.7	2,449	92.2	6.8
Other[5]	6,100	6.2	3,338	54.3	7.1
All nonsurgical	51,540	52.3	33,884	65.5*	3.8
All	98,578	100.0	68,645	69.6	2.6

[1]Example: injury to the ureter or spleen during an operation.
[2]Example: incisional hernia, retained gallstone.
[3]Example: pneumonia or heart attack following surgery.
[4]Example: recurrent disc.
[5]Nonoperative fractures and deliveries, neonatal, anesthesia.
*$p < .05$ for operative vs. nonoperative.

preventable, largely because many were unpredictable allergic reactions or unpreventable effects of chemotherapy.

However, in absolute numbers, that is, in terms of where hospitals might target their efforts, preventable technical complications of surgery (10,981) and wound infections (9,659) were most common. Because of the large number of drug-related AEs (19,130), preventable drug-related AEs ranked third in total number (8,644) despite their relatively low preventability rate. Diagnostic mishaps accounted for almost as many (7,866), even though they represented only 8% of all AEs.

Death and Disability.   Patients with a preventable AE were more likely to suffer prolonged disability or death (27%) than those with nonpreventable AEs (19%). Mortality due to the AE was 60% higher among patients with preventable AEs than among those with unpreventable AEs (16.5% vs. 10.6%). Overall, 78% of all fatal AEs were preventable.

TABLE 2.4
Preventability by Specialty

Specialty	AE Rate[1] (%)	Total Number AEs[2]	Preventable		S.E. (%)
			No.	(%)	
General medicine	3.6	37,135	23,766	64.0	4.8
General surgery	7.0	22,324	16,274	72.9	3.4
Orthopedics	4.1	6,746	5,248	77.8[3]	4.2
Obstetrics	1.5	5,013	3,810	76.0	7.6
Urology	4.9	4,819	3,470	72.0	6.9
Thoracic surgery	10.8	3,557	2,225	62.6	7.9
Vascular	16.1	3,187	1,950	61.2	7.2
Neurosurgery	9.9	2,987	2,201	73.7	8.6
Neonatal	0.6	1,713	NS	NS	
Other	3.0	11,097	8,290	74.7	3.8
All	3.7	98,578	67,234	69.7	2.7

[1]Number of Adverse Events per hundred hospitalized patients cared for by that specialty.
[2]Population estimate for patients hospitalized in New York in 1984.
[3]$p < .01$ for orthopedics vs. general medicine, vascular, and thoracic and cardiac surgery.
NS Number of patients is too small to provide accurate estimates.

Specialty Variations.   Despite the fact that AE rates varied by more than ten-fold among specialties (Table 2.4), there was no significant difference among specialties in the percentage of AEs that were preventable, with the exception of orthopedics ($p < .01$). In fact, the two specialties with the highest AE rates, vascular surgery and cardio-thoracic surgery, had the lowest percentages of AEs that were preventable.

However, the greatest *number* of preventable AEs were accounted for by the two broad specialties, general medicine and general surgery. These two specialties alone are responsible for care that results in 60% of all AEs, and, interestingly, also accounted for 60% of the preventable AEs (40,040/ 67,234).

Site of Care.   There was no significant difference in overall preventability between adverse events that occurred within the hospital (70.7%) and those that occurred outside (65.7%; see Table 2.5). Inside the hospital, the emergency room was the site where adverse events were most likely to be preventable (93%, $p < .05$). Outside the hospital, AEs were most likely to be preventable if they resulted from care in a freestanding ambulatory surgical unit (95%, $p < .001$).

TABLE 2.5
Preventability by Site of Care

Site	Adverse Events		Preventable		
	No.	(%)	No.	(%)	S.E. (%)
Inside Hospital					
Operating room	40,438	41.0	28,879	71.4	2.6
Patient's room	26,097	26.5	17,676	67.7	4.4
Emergency room	2,860	2.9	2,669	93.3[1]	4.7
Labor and delivery	2,750	2.8	2,164	78.7	6.4
Intensive care unit	2,618	2.7	1,840	70.3	11.3
Radiology	2,012	2.0	1,218	60.5	11.0
Other[2]	3,185	3.4	2,015	63.3	9.0
All	79,959	81.2	56,461	70.7	2.8
Outside Hospital					
Physicians office	7,581	7.7	4,735	62.5	5.1
Home	2,662	2.7	1,600	60.1	9.3
Ambulatory surgical unit	1,337	1.4	1,273	95.2[3]	4.0
Other[2]	2,007	2.0	1,310	65.3	10.9
All	13,587	13.8	8,918	65.7	4.6
Not Classified	5,032	5.1	3,266	64.9	—
Total	98,578	100.0	68,645	69.6	2.8

[1]$p < .05$ for emergency room vs. all other inside hospital.
[2]Sites with fewer than 1% of cases.
[3]$p < .001$ for ambulatory surgical unit vs. all other outside hospital.

COMMENTARY

If these results are typical of those in hospitals throughout the country, we estimate that preventable injuries afflict 938,000 hospitalized patients annually. The AEs we identified are complications that are familiar to doctors: wound infections, postoperative bleeding, urinary infections, overdoses of drugs, bleeding from anticoagulants, insulin reactions, and so on. Few are esoteric or caused by rare or unusual circumstances. Because they are common and also have high preventability rates, technical complications, wound infections, and diagnostic mishaps are the types of adverse events that account for the major share of preventable AEs. On the other hand, even though they are often unpreventable, drug-related AEs are also a leading cause of preventable AEs because of their high frequency.

We found that 78% of fatal AEs were preventable, but this does not mean that 78% of the deaths caused by these AEs were preventable. Although the AE was judged to be the proximal cause of death in these cases, often the patients were severely ill and would have succumbed even if the AE had not occurred. We estimate that about half of the deaths due to iatrogenic AEs were preventable. This means that approximately 7% of AEs were fatal and preventable, or about 7,000 deaths in New York. Assuming a similar rate, the total number of preventable deaths for the United States would be approximately 100,000. This is twice the annual highway death rate.

Why Do Accidental Injuries Occur?

All of the reasons for accidental injury to patients in hospitals are not known, of course, but certainly the complexity of modern medical care is a major risk factor. The higher rates of AEs in the highly technical surgical specialties of vascular surgery, cardiac surgery, and neurosurgery attest to this conclusion. We perform many interventions during hospital care. Each of them presents many opportunities for error. Indeed, the wonder is that there are not many more injuries.

Most AEs result from errors in management rather than from patient factors such as allergic reactions. Many physicians (and most lawyers) tend to consider that all errors result from negligence. Not only is such a judgment unjustifiably harsh, but this kind of thinking can be a substantial barrier to efforts to reduce errors. Although some errors are egregious and can be legitimately considered negligent, the vast majority are not. Minor slips or momentary lapses are far more common, such as writing the wrong dosage for a drug or forgetting to obtain the results of a laboratory test. These kinds of slips and mistakes occur frequently to everyone in everyday life. And, although they are arguably among the most careful people in our society, doctors, nurses, and pharmacists also make mistakes.

Unfortunately, even a very small error rate—near perfection—can have serious consequences in a modern hospital. The arithmetic is staggering. For example, in a teaching hospital, the average patient receives more than 30 different medications during a hospitalization. Thus, an average-sized (600-bed) teaching hospital may administer more than 4 million doses of drugs a year. Each dose offers several opportunities for error. If the medication ordering, dispensing, and administration system were 99.9% error free, there would still be over 4,000 errors a year in that hospital, and if only 1% of these result in an AE, this commendably low error rate would cause 40 AEs from medications alone. Of course, even doctors, nurses, and pharmacists do not function at the 99.9% level, so actual AE rates in practice are undoubtedly higher than in this hypothetical example.

How Can Injuries Be Prevented?

Serious efforts at injury prevention in the past have been prompted by concerns about malpractice liability. This emphasis is misplaced for the simple reason that three fourths of AEs are not due to negligence. As a result of the focus on liability, many hospital programs are oriented around "damage control," aimed at reducing practices that may result in suits. Efforts are directed at risk management rather than at overall patient safety. A first step in prevention of AEs would be to make injury prevention a high priority for all hospital staff.

Next, hospitals need to know the extent of accidental injury in their own institutions. Most hospitals rely on self-reporting systems, such as incident reports. There is ample evidence that even when reporting is required, such systems have low yields in comparison to active investigations. A first principle of continuous quality improvement is to obtain data on variance, that is, the types and rates of errors and AEs. Few, if any, hospitals currently obtain those data. To do something about errors and injuries, hospitals need to establish data collection methods that will accurately discover and describe the errors that occur.

Third, hospitals need to rethink the manner in which they deal with human mistakes. Hundreds of mistakes occur every day in a major hospital. Too often in the past, emphasis has been placed on identifying who made the error rather than on why the error occurred and how it can be prevented in the future. Identifying "guilty" persons has not proven to be an effective method for improving quality in other industries, such as the airline industry. Instead, considering errors as evidence of systems failures and directing the efforts of all affected parties to develop ways to minimize these failures has proved more effective (Perrow, 1984; Reason, 1990; Ricketson, Brown, & Graham, 1986).

If errors are recognized as inevitable, as they surely are in any human endeavor, and common, as they must be in a highly complex system such as a hospital, then depersonalizing errors and looking for systems solutions is probably the most efficacious way to reduce iatrogenic injury. "Human error" may appear to be the "cause" of a patient getting one drug instead of another, but the fact that the containers appear identical and are stored side by side is a systems design fault that is easy to correct. Such efforts may prove more successful in preventing AEs than education or discipline of the offending party.

A systems approach to prevention of AEs can also result in the development of new methods to protect patients from injury even when errors occur. If the system can "absorb" errors, that is, automatically correct for them, then inevitable errors need not inevitably cause injury to patients. The redundancy of instruments and warning systems in airlines is based on this concept. It could also work in medicine.

Finally, none of the material presented here should be interpreted as an argument for decreasing education and training efforts to prevent errors. Indeed, the high fraction of technical errors that were found in the Medical Practice Study suggests that there may be major deficiencies in the methods by which doctors are taught the technical aspects of surgery and the performance of nonsurgical procedures. There has always been a strange incongruity between the value placed on technical skill by the specialists who possess them (both surgeons and nonsurgeons) and the casualness with which lesser skills, such as insertion of a chest tube or central venous line, are sometimes taught to those who follow. Much, probably too much, of resident education is still carried out by residents, as embodied in the hallowed, and still oft-repeated, aphorism of "see one, do one, teach one."

One may question why simple measures such as these have not been previously adopted. The reason, we believe, is not callous indifference, but that iatrogenic injury has not been recognized by doctors, nurses, administrators, or hospital and equipment designers as a problem of major significance. With the evidence now at hand that nearly 4% of patients suffer a disabling AE while in hospital, and that more than two thirds of those are preventable, the time has come in medicine to make safety a primary concern.

REFERENCES

Barr, D. P. (1956). Hazards of modern diagnosis and therapy—The price we pay. *Journal of the American Medical Association, 159*, 1432–1436.

Bedell, S. E., Deitz, D. C., Leeman, D., & Delbanco, T. L. (1991). Incidence and characteristics of preventable iatrogenic cardiac arrests. *Journal of the American Medical Association, 265*, 2815–2820.

Brennan, T. A., Leape, L. L., Laird, N. M., Hebert, L., Localio, A. R., Lawthers, A. G., Newhouse, J. P., Weiler, P. C., & Hiatt, H. H. (1991). Incidence of adverse events and negligence in hospitalized patients: Results from the Harvard Medical Practice Study I. *New England Journal of Medicine, 324,* 370–376.

Couch, N. P., Tilney, N. L., Rayner, A. A., & Moore, F. D. (1981). The high cost of low-frequency events: The anatomy and economics of surgical mishaps. *New England Journal of Medicine, 304,* 634–637.

Dubois, R. W., & Brook, R. H. (1988). Preventable deaths: Who, how often, and why? *Annals of Internal Medicine, 109*, 582–589.

Eichhorn, J. H. (1989). Prevention of intraoperative anesthesia accidents and related severe injury through safety monitoring. *Anesthesiology, 70*, 572–577.

Friedman, M. (1982). Iatrogenic disease: Addressing a growing epidemic. *Postgraduate Medicine, 71*, 128–129.

Leape, L. L., Brennan, T. A., Laird, N. M., Lawthers, A. G., Localio, A. R., Barnes, B. A., Hebert, L., Newhouse, J. P., Weiler, P. C., & Hiatt, H. H. (1991). The nature of adverse events in hospitalized patients: Results from the Harvard Medical Practice Study II. *New England Journal of Medicine, 324*, 377–384.

Leape, L. L., Lawthers, A. G., Brennan, T. A., & Johnson, W. G. (1993, May). The preventability of medical injury. *Quality Review Bulletin*, 144–149.

McLamb, J. T., & Huntley, R. R. (1967). The hazards of hospitalization. *Southern Medical Journal, 60*, 469–472.

Medical Practice Study. (1990). *Patients, doctors, and lawyers: Medical injury, malpractice litigation, and patient compensation in New York.*

Mills, D. H., Boyden, J. S., Rubsamen, D. S., & Engle, H. L. (1977). *Medical insurance feasibility study*. Los Angeles: California Medical Association and California Hospital Association.

Perrow, C. (1984). *Normal accidents: Living with high-risk technologies*. New York: Basic Books.

Reason, J. (1990). The contribution of latent human failures to the breakdown of complex systems. *Philosophical Transactions of the Royal Society of London B, 327*, 475–484.

Ricketson, D. S., Brown, W. R., & Graham, K. N. (1986). 3W approach to the investigation, analysis, and prevention of human-error aircraft accidents. *ASEM*, 1036–1042.

Shimmel, E. M. (1964). The hazards of hospitalization. *Annals of Internal Medicine, 60*, 100–110.

Steel, K., Gertman, P. M., Crescenzi, C., & Anderson, J. (1981). Iatrogenic illness on a general medical service at a university hospital. *New England Journal of Medicine, 304*, 638–642.

Weiler, P. C., Hiatt, H. H., Newhouse, J. P., Johnson, W. G., Brennan, T. A., & Leape, L. L. (1993). *A measure of malpractice*. Cambridge, MA: Harvard University Press.

3 Life-Threatening and Fatal Therapeutic Misadventures

Joshua A. Perper
University of Pittsburgh

There are some patients whom we cannot help: there are none whom we cannot harm.

—Bloomfield, quoted in Lambert (1978)

OVERVIEW

The available literature indicates that therapeutic misadventures affect a substantial percentage of hospitalized patients and often result in permanent disability and death. Almost all studies suggest that for a variety of reasons, including fear of litigation, underreporting of such incidents is highly prevalent.

The patients at risk are primarily elderly people in poor health and hospitalized over longer than average periods of time. Though physicians are involved in the majority of the incidents, auxiliary medical personnel are responsible for the adverse event in a substantial proportion of cases.

It is vitally important to increase both the reporting of therapeutic misadventures from the health-care providers as well as to disseminate among the medical community adequate information regarding the nature and possible prevention of adverse peritherapeutic events. Such increased reporting and educational activities may be effectively achieved through confidential professional panels and through appropriate enacted legislation.

INTRODUCTION

In the course of history, people have come to an early realization that the administration of medical care is not free of risk for life and limb. One of the first legal codes, that of the Babylonian King Hammurabi in the second century B.C., held the provider of care strictly liable for death or injury of a patient, and provided a specific menu of quite severe penalties. A surgeon who caused the death of a patient was likely to have his fingers cut off, and a nurse who mistakenly exchanged two infants had to sacrifice her breasts.

Perhaps a similar realization of the physician's limitations prompted the writers of the Talmud to issue the uncomplimentary opinion that "the best of physicians is headed for Hell."

Modern times have been more benign to the shortcomings of the provider of medical care, as society recognizes that peritherapeutic injuries or deaths are not necessarily a result of negligence and may occur without any fault on the part of the provider of care (Lambert, 1978). The physicians or other medical providers are held liable only when an injury occurs as a result of a medical intervention, when there was a legal duty to provide medical care, and when the diagnostic or therapeutic action was below the accepted medical standards.

In recent years, a plethora of somewhat confusing terms have been used to designate injuries related to the rendering of medical care. Those terms include iatrogenic diseases or injuries, therapeutic misadventures, nosocomial diseases, side effects, medical hazards, and adverse (medical) events. In addition, there are terms such as iatrogenic misadventures, peritherapeutic misadventures, peritherapeutic accidents, medical accidents, complications of medical care, and diseases of medical progress.

Some of the terms have been preferentially used in a more general sense, whereas others have been more restricted. The following classification is designed to define more clearly the major types of injuries or illnesses associated with medical diagnosis or treatment in accordance with the generally accepted nomenclature.

IATROGENIC DISEASES/INJURIES

Such conditions include iatrogenic illnesses due to conceptual errors, side effects of procedures or medication, nosocomial diseases, and therapeutic misadventures or peritherapeutic accidents.

Iatrogenic injuries or illnesses are adverse effects resulting wholly or in part from medical procedures or medication, which are not a direct or indirect complication of the patient's primary condition or disease.

This chapter is primarily concerned with the category of iatrogenic injuries designated as therapeutic misadventures, and in particular the fatal therapeutic misadventures. However, the following discussion of the other categories might be helpful to place the therapeutic misadventures in proper context.

Conceptual Errors Resulting in Iatrogenic Disease or Injury

A major group of iatrogenic injuries is the result of conceptual errors or an erroneous concept in selecting a diagnostic or therapeutic procedure. The erroneous concept assumes that an actual or theoretical medical benefit justifies a specific therapeutic approach. There, the error is failing to foresee the nature or magnitude of possible injurious effects. The famous biologist T. H. Huxley adequately defined such frustrating situations as demonstrating "a beautiful theory killed by nasty, ugly little facts." An injurious procedure due to a conceptual error is technically performed correctly according to the initial intent and in line with the general prevailing standards of medical care at that particular time. The procedure ultimately results in a cluster of injuries or fatalities that eventually prompt its discontinuance.

The failure to consider the injurious aspects of a medical procedure or medication may be a result of the intrinsic inability of medical science at a particular time to predict the adverse event, a failure to design a proper procedure for the detection of possible adverse effects of a new procedure or medication, or a predictive failure because of an excessively long interval between the rendered medical care and the injurious result.

There are many examples of iatrogenic injuries due to conceptual errors. Such examples include the antiquity-honored use of bloodletting as a universal remedy, the middle-age and early modern times use of enema as a universal therapeutic procedure—based on the faulty theory of auto-intoxication—and the failure to observe antisepsis during surgical procedures based on ignorance of the existence of infectious microorganisms. Other examples are the indiscriminate practice of tonsillectomy—based on the failure to understand the immunoprotective role of lymphatic tissue—the radiation of the thymus in children based on the faulty concept of status thymo-lymphaticus, and failure to appreciate the tumorigenic effect of irradiation on the thyroid. More recent examples of iatrogenic conditions resulting from erroneous therapeutic concepts include the exposure of premature infants to excessively high oxygen concentration, which led to retrolental fibroplasia and blindness; the use of insulin coma in the treatment of schizophrenics while minimizing the hazards of critical hypoglycemia; and the freezing of gastric ulcers without adequate controls indicating the efficacy of the procedure.

The difference between iatrogenic illnesses due to conceptual errors and side effects of a diagnostic or therapeutic medical procedure is that in the former, the selected medical approach is not indicated, unsound, or inappropriate, and the complicated injury or illness is totally unexpected. In medical side effects, the selection may be totally justified, and some or all of the side effects accepted as reasonable therapeutic trade-offs.

The difference between iatrogenic illnesses due to conceptual errors and therapeutic misadventures is that in the former, as opposed to the latter, the selection of the medical care, though wrong, is free of technical mishaps. As long as conceptual errors are unwittingly accepted in the medical community as standard medical practice, related injuries cannot be considered an outcome of malpractice.

Side Effects of Medication or Therapy

There are no medications or procedures that are totally free of potentially adverse effects. Side effects, however, vary greatly in nature, severity, and frequency with the particular therapeutic approach. The role of the medical care provider is to balance the therapeutic benefit against the potential injury. "If the price is right" and the primary disease significant enough, a competent physician may be justifiably willing to accept the possible occurrence of substantial discomfort or a side-effect-related disability. This may occur even to the point of insisting on the continuation of the therapy in spite of persistent side effects. For example, with regard to radiation treatments of neoplasms or treatment with radiomimetic drugs, the therapist is well prepared to accept secondary side effects of anorexia, nausea and vomiting, extensive loss of hair, and marked depression of bone marrow activity with anemia and granulocytopenia.

Occasionally, the side effects are delayed and unpredictable, and may result in severe disability or death. The tragic examples of congenital fetal malformations associated with the intake of the sedative Thalidomide by the mother and the development of clear cell carcinoma of the vagina in daughters of women who took DES (Diethylstilbestrol) during pregnancy are only two of the more well-known examples.

Nosocomial Diseases

Nosocomial diseases are diseases that develop during the hospitalization of patients. Those diseases most often result from exposure of patients, particularly those susceptible to infections, to an injurious agent more likely to be found in an institutional medical care environment.

Examples of nosocomial illnesses are infections by drug-resistant hospital strains, urethral infections in catheterized patients, and infections in burn victims.

Often the source of infection is insidious and difficult to identify. For example, in some burn victims kept in a totally aseptic hospital environment, the nosocomial infection has been traced to bacteria present in the unremoved stalk remnants of vegetables served to the patient.

Therapeutic Misadventures or Peritherapeutic Accidental Events and Related Fatalities

A therapeutic misadventure or a peritherapeutic adverse event is an unexpected or unexplained medical care-related injury or adverse outcome, which is not an inherent disability, side effect, or unpreventable natural complication of the involved procedure or medication, or a natural complication of the patient's initial conditions. Therapeutic misadventures are adverse effects caused by diagnostic or therapeutic procedures, excessive or inappropriately administered medication, allergy or idiosyncracy to medication, and failure of equipment.

Under this definition, fatal therapeutic misadventures are considered as such when they are caused, wholly or in part, by a diagnostic or therapeutic manipulation that results in mechanical injury; by an incorrect, mistaken, or substandard medical procedure; by an inappropriate, mistaken, or overdose of medication; by incorrect use of medical equipment; or by the use of inappropriate, malfunctioning, defective, or inadequate medical equipment. Fatal therapeutic misadventures are also considered to occur when the death occurs during or immediately following the medical procedure and is not reasonably explained by the patient's prior condition.

On the other hand, deaths resulting from complications and/or developments of the injury or disease for which the patient sought the initial care and inherent side effects or natural complications of the diagnostic or therapeutic procedure should not be considered as therapeutic misadventures. Examples of such adverse nonaccidental events include cardiac arrests occurring during a monitored stress exercise, pneumonia developing after a surgical procedure, or agranulocytosis following treatment with chemotherapeutic agents. In contrast, therapeutic misadventures have clear accidental connotations and are subject to a medico-legal investigation.

THE INVESTIGATION OF FATAL THERAPEUTIC MISADVENTURES

The accidental nature of fatal therapeutic misadventures implicitly mandates their reporting to the local coroner or medical examiner. For a variety of reasons, substantial underreporting of peritherapeutic misadventures is the rule rather than the exception. Beyond underreporting, an additional difficulty is posed by some coroners or medical examiners who exempt all peritherapeutic deaths from the accidental category and therefore abstain from investigating such cases. This approach is based on the erroneous belief that accidental peritherapeutic complications of natural diseases lose their accidental coloration through the magic of the therapeutic involvement and become natural side effects that escape the forensic net. In other words, the very fact that the peritherapeutic accident occurs in the process of treating a natural condition or an injury transforms it into a general peritherapeutic complication free of accidental connotations.

Additional factors that may prompt a disinterest in investigating therapeutic misadventures include an overprotective attitude toward medical care providers by physicians heading medico-legal offices, a belief that because the rendering of care is basically an altruistic act it should not be burdened with the label of accident and possible related malpractice litigation, an effort to avoid a confrontation with individual physicians and/or organized medicine, and a lack of financial or professional resources to perform an adequate forensic investigation and evaluation of adverse medical events.

Initial Investigation

The investigation of therapeutic misadventures is one of the most difficult responsibilities of a medico-legal office. The initial investigation of a possible therapeutic misadventure must consider the reliability of the reporting source and whether the alleged incident is of an accidental nature. Occasionally families or their legal representatives may request, under the guise of a therapeutic misadventure, the medico-legal investigation of a case in which they suspect negligence.

Medical negligence cases are not necessarily accidental therapeutic misadventures. Those cases may involve such issues as misdiagnosis, delayed diagnosis, selection of inappropriate diagnostic or therapeutic procedures, failure to perform certain diagnostic tests or to adequately carry out certain diagnostic or therapeutic procedures, and inadequate or insufficient medication. Such issues, which are the proper subject of civil litigation, are not necessarily medical accidents under the jurisdiction of the coroner or medical examiner, however. Therefore, allegations of therapeutic misadventures should in such cases be properly rejected unless there is credible or at least reasonable evidence that a medical accident might have occurred.

It is equally common that some providers of care who report a fatality involving a suspicious peritherapeutic incident try at the same time to minimize the significance of the incident or to argue that the occurrence of a medical accident was extremely unlikely. In this way the medical informer attempts to have the best of both worlds: The reporting is legally made, but the coroner or medical examiner may be convinced to decline jurisdiction.

Once a determination has been made that a bona fide suspicious case of a fatal therapeutic misadventure is to be investigated, prompt preliminary interviews with the principals involved and the primary physician should be scheduled. Such "warm" interviews are very important as they are likely to elicit a recollection of recent, undistorted information. Also, it is strongly recommended that immediate steps be taken to secure any biological specimens that are in the custody of the provider of care, such as blood samples and frozen or formalin fixed tissue.

Pertinent medical equipment and medication such as anesthesia apparatus in anesthesia-related deaths, containers of intravenous blood or other infusions, intravenous tubing, syringes of recently used medications, and bottles with medication are to be obtained in appropriate cases. Hospital personnel may discard such evidence. Therefore, the importance of promptness in subpoenaing such evidence cannot be overemphasized. It is good practice to establish in advance a therapeutic misadventure reporting protocol for the medical provider. That protocol should specify the type of evidence that is likely to be requested in suspected therapeutic misadventures. For example, hospitals should be requested not to remove any intravenous lines, tracheotomy tubes, or any other invasive devices from the body of the deceased; to preserve all biological samples; and to make available any extracorporeal devices or equipment that may be related to the death.

Finally, prior to the autopsy, the full medical records should be subpoenaed for review. This is important because the nurses' notes may contain more information about the nature of the incident than is contained in the usually terse notes of the physician's follow-up. Furthermore, the incident report and all internal investigation notes should be specifically requested. Many health facilities regard these confidential records as not being part of the file proper and do not usually include them in the submitted records.

The Autopsy

The autopsy should obviously address the question of whether the medical incident was related to the patient's death. Such focusing may require an examination of invasive devices in situ. This includes special examination for air embolism, special toxicological or immuno-serological studies, and special histochemical stains. Full and thorough documentation of gross, microscopic, toxicological, and other findings is invariably required because of possible civil litigation.

The Inquest

The usually complex nature of many therapeutic misadventures and the details of the incident, which are often murky or inadequately reflected in the medical records, require an additional evaluation. That evaluation follows the completion of the autopsy and the relevant auxiliary tests and can be made through the vehicle of an informal or formal inquest. Most coroners and many medical examiners have a statutory right to subpoena witnesses and documents and to take testimony under oath during informal and/or formal inquest proceedings.

Though the coroner's office in Allegheny County, Pennsylvania has the legal authority to conduct both informal investigations and public formal inquests, a policy was adopted of conducting nonpublic inquests rather than public inquests in most cases in which a therapeutic misadventure is suspected.

The testimony in such cases is taken under oath and recorded stenographically. The witnesses are allowed to bring their own attorneys to the proceedings. The reason for such in-chamber procedures is that it is not uncommon that an alleged therapeutic misadventure cannot be substantiated. A public inquest in such cases may unfairly affect the reputation of the medical institution or provider of care. The confidential nature of the investigation also prompts more reporting and cooperation from the health-care providers. Furthermore, the stenographic record of the inquest is provided to the bereaved family or representatives of the estate. The stenographic record offers the bereaved family a credible picture of the events preceding and following the misadventure incidents and permits them to make an informed decision about whether or not civil action should be contemplated.

If the initial investigation indicates gross or criminal medical negligence, a public inquest is scheduled rather than an in-chamber inquest. If it is determined that a direct or indirect primary or secondary complication of a therapeutic misadventure caused the death of the individual, then the death certificate is completed to indicate whether such cause of death was primary or contributory, and the manner of death is listed as "accident."

REPORTING OF THERAPEUTIC MISADVENTURES

The published literature is unanimous in that therapeutic misadventures are notoriously underreported. The major reason for the underreporting is the adverse effect of the incident on the reputation of the medical care providers and the potential threat of malpractice litigation. The longer the time interval is between the incident and the death, the less the chance of reporting the incident. This is either because of a failure to relate the final complication to the medical misadventure or because a transfer to another medical facility may obscure sight of the prior medical accident. Also, the fact that most victims of therapeutic misadventures are old and very ill facilitates the shifting of responsibility for the death from an accidental medical event to natural or other unrelated causes.

Nevertheless, reporting is usually forthcoming when the coroner or medical examiner is known to pursue the investigation of therapeutic misadventures vigorously, when the families are aware of the incident, or when the incident is of such magnitude that it cannot be concealed. The reporting occasionally originates outside official hospital channels and may come from family, nursing or technical personnel, or other patients.

Epidemiologic Data on Therapeutic Misadventures

Therapeutic misadventures are reported in the literature as isolated case reports, reviews of adverse events associated with various medical procedures, and population studies.

Population studies may focus on a population of patients in a particular hospital, groups of patients with specific pathology or who are subjected to specific medical procedures, random samples from the patient population of county or state hospitals, and peritherapeutic fatalities reported to coroners or medical examiners.

Case Reports of Therapeutic Misadventures

Examples of isolated reports of therapeutic misadventures include reports of traumatic injuries during invasive manipulations of hollow organs, such as cardiac injuries during valve replacement surgery (Hawley, Kennedy, Pless, Gauger, & Waller, 1988); ureteral injuries during manipulative or invasive procedures (Woodland, 1992); unusual surgical or medical complications, such as mediastinal emphysema secondary to dental drilling (Hunt & Sahler, 1968); preventable postoperative, life-threatening electrolyte imbalances (Arieff, 1986); anesthesia-related complications or deaths (Sperry & Smialek, 1986); medical equipment-related injuries (Jost et al., 1989); and injuries associated with inappropriate or excessive medication (Hejka, Poquette, Wiebe, & Huntington, 1990).

THERAPEUTIC MISADVENTURES ASSOCIATED WITH SPECIFIC MEDICAL PROCEDURES

There are quite a number of studies evaluating therapeutic misadventures associated with specific medical procedures. Limitations of space permit only the presentation of a few salient examples.

A 1988 study done by Natali in France reviewed 277 vascular iatrogenic injuries that occurred over a period of 24 years (Natali, 1989). Most of the incidents (121 occurrences or 44.8%) were due to invasive procedures. These procedures included injection, perfusion, catheterization, and arteriography. Of the remaining injuries, orthopedic surgery injuries accounted for 69 incidents (24.9%), general surgery for 14 (5%), and surgery for sclerosis of varicose veins for 71 cases (25.6%). In 82 instances (29.6%) the nature of the iatrogenic injury was defined as being very severe. The authors emphasized that most of the injuries were due to imperfect techniques or professional errors.

Linden, Paul, and Dressler (1992), reported on the incidence and characteristics of transfusion errors that occurred in New York State over a period of 22 months, from January 1990 through October 1991. Among 1,784,600 transfusions of red cells there were 92 cases of erroneous transfusion (1/19,000 transfusions), including 54 ABO-incompatible transfusions (1/33,000 transfusions) and three fatalities (1/600,000 transfusions). In most cases, patients with an ordered transfusion received blood of an incorrect group, and in several cases patients with no ordered transfusions received blood intended for another patient. The majority of the reported errors occurred outside of the blood bank. Forty-three percent of the incidents arose solely from failure to identify the patient and/or the unit prior to transfusion, and 11% from phlebotomist (blood-drawing technician) errors. The blood bank alone was responsible for 25% of the errors (wrong blood group was used or cross-matched). The study (Linden et al., 1992) also demonstrated a substantial underreporting of transfusion-related misadventures, particularly those involving phlebotomy (blood drawing) and ordering. When corrected for underreporting, the estimated true risk for transfusion errors rose to 1 per 12,000. The authors emphasized that a national application of the New York State data resulted in an estimate of 800 to 900 projected red-cell-associated catastrophic errors annually in the United States.

Some of the studies have concentrated on groups of patients who by virtue of their age, illness, or injuries are particularly at risk for therapeutic misadventures. Davis et al. (1991) analyzed the magnitude and significance of critical errors on preventable mortality and morbidity in a regionalized system of trauma care in a group of 12,910 trauma patients admitted to six trauma centers over a 3-year period. Critical-care errors were found in 151 (23%) of the patients. Such critical-care errors were identified as the cause of death in 30 (48%) of the 62 preventable deaths. Accidental toxic exposure may also occur within health-care facilities. Scalise, Harchelroad, Dean, and Krenzelok (1989) described six categories of such therapeutic misadventures that were reported to a poison center within a 12-month period. Those categories were right patient/wrong medication, 18%; right patient/right medication/wrong dose or route, 16%; lack of patient education, 2%; proximity of potentially harmful substances to confused persons, 54%; and incorrect equipment management, 4%.

In a number of studies, the database of adverse peritherapeutic events was drawn from reports of malpractice claims related to a particular medical specialty or procedure (Ahamed & Haas, 1992; Kravitz, Rolph, & McGuigan, 1991; Rosenblatt & Hurst, 1989).

THERAPEUTIC MISADVENTURES IN SPECIFIC DEPARTMENTS AND HOSPITALS

Data on the characteristics and incidence of therapeutic misadventures in specific hospital populations, though obviously providing limited information, are easily available for both retrospective and prospective studies. Even so, the number of such studies both in the United States and abroad is surprisingly small.

In 1963, Schimmel conducted a prospective study of the hazards of hospitalization in 1,014 patients admitted to Yale University Medical Service of Grace New Haven Community Hospital in Connecticut over an 8-month period (Schimmel, 1964). Unfortunately, the study specifically excluded any adverse reactions resulting from inadvertent errors by physicians or nurses, or postoperative complications. During the study period, the hospital staff reported 240 episodes in 198 different patients; that is, 20% of the hospitalized patients had one or more such adverse events. The episodes were classified as adverse reactions to diagnostic procedures, therapeutic drugs, transfusions, other therapeutic procedures, acquired infections, and miscellaneous hospital hazards. The most common adverse events, which accounted for 119 episodes (49.6%), were reactions to therapeutic drugs. There were 16 deaths that were determined to be related to adverse events, "whose precise causal role was difficult to evaluate" (Schimmel, 1964, p. 105).

In spite of the provisos, most of these deaths in the Schimmel (1964) study clearly appeared to be due to therapeutic misadventures. Such deaths included the case of a middle-aged woman with cirrhosis who died with mediastinal emphysema (air infiltrating soft tissues of the chest) following a minor laceration of the esophagus during diagnostic esophagoscopy and a patient who was treated with heparin (a blood thinner) and died from massive retroperitoneal bleeding arising in a previously undiagnosed malignant kidney tumor. There was a case of a gastric esophageal balloon that ruptured, producing asphyxia, as well as nine fatalities associated with drug administrations including three involving digitalis (overdoses of cardiac medication). Prolonged hospitalization (averaging more 1½ weeks) and increased severity of primary illness were identified as major risk factors of the adverse events.

A study by Steel, Gertman, Crescenzi, and Anderson (1981) prospectively followed 815 patients who were admitted consecutively over a period of 5 months to the general medical service of a university hospital. The Steel group defined an iatrogenic illness as any illness that resulted from a diagnostic or therapeutic procedure or any form of therapy. Their definition of iatrogenic illness also included harmful occurrences (such as a fall) that were not a natural consequence of the patient's disease. The study found that 290 patients (36%) developed an iatrogenic injury. The incidents were major or life threatening in 76 cases (9%), and fatal in 15 (2%). Thirty of the 290 patients who experienced iatrogenic events died, as compared to only 33 of the 525 with no complications ($p < .05$).

Most hospital interventions leading to iatrogenic complications in the Steel study were drug related (208 cases) followed by diagnostic and therapeutic procedures (175 cases) and miscellaneous (including falls, 114 cases). However, in analyzing the percentage of major complications within each one of these groups, diagnostic and therapeutic procedures took the lead with 28% major complications, followed by miscellaneous with 21%, and drugs, with 19%. The study found that the risk factors for major peritherapeutic misadventures were older age, severity of primary illness, polydrug exposure, and increased hospital stay (19.3 days average stay).

In a 1986 prospective study of 1,176 consecutive patients admitted to a department of internal medicine in Barcelona, Spain, De La Sierra et al. (1989) reported that a total of 295 (25.1%) of these patients developed 367 episodes of iatrogenic illness. The definition of iatrogenic illness was that cited in the previously discussed study by Steele et al. (1981): "an adverse situation due to any diagnostic or therapeutic procedure, as well as those harmful events occurring during hospitalization that are not a direct consequence of the disease of the patient, but do have a specific etiology" (De La Sierra et al., 1989, p. 268).

Though most of the adverse peritherapeutic misadventures were relatively minor, 19 patients developed life-threatening events, including two who died. Identified risk factors were being female, old, or in poor general status on admission; having a hospital stay of longer than 12 days; and having intravenous catheterization or intravenously administered antibiotics and anticoagulants. The most common incidents included intravenous catheters, 79%; drugs, 9.5%; falls from bed, 5.4%; diagnostic procedures, 3.3%; and urinary catheterization, 1.6%.

In a study of 295 patients admitted to a medical intensive care unit over a period of 10 months, Rubins and Moskowitz (1990) reported that 42 patients (14%) experienced one or more care-related complications during their stay. The patients who experienced the adverse events tended to be older, were more acutely ill, had a significantly longer length of stay, and had a much higher hospital mortality rate (67% vs. 27%).

STATEWIDE OR NATIONAL STUDIES OF
THERAPEUTIC MISADVENTURES

A bibliographic search for comprehensive nationwide or statewide reporting of hospitalized patients with therapeutic misadventures revealed only two sources of data: the 1974 California Medical Insurance Feasibility Study (Mills, 1977) and the excellent studies published by the Harvard Medical Practice Study group (Hiatt et al., 1989).

The California study found an incidence of 4.65 injuries per 100 hospitalizations, with only 17% of those due to negligence.

The more comprehensive Harvard Medical Practice Study reported much more dramatic findings. The Harvard group reviewed 30,121 randomly selected records from 51 randomly selected acute-care, nonpsychiatric hospitals in New York State in 1984 (Brennan, Leape, et al., 1991). This represented a random sample of 2,671,863 patients discharged from New York hospitals in 1984.

The Harvard group identified adverse events in 3.7% of the hospitalizations and 27.6% of the adverse events were due to negligence. Although 70.5% of the adverse events were disabling for less than 6 months, 2.5% caused permanent disabling injuries, and 13.6% led to death. Risk factors for an increased advent of adverse events included age over 65 years. People over 65 years of age were also more at risk for negligent adverse events.

Leape et al. (1991) reported that drug complications were the most frequent type of adverse events and accounted for 19% of all incidents, followed by wound infections with 14%, and technical complications with 13%.

Nonsurgical events were more likely to be due to negligence (37%) than intrasurgical events. Diagnostic mishaps accounting for the highest percentage of negligent adverse events were: diagnostic mishaps (75%), noninvasive therapeutic mishaps (errors of omission, 77%), and emergency room incidents (70%). Errors in management occurred in 58% of adverse events with almost half of the errors due to negligence. Certain specialties (neurosurgery, cardiac and thoracic surgery, and vascular surgery) were found to have higher rates of adverse events, but lower rates of negligence (Brennan, Leape, et al., 1991).

Univariate analyses revealed that primary teaching institutions had significantly higher rates of adverse events (41%) than rural hospitals (1%; Brennan, Hebert, et al., 1991). The percentage of adverse events due to negligence was, however, lower in primary teaching hospitals (10.7%) and for-profit hospitals (9.5%) and was significantly higher in nonteaching hospitals. The percentage of negligent adverse events was highest (37%) in hospitals with predominantly minority patients (more than 80%). These results suggested that certain types of hospitals have significantly higher rates of injuries due to substandard medical care.

POPULATION STUDIES OF FATAL THERAPEUTIC MISADVENTURES

A review of the pertinent literature elicited only two studies that looked in detail at fatal therapeutic misadventures. One is an 11-year study (1973–1983) from the Montgomery County Coroner's Office in Ohio (Murphy, 1986). The other is a 10-year study (1982–1992) from the Office of the Coroner of Allegheny County (Pittsburgh) in Pennsylvania (Perper, Kuller, & Shim, 1993).

The Ohio study indicated that during the entire 11-year research period, 44 fatalities reported to the Montgomery County Coroner's Office were identified as caused by therapeutic misadventures. Unfortunately, the study does not indicate the total number of admissions to the county's hospitals or the hospitals' total bed capacity, so a misadventure-to-admission ratio cannot be determined. However, the study mentioned that the total population of the county was 743,600 inhabitants; the yearly average of cases reported to the coroner's office was about 2,000; and the number of cases brought to that office for examination was 809 cases. The 44 cases represented an incidence of 0.46% out of the 9,497 cases examined during the 11-year period.

There were marked differences in the frequency of the various medical categories involved in therapeutic misadventures. Surgical events (e.g., vascular trauma, complications of tracheotomy) accounted for 36% of the cases, anesthesia for 30%, therapeutic procedures for 18%, diagnostic procedures for 14%, and drug reaction for 2%.

The Allegheny County (Pittsburgh) study (Perper et al., 1993) identified during a 10-year study period a total of 63 fatalities due to therapeutic misadventures. This is from a county with a population of 1,400,000 and a total hospital capacity of 9,718 beds. The annual average rate of fatal therapeutic misadventures was 2.2 per 100,000 hospital admissions.

Among the 63 fatal therapeutic misadventures in the Allegheny County study, women accounted for 39 (62.9%) and men for 24 (38.1%). African Americans, who accounted for only 10.5% of the county's population, were overrepresented with 17 deaths (27.0%). More than half of the fatalities (33 or 52.4%) occurred in people 65 years of age and older, many of whom had critical clinical conditions. The survival time from the occurrence of the incident to death was variable, 60% survived 24 hours or less, 17.5% survived between 24 hours and a week, and 22.2% survived more than a week.

In the Pittsburgh study, most of the medical interventions resulting in death (76.2%) were therapeutic, and less than one fourth (23.8%) were diagnostic. The majority of fatalities (47.6%) were caused by traumatic medical injuries, primarily perforations and bleeding caused by intravascular devices. Most of the remainder were due to overdose of medication (17.5%), obstruction of airways (14.3%), anaphylactic reactions (12.7%), and anesthesia (3.2%). Almost all of the incidents (92.1%) occurred in hospitals, with only three occurring in nursing homes and two occurring at the patient's home.

The Harvard study reported differences in the frequency of fatal misadventures between university-related and community-based hospitals, with the former having double the rate of misadventure fatalities than the latter. In the Pittsburgh study all university-related hospitals reported at least one fatal case. However, six nonuniversity hospitals, five with fewer than 400 beds, reported no misadventure cases during the 10-year study. Within hospitals, the incidents occurred most frequently in operating rooms (57.1%) and hospital wards (27.0%), followed by X-ray rooms, catheterization rooms, and intensive care units.

Among the 63 fatal therapeutic misadventures, 29 (46%) were attributed to negligent conduct; the remainder were not negligent or questionable. Among the cases showing clear evidence of negligence, 13 (44.8%) were related to medication, 6 (20.7%) were related to traumatic misadventures, 5 (17.2%) were related to airway obstruction, and 5 (17.2%) were related to other causes.

FATAL THERAPEUTIC MISADVENTURES AND NEGLIGENCE

Although a substantial number of therapeutic misadventures are due to negligence or malpractice, the two entities are not identical. Therapeutic misadventures may well be accidents that are beyond the control of the provider of care. That is not so of negligence.

Furthermore, it should be emphasized that although a therapeutic misadventure is often due to error, unless the error manifests as a substandard medical action, no case for negligence can be made. Though this may be a reasonable argument in some of the nonfatal therapeutic misadventures, in fatal therapeutic misadventures it would be difficult if not impossible to demonstrate that an error that led to the patient's death was due to anything other than substandard quality of care.

Not infrequently, cases of death due to negligence may lack the characteristics of an accident and may be a result of a judgment failure in the diagnostic or therapeutic process. Examples include a misdiagnosis in spite of adequate data, failure to select appropriate diagnostic tests or therapeutic procedures, and delay in diagnosis or treatment. This is precisely the reason why, as previously discussed, deaths associated with medical negligence do not automatically fall under the jurisdiction of the coroner or medical examiner.

It should also be recalled that only a small percentage of negligent therapeutic misadventures end in civil litigation. Localio et al. (1991), investigators with the Harvard study, reported that out of their 280 patients who had adverse events caused by medical negligence, only eight filed malpractice claims. The total number of malpractice claims among the 30,195 patient records that were reviewed in the study constituted only 47 suits.

The study indicated that the files of the Office of Professional Medical Conduct of the New York Department of Health reflected a higher fraction of medical negligence cases resulting in claims (close to 2%). The explanation for this higher percentage as compared with the Harvard Study population was due to the fact that additional claims were made in situations in which no malpractice did in fact exist.

Localio et al. (1991) concluded that "medical-malpractice litigation infrequently compensates patients injured by medical negligence and rarely identifies and holds providers accountable for substandard care" (p. 245). The primary reason for the low rate of justifiable malpractice suits may well be a lack of awareness by the patients or their families as to the negligent medical care, either directly or by not being notified by the health-care provider. Such lack of notification is not surprising in view of the literature reports previously quoted.

Other factors may also be involved in the low rate of malpractice suits, such as receipt of adequate health or disability insurance benefits, the will to preserve a good patient–physician relationship, or refusal by lawyers to accept on the basis of contingency fees cases that are likely to result in small monetary awards.

Furthermore, patients who sustained only minor injury may well choose to forego litigation. Obviously such reasoning does not apply to therapeutic misadventures that result in severe injuries or death. Unfortunately, the Localio study did not specifically address the question of frequency of malpractice litigation in such cases.

PREVENTION OF THERAPEUTIC MISADVENTURES

A successful prevention program for therapeutic misadventures requires improved reporting, quality care monitoring, identification of risk factors for the patient, identification of risk factors for the care provider, identification of risks in the medical environment, and effective educational and preventive programs.

Reporting and Quality Care Monitoring

As previously noted, the medical literature is unanimous in concluding that there is substantial underreporting of therapeutic misadventures.

The major culprit for the underreporting of therapeutic misadventures seems to be the secretive method by which adverse incidents in hospitals are internally handled. In most states, including Pennsylvania, there are no mandatory statutory provisions requiring the reporting of nonfatal medical misadventures that occur in health-care facilities, regardless of how severe such incidents may be. The incident reports that are filed in such cases are neither included nor mentioned in the patient's regular records and therefore may easily escape external scrutiny. Though many states require that all evaluations and treatments be accurately recorded in the patients' charts, it may not actually occur. Beyond a general duty to provide adequate guidelines and to activate quality control committees, the hospital administration is often not specifically directed to monitor the accuracy and completeness of the medical records and to provide mechanisms for correction of inadequate or inaccurate entries. One should not be surprised, therefore, by the conclusions of the Pittsburgh study that even fatal therapeutic misadventures may remain undisclosed and unreported to the local coroner or medical examiner.

When comparing the results of the Harvard studies of the New York State hospital records with the Ohio and Pittsburgh studies of fatal therapeutic misadventures reported to a medico-legal system, it becomes clear that whereas in the Harvard study, diagnostic errors leading to therapeutic misadventure were quite common, there was only one such incident reported in the Pittsburgh study. The reason is that errors in diagnosis are usually not reported in the Pittsburgh sites, and even if reported they are likely to be refused jurisdiction by the coroner or medical examiner on the grounds that such events lack an accidental coloration and are primarily errors in medical judgment open only to civil litigation. The fatalities accepted under the medico-legal jurisdiction are, as a rule, mostly due to errors in the performance of a medical or surgical procedure, overdoses of medications, or wrongfully switched medications. Even such adverse events are often not officially reported by hospitals, and as the Pittsburgh study indicates, reporting from other sources (family, anonymous hospital personnel, attorneys) is not unusual. Underreporting is more likely to occur in cases with a longer survival interval or when a patient is transferred from one medical facility to another. For institutional medical care facilities other than hospitals (e.g., nursing homes), reporting of therapeutic misadventures is usually very low, or even nonexistent.

It is unquestionable that the willful nondisclosure of fatal or severely disabling medical misadventures constitutes both a violation of medical ethics and of the law. The underreporting of fatal therapeutic misadventures is particularly onerous. It is bound to result in erroneous certification of the cause and manner of death, failure to identify highly dangerous medical procedures and to inform others about such risks, failure to identify highly incompetent health-care providers, and denial of accidental death benefits and the right of the estate to sue for a wrongful death.

The current situation is clearly unacceptable and therefore urgently mandates a substantial improvement in the reporting of serious and fatal therapeutic misadventures. Most pertinent studies fail to offer specific solutions and content themselves with a general exhortation to increase quality care through increased monitoring of peritherapeutic incidents and implementation of appropriate corrective measures.

Quality Control Agencies and Therapeutic Misadventures

The proliferation of quality care agencies in the United States did not substantially increase the reporting of therapeutic misadventures.

The major monitors of quality of care are the quality control boards of hospitals, state boards of medicine, insurance companies, the National Data Bank, the Food and Drug Administration (FDA), and medico-legal investigative offices (coroner or medical examiner). These are clearly disparate bodies with very different responsibilities and interests. Some are passive repositories of data; others are active participants in the preventive or corrective effort. The limitations of the quality control boards of hospitals have been mentioned previously in this chapter. The state boards of medicine have a general duty of monitoring the quality of medical practice; however, most state boards become actively involved only after complaints of reckless medical care against specific physicians.

The state boards of medicine receive from insurance companies periodic "Medical Malpractice Payment Reports," which detail incidents of injury to patients for which monetary settlements were made or court awards were given. Such reports could constitute an excellent source for the identification of therapeutic misadventures. Unfortunately, few state medical boards have the resources or the interest to use such a large and comprehensive body of data. Furthermore, the resolution of the malpractice suits may take years, and by the time the case is reported to the state board, it may be stale both from a legal and a medical viewpoint. The data from insurance companies are highly confidential and are therefore virtually useless from a public standpoint except for the mandatory reports already mentioned. State laws are often silent as to the reporting of therapeutic misadventures.

The recently enacted National Data Bank, mandated by federal law (P.L. 99-660, P.L. 100-93), keeps records of various penalties imposed on specific physicians for professional or ethical violations. These violations include medical malpractice payments, adverse licensure actions, adverse actions on clinical privileges, and adverse actions on professional society memberships. Hospitals, other health-care entities, professional medical societies, and insurance companies must submit reports identifying themselves as the reporting entity and the involved medical practitioners. They also must provide descriptive information on the adverse action taken or malpractice payment made. The Data Bank is confidential and releases information only to specified authorized agents. Unfortunately, the Data Bank does not compile statistical information and does not release periodic reports on the national quality of medical care.

Since 1984, the FDA has required the manufacturers and importers of medical devices to report any related serious injuries or deaths. However, later Congressional hearings and reports by the General Accounting Office (GAO) and the Office of Technology Assessment (OTA) have brought to light the fact that the reporting is largely unsatisfactory. In 1986, the GAO found that "less than one per cent of device problems occurring in hospitals were reported to the FDA and that the more serious problem with the device, the less likely it was to be reported" (U.S. Department of Health & Human Services, 1991; U.S. General Accounting Office, 1986).

These findings prompted Congress to enact the Safe Medical Device Act (SMDA) of 1990, which became effective on November 28, 1991. The SMDA mandates both the medical device industry and the users of medical devices to report device-related illness or injuries and includes severe penalties for violators. The Pittsburgh study group adopted a similar stance in suggesting legislation mandating the reporting of therapeutic misadventures that result in severe disability, coma, or death, and providing substantial penalties for violators. Such legislation has been indeed advocated in Pennsylvania by the Allegheny County (Pittsburgh) Coroner's Office, and is currently under consideration in the Pennsylvania House of Representatives. Opposition to that legislation is primarily represented by the Pennsylvania Medical Society and to a lesser degree by the Pennsylvania Hospitals Association.

Identification of the Patients at Risk

All studies, both hospital and coroner's office based, mention that the major risk factors of patients to therapeutic misadventures are older age (over 65), being female, serious or critical illnesses, and longer hospitalization (longer than 7–10 days). Also more at risk are patients who undergo invasive procedures, in particular procedures with intravascular devices, or those who undergo penetration of hollow organs.

The increased risk to the elderly may be largely explained on the basis of biological frailty and decreased resistance to injury and disease. However, one cannot exclude the possibility that their care is perhaps less attentive or more indifferent.

Identification of Risk Factors for the Health-care Provider

Most studies do not specify the characteristics of the providers of care involved in therapeutic misadventures. The Pittsburgh study found that among the individuals who initiated the fatal therapeutic misadventures, 68.3% were physicians, 27% were nursing staff, and 4.8% were the patients themselves.

Interesting questions about the personal or professional traits of the health-care providers involved in therapeutic misadventures still remain unanswered. How do factors such as their personalities, family status, health condition, graduating medical school, residency program, medical experience, and tendency for recurring episodes relate to an increased risk of getting involved in a therapeutic misadventure incident in general, and in a negligent adverse event in particular?

The following list refers to some of the characteristics that may be observed in some of the care providers involved in therapeutic misadventures.

1. The inexperienced or ignorant
2. The reckless or risk taker
3. The absent minded
4. The deaf to communication or argument
5. The impatient
6. The incompetent slow-poke
7. The procrastinator
8. The alcoholic or drug user
9. The overtired
10. The reluctant to seek advice

The list is written in a somewhat light vein, but all prototypes were sketched from observing a much more sobering reality. The following incidents may exemplify some of the professional or personality characteristics that are likely to result in therapeutic misadventures, often with a strong negligence character.

In one case, a registered nurse, 6 months out of school, erroneously gave a patient an intravenous infusion of a feeding liquid mixture that is supposed to be given only orally or through a gastrointestinal tube. The patient died within a couple of minutes with innumerable emboli of fat and vegetable fragments, which clogged the capillaries of the lungs and brain.

In another case the patient, an old woman, was known to be allergic to penicillin and a warning was recorded in her chart. By error, a house physician ordered a shot of penicillin. Shortly thereafter, the doctor realized his error and canceled his instructions. Unfortunately by that time, the order was taken by a floor nurse. On arriving in the patient's room with the fateful loaded syringe, the nurse was asked by the patient whether she was to be given penicillin. The patient told the nurse that if she was to receive penicillin she would die. The nurse, believing that "the doctor knows better," injected the penicillin in spite of the patient's protests. The patient died within minutes of anaphylactic shock.

It should be emphasized that in some cases the therapeutic misadventure is virtually unpreventable and occurs by no fault on the part of the health-care provider. For example, during a coronary artery catheterization, a fracture of an arteriosclerotic plaque with atheromatous embolization, a local mural dissection, or a perforation of a severely arteriosclerotic vessel may be absolutely unpreventable. Under such circumstances, if the procedure was medically indicated, no reasonable case for negligence can be made. Similarly, therapeutic misadventures due to defective medical products cannot be blamed on the provider of care unless the latter adds contributory negligence.

Identification of Risks in the Hospital Environment

Some of the major risks in the hospital environment include the necessary regimentation in the distribution of medicines, the use of numerous medical devices, and the impersonal relationship between certain health-care specialists and the patient. As a result, unless strict checking procedures and periodic quality control are in place, serious errors may occur in identifying the patient who is to be subjected to an indicated medical procedure. Withholding therapy from the appropriate patient or administering therapy to the wrong patient may result in serious injury or death.

Table 3.1 illustrates three cases in which the misidentification of patients scheduled for blood transfusions resulted in death.

It should be emphasized that misidentification errors may well be caused by physicians and not only by auxiliary health-care personnel. There are sporadic incidents of physicians confusing one patient with another and ordering the wrong medication, and operating on the wrong person or on the wrong side of the patient's body.

Sometimes the error of the provider of care is facilitated by inadequate labeling of medication or defective or outdated medical equipment. For example, in the Pittsburgh study, two episodes of death due to overdoses of Lidocaine were due to the fact that the involved nurses mistakenly used the medication bottle with the higher dose, which had the identical packaging appearance as the lower dose. The only difference in labeling was the different concentration. Obviously, adequate preventive strategies can be easily developed to avoid such accidents.

TABLE 3.1
Erroneous Fatal Blood Transfusions

Nature of Incident	Health-care Provider Responsible for the Error	Nature of Error	Explanation of Error
52-year-old woman in renal failure due to diabetes given type A blood instead of type O	Registered Nurse	Given blood intended for another patient	Nurse checked requisition form against blood bag label but did not check patient identification
86-year-old woman electively scheduled for surgery for cancer of the uterus received mismatched blood	Phlebotomist	Blood for cross-matching instead of being taken from patient was taken from two other persons on two different occasions	Overtired worker (worked two shifts) Passed polygraph test
76-year-old woman with massive myocardial infarction in cardiogenic shock received two units of type A blood instead of type B	Laboratory technician	Wrong blood was used for cross-matching	Overworked technician cross-matched the blood of three patients and mixed up the samples

Education and Preventive Programs

The dissemination of detailed information among health-care providers as to the nature and pathogenesis of therapeutic misadventures, particularly of those serious enough to result in disability or death, is essential to any effective preventive effort. Unfortunately at the present time, the secrecy with which such incidents are handled within the medical care institution obviously impedes the educational effort. As a result of this communication failure, the same type of adverse event may reoccur repeatedly in different health-care facilities.

When for one reason or another a particularly dramatic peritherapeutic accident becomes publicized in the media, it is not unusual to see hospital representatives come forward with most unreasonable explanations that deny the advent of that therapeutic misadventure in their facility. For example, in the case of a newborn who developed severe birth anoxia and subsequently died because the mother was erroneously subjected during delivery to a large overdose of pitocin, a well-qualified medical expert stated to the press that the death could not be due to pitocin because pitocin is not a drug but an internal hormone, and as such is innocuous.

In another instance, an elderly woman sustained an intestinal perforation in a vehicular accident due to a tight safety belt. The emergency room examination, testing, and record were clearly deficient and the patient was released prematurely to her home where she collapsed 2 days later with peritonitis. A laparotomy revealed a single intestinal perforation with adjacent hemorrhage and otherwise a normal structure. The medical administrator of the involved facility announced unabashedly to the media that the perforation was not due to the accident but to a spontaneous bursting of the intestines due to overeating in a patient who had some abdominal adhesions from prior surgery.

IMPROVING REPORTING

A number of strategies may be devised in order to deal with the reluctance to report serious or fatal therapeutic misadventures. One strategy is the demonstration of a persistent and clear interest by responsible public agencies in the investigation of accidental peritherapeutic deaths. For example, the Pittsburgh study indicates that the progressively increased reporting of total therapeutic misadventures in the study area was at least in part attributable to the awareness of the local providers of care to the manifested interest and scrutiny by the coroner's office, which is charged under the law with the investigation of unexplained, suspicious, or accidental deaths.

A second strategy is the formation of a therapeutic misadventures panel representing all regional health-care institutions. The Allegheny County Coroner's Office has created such a panel, which is open to representatives of all hospitals in the area. Following the completion of the investigation by the coroner's office, the panel periodically receives a detailed report on the therapeutic misadventures that occurred over a specific period of time. The deliberations of the panel are confidential. The reports do not disclose the names of the patients, of the facility where the incident occurred, or of the providers of care involved. The panel discusses possible alternatives for preventing such episodes in the future, and following its deliberations, a list of alternative strategies for preventing different types of therapeutic misadventures is mailed to all health-care institutions in the area. Similar panels may be established under the aegis of other agencies or associations and may evaluate an even wider range of peritherapeutic incidents.

A third strategy is statutorily mandated notification of serious or fatal therapeutic misadventures by hospitals and individual health-care providers to the state board of health or medicine. The statutory directives mandating health-care providers to notify state health agencies of serious or fatal therapeutic misadventures should be coupled with provisions requiring the state board of health or medicine to compile annual or semiannual reports of all such incidents. The reports should also indicate how the incidents occurred, what the possible reasons are, and how they may be prevented.

REFERENCES

Ahamed, S., & Haas, G. (1992). Analysis of lawsuits against general surgeons in Connecticut during the years 1985–1990. *Connecticut Medicine, 56*(3), 139–141.

Arieff, A. I. (1986). Hyponatremia, convulsions, respiratory arrest, and permanent brain damage after elective surgery in healthy women. *New England Journal of Medicine, 314*, 1529–1535.

Brennan, T. A., Hebert, L. E., Laird, N. M., Lawthers, A., Thorpe, K. E., Leape, L. L., Localio, A. R., Lipsitz, S. R., Newhouse, J. P., Weiler, P. C., & Hiatt, H. H. (1991). Hospital characteristics associated with adverse events and substandard care. *Journal of the American Medical Association, 265*(24), 3265–3269.

Brennan, T. A., Leape, L. L., Laird, N. M., Hebert, L., Localio, A. R., Lawthers, A. G., Newhouse, J. P., Weiler, P. C., & Hiatt, H. H. (1991). Incidence of adverse events and negligence in hospitalized patients, results of the Harvard Medical Practice Study I. *New England Journal of Medicine, 324*, 370–376.

Davis, J. W., Hoyt, D. B., McArdle, M. S., Mackersie, R. C., Shackford, S. R., & Eastman, A. B. (1991). The significance of critical care errors in causing preventable death in trauma patients in a trauma system. *Journal of Trauma, 31*(6), 813–819.

De La Sierra, A., Cardellach, F., Cobo, E., Bové, A., Roigé, M., Santos, J. M., Ingelmo, M., & Urbano-Márquez, A. (1989). Iatrogenic illness in a department of general internal medicine—A prospective study. *The Mount Sinai Journal of Medicine, 56*(4), 267–271.

Hawley D. A., Kennedy, J. C., Pless, J. E., Gauger, D. W., & Waller, B. F. (1988). Cardiac injury during valve replacement surgery. *Journal of Forensic Sciences, 33*(1), 276–282.

Hejka, A. G., Poquette, M., Wiebe, D. A., & Huntington, R. W. (1990). Fatal intravenous injection of mono-octanoin. *American Journal of Forensic Medicine and Pathology, 11*, 165–170.

Hiatt, H. H., Barnes, B. A., Brennan, T. A., Laird, N. M., Lawthers, A. G., Leape, L. L., Localio, A. R., Newhouse, J. P., Peterson, L. M., Thorpe, K. E., Weiler, P. C.,

& Johnson, W. G. (1989). Special report—A study of medical injury and medical malpractice—An Overview. *New England Journal of Medicine, 321*(7), 480–484.

Hunt, R. B., & Sahler, O. D. (1968). Mediastinal emphysema produced by air turnine dental drills. *Journal of the American Medical Association, 205*(4), 101–102.

Jost, S., Simon, R., Amende, I., Herrman, G., Reil, G. H., & Lichtlen, P. R. (1989). Transluminal balloon embolization of an inadvertent aorto-to-coronary venous bypass to the anterior cardiac vein. *Journal Catheterization and Cardiovascular Diagnosis, 17*(1), 28–30.

Kravitz, R. L., Rolph, J. E., & McGuigan, K. (1991). Malpractice claims data as a quality improvement tool, I. Epidemiology of error in four specialties. *Journal of the American Medical Association, 266*(15), 2087–2092.

Lambert, E. C. (1978). *Modern medical mistakes.* Bloomington: Indiana University Press.

Leape, L. L., Brennan, T. A., Laird, N., Lawthers, A. G., Localio, A. R., Barnes, B. A., Hebert, L., Newhouse, J. P., Weiler, P. C., & Hiatt, H. (1991). The nature of adverse events in hospitalized patients, results of the Harvard Medical Practice Study II. *New England Journal of Medicine, 324,* 377–384.

Linden, J. V., Paul, B., & Dressler, K. P. (1992). A report of transfusion errors in New York State, *Transfusion, 32,* 601–606.

Localio, A. R., Lawthers, A. G., Brennan, T. A., Laird, N. M., Hebert, L. E., Peterson, L. M., Newhouse, J. P., Weiler, P. C., & Hiatt, H. H. (1991). Relation between malpractice claims and adverse events due to negligence, results of the Harvard Medical Practice Study III. *New England Journal of Medicine, 325,* 245–251.

Mills, D. H. (Ed.). (1977). Report on the medical insurance feasibility study. San Francisco: California Medical Association.

Murphy, G. K. (1986). Therapeutic misadventure—An 11-year study from a metropolitan coroner's office. *American Journal of Forensic Medicine and Pathology, 7*(2), 115–119.

Natali, J. (1989). Iatrogenic vascular lesions. *Bulletin Acadamie Nationale de Medecine, 173*(6), 753–765.

Perper, J. A., Kuller, L. H., & Shim, Y. K. (1993). Detection of fatal therapeutic misadventures by an urban medico-legal system. *Journal of Forensic Sciences, 38*(2), 327–338.

Rosenblatt, R. A., & Hurst, A. (1989). An analysis of closed obstetric malpractice claims. *Obstetrics and Gynecology, 74*(5), 710–714.

Rubins, H. B., & Moskowitz, M. A. (1990). Complications of care in a medical intensive care unit. *Journal of General Internal Medicine, 5*(2), 104–109.

Scalise, J., Harchelroad, F., Dean, B., & Krenzelok, E. P. (1989, April). Poison center utilization in nosocomial toxicologic exposures: A prospective study. *Veterinary and Human Toxicology, 31*(2), 158–161.

Schimmel, E. M. (1964). The hazards of hospitalization. *Annals of Internal Medicine, 60,* 100–109.

Sperry, K., & Smialek, J. E. (1986). The investigation of an unusual asphyxial death in a hospital. *Journal of the American Medical Association, 255,* 2472–2474.

Steel, K., Gertman, P., Crescenzi, C., & Anderson, J. (1981). Iatrogenic illness on a general medical service at a university hospital. *The New England Journal of Medicine, 304*(11), 638–642.

U.S. Department of Health and Human Services. (1991). Medical device reporting for user facilities, HHS. *FDA Publication,* 92–4247.

U.S. General Accounting Office. (1986). *Early warning of problems is hampered by severe under-reporting* (Rep. GAO/PEMD, 87-1. Washington, DC: Author.

Woodland, M. B. (1992). Ureter injury during laparascopy-assisted vaginal hysterectomy with the endoscopic linear stapler. *American Journal of Obstetrics & Gynecology, 167*(3), 756–757.

4 Human Errors: Their Causes and Reduction

Harold Van Cott
Van Cott and Associates

"Human Error: Avoidable Mistakes Kill 100,000 Patients a Year" (Russell, 1992): Headlines about human error in hospitals, railroads, and chemical and nuclear power plants have become everyday breakfast-table news, but they are neither inevitable nor unavoidable.

The goal of this chapter is to examine the nature and extent of human error in health care and to introduce some of the lessons that have been learned from experience with other systems and from psychological theory and research on human error.

This chapter must be prefaced with a comment about the perspective from which it was written. It was not written by a health-care professional but by a human-factors practitioner with experience in human error acquired from work in power plants, refineries, aircraft and space vehicles, submarines, and with consumer products. It is practical knowledge learned in trying to develop workable ways to improve the reliability and safety of human–machine systems.

THE CHARACTERISTICS OF SYSTEMS

Despite many differences in function and form, all systems have common characteristics. They involve technology: the tools and machines that serve human needs; systems have an interface: the means or affordances by which users interact with the system; they involve people: in every system but the most highly automated, there are one or more people who operate and maintain it.

People are a constant across the spectrum of built systems. As a constant they behave in much the same way in every system, performing needed jobs, usually faithfully, but not always faultlessly. People can make errors that can have serious, often tragic, and costly outcomes. It is therefore important to understand the nature of human error and the mechanisms that cause it if measures are to be taken to reduce its frequency and consequences.

Slowly, painfully so, some answers are beginning to take shape. Some come from detailed studies of actual accidents and careful reconstructions of their precursor events (Rogovin, 1975; U.S. Nuclear Regulatory Commission, 1985). Some come from information from incident-reporting systems that are used to compile actuarial information on near accidents and accidents, and their assumed causes. An example is the National Electronic Surveillance Systems operated by the Consumer Product Safety Commission, which collects information on product-related injuries admitted to emergency rooms. Some come from recent research on theory by cognitive psychologists (Reason, 1990). All point to human error as a major contributor to the great majority of accidents and near accidents in the massively large sociotechnical systems on which contemporary life has become so dependent.

THE EXTENT OF HUMAN ERROR

Precise data on the extent of human error as a percentage of all system failures are hard to come by and vary from system to system, depending on how error is defined, classified, and reported. Meshkati (personal communication, 1993) collected as many reports on the incidence of human error in different systems as he could find. He encountered cases as high as 90% for air traffic control and as low as 19% in the petroleum industry. Rates for some other systems are as follows:

- 85% for automobiles.
- 70% for U.S. nuclear power plants.
- 65% for worldwide jet cargo transport.
- 31% for petrochemical plants.

The average across all systems is about 60%. The remaining 40% is attributable to material, electrical, and mechanical failures.

Why is human error so prevalent? Is it at the root of the majority of accidents or a contributing factor? Why does it occur? Is it caused by irrational tendencies, perversity, malevolence, or by more mundane conditions such as forgetfulness, inattention, or carelessness? Is it human nature to err, as Alexander Pope the poet claimed? If so, does error have its origins in fundamentally useful, adaptive psychological processes (Reason, 1990)? Does it represent noise in an imperfect biological human system? Or, could it be that human error is the result of mismatches between the way things are designed and the way people perceive, think, and act?

HEALTH-CARE DELIVERY AS A SYSTEM

Systems thinking—and systems analysis and design tools—have been around for a long time, but health-care delivery is usually *not* thought of as a system. Yet, of all sociotechnical systems, it surely is the largest, most complex, most costly and, in at least one respect, the most unique.

What is the health-care system like? It is an enormous number of diverse and semiautonomous elements: ambulance services, emergency care, diagnostic and treatment systems, outpatient clinics, medical devices, home care instruments, patient-monitoring equipment, testing laboratories, and many others. All of these elements are loosely coupled in an intricate network of individuals and teams of people, procedures, regulations, communications, equipment, and devices that function in a variable and uncertain environment with diffused, decentralized management control.

If there is one characteristic of the health-care system that distinguishes it from others, it is its uniqueness as a sociotechnical system. Each of its many component subsystems—hospitals, emergency care, pharmacies, clinics, laboratories, and others—represents a distinct culture with its own unique goals, values, beliefs, and norms of behavior. Each is managed separately from the others. Coordination among the subsystems is accomplished by informal networking, custom, and regulation.

In contrast with the health-care delivery system, change is effected in centralized systems through the authority of a hierarchical, vertical management structure. The process is relatively quick and reasonably efficient. Change in the health-care system is accomplished laterally across several subsystems in which responsibility and decision making are distributed across many people and units. In such a diffuse system, change is a slow, often difficult process with more opportunities for error and more unpredictable outcomes than in a single, hierarchical system (National Research Council, 1990). As a further drawback, the health-care system must cope with very rapid advances in medical technology and practice. It must also cope with legal and economic constraints. Any program aimed at the reduction of human error in the health-care system must be designed with an understanding of characteristics such as these if desired cultural learning and change are to be fostered.

ON HUMANS IN THE HEALTH-CARE SYSTEM

The humans in the health-care delivery system are its most ubiquitous and important element. Unlike the manufactured components of a system, people differ greatly from one another on every physical and mental attribute that one can mention. One person does not necessarily behave in the same way as another person in the same situation.

Notwithstanding the differences among individuals and individual performance, the reliability of the health-care delivery system rests on people—hundreds of thousands of them. Among them are doctors, administrators, nurses, technicians, aids, orderlies, pharmacists, accountants, engineers, and maintenance technicians. These people are the nodes that link the components of health care; they are critical to its reliability and safety. Unlike other systems in which technology is the center of the system and humans serve as equipment monitors and supervisors, the health-care system is people-centered and people-driven. Pieces of equipment, such as X-rays, glucose monitors, and dialysis machines are linked together with procedures and policies by people to serve other people as patients.

ON HUMAN ERROR IN HEALTH CARE

Malfunctions and accidents in many systems are dramatic, often terrifying events, with catastrophic consequences: explosion, fire, toxic releases. These events cause property damage, injury, and often loss of human life. Chernobyl, Vicennes, and the Exxon Valdez were megacatastrophic system failures. The accidents that occur in health-care delivery are less visible and dramatic than those in other systems.

Although accidents and adverse events in health care are numerous, they are insidious. Except for celebrated malpractice claims, they usually go unnoticed in the public media and underreported in compilations of accident statistics. These microcatastrophes occur throughout the health-care system. They are often one of a kind, and there are many different kinds. The causes are not always easy to isolate and usually have multiple origins.

Because the system is so loosely coupled, errors originating with the health-care system from human and equipment failures do not propagate throughout it or culminate in major catastrophes such as the Challenger tragedy. Rather, there is an injury here, a mistake there, an accident here, a death there. These microevents add up to over 100,000 a year (Russell, 1992).

As a sociotechnical system, health-care delivery is subject to extrinsic as well as intrinsic events: societal pressures, legal and regulatory rules. The extent and nature of these additional extrinsic forces and their impact on the reliability of health care is difficult to assess but must be considered, lest overemphasis be given to the role of the technical components of health care.

Good data—even crude data—on the nature and extent of the errors that occur in health-care delivery do not exist. There is no generally accepted way to classify and tally them. The data that exist are fragmented, coming from a few reporting systems such as the product-related emergency room admissions reported by the National Electronic Injury Surveillance System of the Consumer Product Safety Commission and scattered studies, audits, and malpractice files.

Most studies of human error in health care have centered on hospitals and hospitalized patients. One of the most comprehensive (Brennan et al., 1991; Leape et al., 1991) was the Medical Practice Study conducted by the Harvard School of Public Health. The investigators reviewed 30,121 randomly selected records from 51 randomly selected acute-care, nonpsychiatric hospitals in New York State. Adverse events were found in 3.7% of hospitalizations. Of these, 70.5% of the events led to disabilities of up to 6-months' duration; 2.6% caused permanent disability; and 13.6% led to death. The most frequent class of errors (44.4%) were technical errors or flaws in an operation, procedure, or test.

HUMAN ERROR CASES

It is instructive to examine the types of error that occur with great frequency in incident reports, research studies, and malpractice claims. Of special interest are those errors that could have been prevented had consideration been given to the design of equipment, procedures, and training.

Error in the Administration of Medications

On June 23, 1990, the *New York Times* reported the death of three infants in a pharmacy mix-up at Philadelphia's Albert Einstein Medical Center ("Three Infants Die," 1990). Giving medication to the wrong patient or to the right patient in an incorrect dosage or at the wrong time is commonplace in hospitals, nursing homes, and other health-care settings (Cardinale, 1990; Carey, 1989; Clayton, 1987; Cushing, 1986; Karp, 1988; Robinson, 1987).

Clayton (1987) and others have suggested various ways to reduce medication errors, but it is clear that better data are needed to identify probable causes, frequency, site, time, location, and other factors before priorities can be set and comprehensive measures taken to deal with the problem. The many contributing factors, such as excessive nursing or pharmacy staff workloads, poor labeling, illegible prescriptions, improper storage, inventory and control procedures, poor lighting, and other likely causal factors also must be explored systematically.

Errors in the administration of medication are at least as likely to occur in home health-care settings as in hospitals and clinics. Considering that individuals receiving home care may range from infants to the very old and may be affected by language, visual, hearing, and physical disabilities, it is surprising that research on this topic is only infrequently reported in the literature.

Errors in Image Interpretation

The misreading of sonograms, X-ray, CAT scan, electrocardiograms, and other images is a frequent cause of diagnostic errors (Hamer, Morlock, Foley, & Ros, 1987; Lane, 1990; Morrison & Swann, 1990; Vincent, Driscoll, Audley, & Grant, 1988). The problem is often associated with inexperienced, junior doctors (Morrison & Swann, 1990; Vincent et al., 1988).

Research is being conducted to explore aspects of the interpretation process that may shape image interpretation performance (Metz & Shen, 1992; Rolland, Barrett, & Seeley, 1991). Signal detection theory, a theory of judgment as to what is seen or not seen based on the observer's expectations, has been used to evaluate diagnostic systems (Swets 1992; Swets & Pickett, 1983).

Errors in Medical Technology Use

As medical technology has become more sophisticated and complex, errors in its operation have increased (Cooper, Newbower, & Kitz, 1984). The problems are the result of many factors. Devices have not been standardized; they look and work differently. Equipment has not been designed in accordance with human-factors design practices; it does not work the way users expect it to work. Procedures for operating and maintaining equipment are often incomplete and difficult to comprehend. Users are not given adequate, task-specific training and opportunities for supervised practice.

Problems such as these have been encountered previously in other systems, but the lessons learned in solving them have not transferred to the health-care field. For example, many systems employ audible annunciator alarms to signal operators of equipment or process problems. These alarms, which can be loud and incessant, can become so annoying that operators disable them or diminish the volume so it is inaudible except locally ("Tracking Medical Errors," 1992). Similar cases of alarm silencing led to power plant and railroad accidents.

Fortunately, the lessons learned from other systems have, in some cases, begun to stimulate efforts to improve the usability of medical devices. For example, Shepard and Brown (1992) categorize device-related failures according to their causes and consequences and from this gained insights into how these device-related problems might be prevented. Smith and colleagues (1992) did an ergonomic analysis (i.e., an analysis of task-related movement patterns) of work stations in a trauma resuscitation room. The findings were used to improve the conditions of work in this crowded, stressful environment.

An analysis of medical device failures by Nobel (1991) found that over half of the alleged failures of medical devices were due to operator, maintenance, service, and actions taken by patients. Nobel concluded that, in addition to lack of training and not following instructions, "deficiencies in the man–machine interface are far from solved for many types of devices" (p. 121).

Errors in Patient Billing

Equifax, a national auditing firm, found errors in 97% of a sample of 13,000 audited hospital bills. One 31-page bill had 91 entries labeled "partial fills" and 18 labeled "pharmacy" (Merken, 1989). No explanation of either type of entry was given on the bill and traceability was impossible. The total bill of $38,000 was for ulcer surgery. Many of the billing mistakes were the result of simple keyboard entry errors. This is a problem found wherever aids to keyboard data entry and display have not been incorporated into design.

Similar reports of errors and patient confusion over hospital bills make frequent news headlines (Crenshaw, 1990; Rosenthal, 1993; Sloane, 1992).

Errors in Laboratory Testing

Health status tests have become a routine part of preventive and clinical medicine. But are they reliable?

New York State laboratory regulators found testing errors in 66% of the laboratories that offered drug-screening services (Squires, 1990). In another study of the reliability of commercial laboratory tests, 5,000 of the nation's top laboratories measured the amount of cholesterol in samples that contained a known cholesterol value of 262.6 milligrams per liter ("Three Infants Die," 1990). The labs submitted reports with cholesterol values ranging from 101 to 524 milligrams.

Errors in Radiotherapy

Case histories in the use of nuclear materials for patient therapy illustrate the opportunities that exist for error in still another segment of health care. Many such incidents are available in the annals of the U.S. Nuclear Regulatory Commission (USNRC), which has regulatory authority over radioactive materials. Two cases obtained by the author from USNRC records illuminate the severity of the problem (U.S. Nuclear Regulatory Commission, 1991).

In one case, a patient was administered 50 millicuries rather than the 3 millicuries prescribed by the physician. In a second case, the wrong patient was administered 100 rad to the brain. Two elderly patients had been brought to the teletherapy room at the same time, and despite the fact that photographs of patients were routinely taken for use in verifying patient identity, the wrong individual was selected for the treatment.

Other Cases of Error

A recent search of the literature for the period 1989–1992 on human error in health care found additional examples:

- Numerous instances in which illegible handwriting caused errors in patient treatment.
- Many cases of errors in the administration of anesthesia.
- Failures to complete patient record forms accurately.
- Mismatching of patient blood types.
- Failures to operate diagnostic devices and other equipment correctly.
- Many errors attributed to "mismanagement" and diagnosis.

Two of the most frequent error-occurrence sites are emergency and operating rooms. Other common sites are clinics, pharmacies, laboratories, and homes. Errors are made by teams as well as individuals. Some errors are to due to failures in understanding speech or written communications that could have been prevented with appropriate training.

The probability of error tends to increase with high workloads and long or rotating work shifts. Stress and fatigue also degrade the performance of medical students, interns, and other hospital staff.

THE NATURE OF HUMAN ERROR

Conventional Wisdom

A lack of scientific knowledge has never prevented popular wisdom from forming on any topic. Human error and its causes are no exception. Conventional wisdom holds the view that human beings are intrinsically unreliable. From this it follows that when something goes wrong, someone must have erred or goofed. It is assumed that mistakes such as these are the result of inattention, laziness, carelessness, and negligence, all manifestations based on the presumption that humans are innately unreliable.

The remedies invoked by stakeholders in popular wisdom to prevent further lapses into unreliability include finding the culprit, assigning blame, and acting to correct future misdeeds. Corrective actions consist of warnings to be more careful in the future, a formal reprimand, additional training, and sanctions such as loss of wages or benefits. Repeat offenders are suspended, sometimes terminated. Whatever corrective action is taken, it is usually initiated shortly after the event. This tendency toward rapid action reduces the organizational frustration and embarrassment that follow an accident.

The cry for "Somebody do something, anything!" may reduce stress, but it also leads to ill-considered and inappropriate actions.

Admiral Hyman Rickover, former head of the Nuclear Navy, used to say "All errors are human" in the choice of materials, in design, in procedures, or in training. Had he thought further he might have realized that the solution was not to indict people but their error-inducing designs, procedures, and training.

Adherence to a philosophy of conventional wisdom in the management of error represents a failure to grasp a fundamental truth. That truth is that the errors that people make are often traceable to extrinsic factors that set the individual up to fail rather than to intrinsic reasons such as forgetting or inattention. Recognizing this truth was one of the most important lessons learned by the nuclear industry as it went about its investigation of the causes of Three Mile Island, Davis Besse, and other U.S. nuclear power plant accidents. Whenever human error is suspected, whether the system is a power plant or a hospital or clinic, it is always sound policy to trace the error to its root causes. To assign blame to an individual who makes an error is no assurance that the same error will not be made again by a different person.

Another axiom of conventional wisdom is that people are intrinsically unreliable. Given that premise, it would also follow that the entire problem of human error could be settled and solved simply by replacing humans with automation. Proponents of this approach have an abiding faith in the ability of engineering ingenuity and automation to save humanity from human error. This is a questionable assumption at best, given the rigid inability of automated systems to cope with infrequent events and variable environments.

Scientific Perspective on Error

Research on human error was regarded by the behavioral psychology that dominated the field from 1912 to 1970 as a phenomenon that involved mental processes that are inaccessible to legitimate scientific scrutiny. Two events began to lower this barrier in the mid-1970s. First, a new cognitive psychology developed scientifically acceptable ways to study human cognition. Today, a respectable body of theory and research provides a good understanding of the mental mechanisms that underlie human performance. Second, a growing wave of industrial disasters in which human error was a prominent factor stimulated industry and government to support research on human error and to develop practical policies and procedures for error management.

Several lessons have been learned from the past 15 years of human error theory, research, and application. First, academic laboratories have provided us with a good understanding of the basic nature of human error (Reason, 1990). Errors are rare events compared with successes or correct actions. The forms that error can take are limited: Errors appear in many different contexts, but only a few behavioral mechanisms seem to be responsible for all of them. Comparable forms of error are found in speech, perception, decision making, problem solving, and action. Errors are less likely to occur when tasks are skill based or automatic, like driving, than when tasks call for a rule or procedure to be followed. Tasks that call for the use of knowledge to solve new problems are the most vulnerable to human error. This is because knowledge, such as a principle, has to be translated into an appropriate course of concrete action. Errors can occur in this translation process.

Another lesson learned is that as a stand-alone system, the individual—unencumbered by technology of any sort—is remarkably reliable. Everyone makes mistakes and slips that cannot be attributed to any extrinsic cause. Failures to act, taking wrong actions, and lapses of memory are common examples of intrinsic errors. These errors can often be prevented or their consequent effects minimized by applying an understanding of cognitive psychology to the way in which work procedures and equipment are designed. When an intrinsic error could lead to serious consequences, it is a sound human-factors principle to minimize its consequences by error-absorbing designs that will buffer the consequence.

Perhaps the best evidence for the intrinsic reliability of the human species comes from longevity and accident statistics. People live a long time—many over 90 years—and perform a seemingly infinite number of tasks in uncertain and often hostile environments. Despite this, injury and accidental death rates are remarkably low. Contrary to media hype, there are no machines that can duplicate human performance with the versatility and reliability achievable by humans.

It is when people and technologies interact that problems arise. Many human errors are triggered by the technology, its environment, and the conditions, conventions, and procedures for its use. People can be "set up" to make errors by the failure of designers, managers, and planners to take into account what is now known about human behavior and how to apply that knowledge to error-reducing designs.

As Casey (1993) pointed out in the preface of his book, *Set Phasers on Stun and Other True Tales of Design, Technology and Human Error*, "New technologies will succeed or fail based on our ability to minimize these incompatibilities between the characteristics of people and the characteristics of the things we create and use" (p. 9).

People in health care have long been aware that human error is a problem in medicine. *Iatrogenics*, coming from the Greek words *iatro* and *genics*, means "doctor-caused." In recent years, the medical community has become acutely conscious of iatrogenic or human-factors problems because of an increase in the frequency of what appear to be systematic human errors in human–machine interactions, use of procedures, and other health-care activities (Gardner-Bonneau, 1993). We can expect to see the partnership formed between medical and human-factors researchers nearly a decade ago to lead to important advances in the future in the understanding of the causes and management of human error in health care.

SUMMARY AND CONCLUSIONS

Human error has become a serious problem in today's complex, high-technology world. It is especially serious in health-care delivery because of the massive size, complexity, and character of that system. Fortunately, many of the things that trigger or initiate human error can be changed, and many of the changes that can be made are within the power of organizational management to implement. Work environments can be improved. Training can be strengthened. Workplaces, instruments, and equipment can be designed according to accepted human-factors design criteria. Work-rest and shift-change cycles can be designed in accordance with current knowledge about the effects of circadian rhythm disturbances on performance. Workers can be assigned to jobs based on a systematic assessment of the match between their skills and abilities and the demands of those jobs. These and other corrective actions can and do reduce human error. Such actions have been invoked successfully in the past decade on many systems, ranging from nuclear power plants and aircraft design to industrial equipment and consumer products.

REFERENCES

3 infants die in hospital pharmacy mix-up. (1990). *New York Times,* p. 9.
Brennan, T. A., Leape, L., Laird, N. M., Hebert, L., Russell, L., Lawthers, A. G., Newhouse, J. P., Weiler, P. C., & Hiatt, H. H. (1991). Incidence of adverse events and negligence in hospitalized patients. *New England Journal of Medicine, 324,* 370–376.

Cardinale, V. (1990). Medication errors linked to heavy pharmacy workloads. *Drug Topics*, *134*, 12.

Carey, C. (1989). Drug label mix-ups lead to recalls, lawsuits. *FDA Consumer*, *23*, 114.

Casey, S. (1993). *Set phasers on stun and other true tales of design, technology, and human error*. Santa Barbara, CA: Aegean.

Clayton, M. (1987). The right way to prevent medication errors. *RN*, *50*, 30.

Cooper, J. B., Newbower, R. S., & Kitz, R. J. (1984). An analysis of major errors and equipment failures in anesthesia management: Considerations for prevention and detection. *Anesthesiology*, *60*, 34–41.

Crenshaw, A. B. (1990, July 19). Decoding complex bills. *New York Times*.

Cushing, M. (1986). Drug errors can be bitter pills. *American Journal of Nursing*, *86*, 895–899.

Gardner-Bonneau, D. J. (1993, July). What is iatrogenics, and why don't ergonomists know? *Ergonomics in Design*, 18–20.

Hamer, M. M., Morlock, F., Foley, H. T., & Ros, P. R. (1987). Medical malpractice in diagnostic radiology: Claims, compensation, and patient injury. *Radiology*, *164*, 263–266.

Karp, D. (1988). Your medication errors can become malpractice traps. *Medical Economics*, *65*, 79–85.

Lane, D. (1990). "A medical error could cost me my life." *Redbook*, *175*, p. 64.

Leape, L., Brennan, T. A., Laird, N. M., Lawthers, A. G., Localio, A. R., Barnes, B. A., Hebert, L., Newhouse, J. P., Weiler, P. C., & Hiatt, H. H. (1991). The nature of adverse events in hospitalized patients. *New England Journal of Medicine*, *324*, 377–384.

Merken, G. (1989, May 10). Decoding hospital bills can make you sick. *Wall Street Journal*.

Metz, C. E., & Shen, J. H. (1992). *Medical decision making*, *12*(1), 60–75.

Morrison, W. G., & Swann, I J. (1990). Electrocardiograph interpretation by junior doctors. *Archives of Emergency Medicine*, *7*, 108–110.

National Research Council. (1990). *Distributed decision making: report of a workshop*. Washington, DC: National Academy Press.

Nobel, J. L. (1991). Medical device failures and adverse effects. *Pediatric Emergency Care*, *7*, 120–123.

Reason, J. (1990). *Human error*. New York: Cambridge University Press.

Robinson, B. (1987). Pharmacists stand watch against prescribing errors. *Drug Topics*, *131*, 40–43.

Rogovin, M. (1975). *Report of the President's Commission at Three Mile Island*. Washington, DC: U.S. Government Printing Office.

Rolland, J. P., Barrett, H. H., & Seeley, G. W. (1991). Ideal versus human observer for long-tailed point spread functions: Does deconvolution help? *Physiology, Medicine, & Biology*, *36*(8), 1091–1109.

Rosenthal, E. (1993, January 27). Confusion and errors are rife in hospital bills. *New York Times*, p. B7.

Russell, C. (1992, February 18). Human error: Avoidable mistakes kill 100,000 patients a year. *Washington Post*, p. WH7.

Shepard, M., & Brown, R. (1992). Utilizing a systems approach to categorize device-related failures and define user and operator errors. *Biomedical Instrumentation Technology*, *26*, 461–475.

Sloane, L. (1992, November 28). Checking bills carefully to catch costly errors. *New York Times.*

Smith, H., Macintosh, P., Sverrisdottir, A., & Robertson, C. (1992). The ergonomic analysis of a trauma resuscitation room. *Health Bulletin, 50,* 252–258.

Squires, S. (1990, March 6). Cholesterol guessing games. *Washington Post.*

Swets, J. (1992) The science of choosing the right decision threshold for high-stakes diagnostics. *American Psychologist, 47,* 522–532.

Swets, J., & Pickett, R. (1983). *Evaluation of diagnostic systems.* New York: Academic Press.

Tracking medical errors, from humans to machines. (1992 March 31). *New York Times,* p. 81.

U.S. Nuclear Regulatory Commission. (1985). *Loss of main and auxiliary feedwater event at the Davis Besse plant on June 9, 1985* (NUREG-1154). Washington, DC: Author.

U.S. Nuclear Regulatory Commission (1991, May) *NMSS Licensee Newsletter.* Washington, DC: Author.

Vincent, C. A., Driscoll, P. A., Audley, R. A., & Grant, D. S. (1988). Accuracy of detection of radiographic abnormalities by junior doctors. *Archives of Emergency Medicine, 5,* 101–109.

5 Error Reduction as a Systems Problem

Neville Moray
University of Illinois at Urbana-Champaign

As other chapters in this book make clear, error is rife in medical systems, many errors are hazardous, and it is clearly desirable to reduce their occurrence and impact. In any industry, organization, or setting where error is common, there are many ways in which error reduction may be approached. One approach assumes that if people are more careful, pay more attention, and in general take more trouble over what they are doing, then errors can be reduced and their effects mitigated. This approach tends to put great emphasis on the psychology of the individual who makes the error and on training, admonition, supervision, and ever-tighter and more detailed rules, with an implication of blame attached to those who make errors. In contrast to this is an approach that sees relatively few errors as being the fault of the human who commits them, and even fewer as being blameworthy. Rather, one sees the *design* of objects, activities, procedures, and patterns of behavior as being the source of errors. This approach, epitomized by writers such as Norman (1981, 1988) and Reason (1990), emphasizes that people of good intention, skilled and experienced, may nonetheless be forced to commit errors by the way in which the design of their environment calls forth their behavior. One need not deny that people make errors because of fatigue, carelessness, or lack of training in order to espouse this approach to error. However, the fundamental claim is that *the systems of which humans are a part call forth errors from humans*, not the other way around. Only as an attribution of last resort, or because of the tendency of a legal system to be less concerned with justice than with economics, does one ascribe blame to an individual who commits an error.

This chapter begins by noting that complex systems such as health-care delivery are composed of a series of hierarchically organized subsystems. These include equipment, individuals, teams and groups, and organizations. The way in which the different levels may cause errors is described from a psychological standpoint. These causes are then discussed in the more general context of a system of constraints, and strategies to reduce errors are seen as a search for ways to alter constraints. Finally, the question is raised whether the psychology of the individual is the best point at which to try to reduce error, or whether a different philosophy is needed, based on the assumption that it is never possible to reduce the probability of error to zero. The chapter concludes with an argument for the use of an interdisciplinary approach to error management within the context of a chosen social philosophy.

There are many cases in which it is easy to observe how a system elicits human error. When North Americans drive a car for the first time in England or Japan, the habits of a lifetime tend to make drivers behave in stereotypical ways. They tend to glance to the right when looking for the image in the rear view mirror, turn to the right to move closer to the near side verge and expect to be overtaken on the left. All these behaviors are incorrect in those countries and predispose the well-intentioned driver to make errors. When people buy a kitchen range in which the relations between the positions of the burners and controls are other than those they have learned habitually to expect, the probability that they will turn on the wrong burner can be increased by a factor of 10 or more (Grandjean, 1980). Many similar examples will be found in Norman (1988).

In the context of medical error, it is easy to find similar cases. The labels on bottles may be very similar, the shapes of bottles may be confusing, the controls on equipment may violate expected stereotypes, or there may be a lack of labeling on connections in equipment that allows potentially lethal interconnections to be made. The layout of operating rooms may require wires to run across the floor in ways that make it likely that people will trip over them or accidentally pull them out of the socket. Notoriously, the handwriting of physicians may verge on the illegible, increasing the likelihood of error in interpreting orders or filling prescriptions.

Errors that arise from such sources are not caused by a lack of good intent, nor commonly from carelessness, and in many cases the nature of the psychological mechanisms that underlie them are well understood. The existence of a phenomenon such as the speed–accuracy trade-off, in which forcing people to act rapidly increases the probability of error, is well known. The very strong effect of stimulus–response stereotypes, strong expectations in a culture about the directions in which knobs turn, switches move, and so on is well documented. The relation between the architectural layout of a room, the equipment therein, and the likelihood that their combination will make people stumble, trip over cables, or otherwise behave clumsily is clear. There are strong models for the effect of display parameters on the probability of correct identification of displayed information.

In all of these cases, it is relatively easy to say how the design of the object or system should be altered to reduce error. Indeed, one can go further. Enough is now known about the relation between ergonomic design and the way that error is caused by ignoring that knowledge to make it certain that almost all errors involving these aspects of a system should be laid directly at the door of designers and manufacturers. It is trivially easy to come by the relevant information that is needed to reduce error in the design of equipment to be used by humans. This is the field of the *ergonomics* of equipment design. Anyone who manufactures or installs equipment that violates the published data on these matters is directly responsible for the vast majority of the errors that arise in its use. Errors arising from the violations of well-known design rules are the responsibility of the designer, not of the user.

On the other hand, one of the major discoveries of recent years has been that even when all that is known about ergonomics is applied to design, the probability of error cannot be reduced to zero. Moreover, further research of this type will never reduce error to zero. There are factors at work in a complex human–machine system that have far greater potency for causing errors than do ergonomic factors. Such factors lead to the notion of *systems design* rather than equipment design and into an area that has been relatively poorly studied.

The most that can be done in this chapter is to indicate some of the factors that are relevant, and what kinds of approaches will tend to produce safer systems. Few designers have been trained to take such factors into account and, more importantly, they *cannot* be taken into account by designers when design is done piecewise. This is usually the case in medical systems. The invention of a new piece of equipment, its realization in terms of physical hardware, and its purchase and installation are all done independently and separately from the design of a system, such as a hospital or other health-care delivery system, as a whole. The situation is little different in practice whether the entire hospital or health-care system is regarded as a single system, or whether some parts are regarded as subsystems of a larger whole. The particular opportunities for error will be different, but the interaction of system and component will be very similar.

TOWARD A SYSTEMS APPROACH

Figure 5.1 is a representation of the causal structure of a complex hierarchical human–machine system based on an analysis by Gaines and Moray (1985). It is quite general, so there are few systems that cannot be mapped

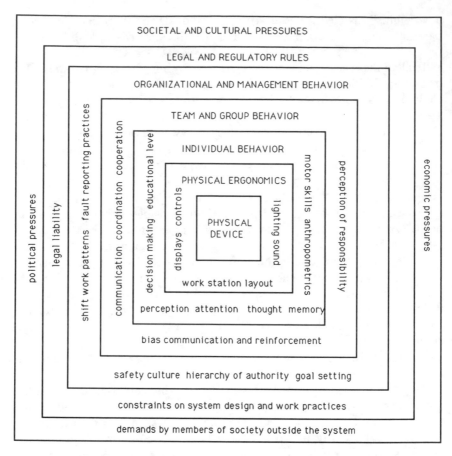

FIG. 5.1.   A generic hierarchical systems oriented approach to design and analysis. The terms in upper case define levels of description. The terms in lower case describe typical variables relevant to each level of description.

onto such a diagram. Note that the term "human–machine system" is interpreted in a far wider sense than is usual. It is commonplace to use the term to talk about such things as a vehicle, an aircraft, or even a factory or power station where the relation between the human and the machine is what engineers call *tightly coupled*. A human–machine system is usually thought of as something where there is a piece of equipment that the operator controls directly or through the use of an automatic controller.

It is not usual to think of a bureaucratic organization as a human–machine system, but it is a goal of this chapter to extend the term to include all such systems. A *system* , in this sense, is any collection of components and the relations between them, whether the components are human or not, when the components have been brought together for a well-defined goal or purpose. Thus, the management of a health-care system includes human components, such as doctors, nurses, and managers; hardware components such as computers and telephones that transmit and store information, paper and magnetic records, drugs, operating theaters, scalpels, and beds; the management *policies* that are adopted; and the financial mechanisms in place to govern the economic control of the system.

Only when the entire *system* is designed correctly will error be minimized. The components of the system must not be merely correctly designed and chosen, but the relations between the components must also be part of the design, as must the rules for its operation. If, for example, standard operating procedures are written without reference to the particular choice and layout of equipment, without reference to the training or social organization of the users, and without reference to maintenance practices and manning levels, then the system will be accident prone (as the accident at Three Mile Island proved). These aspects of system design must be integrated. Error will even then not be eliminated. It will, however, be reduced, and the effects of errors rendered more manageable.

Figure 5.1 is designed so that the causes and effects of errors and the steps needed to reduce and control them are most local in scope in the inner regions of the diagram. As we go outward, the scope of causal variables becomes increasingly global. Local causes of error are usually more readily manageable than global causes, but there is a sense in which global causes, if they can be controlled, have greater payoffs, because the effect of action at a global level will be far more pervasive throughout the system than will local intervention.

Physical Ergonomics

At the center of the system lies the physical design of equipment and the immediate work environment. In the context of health-care systems, this includes such things as:

- The legibility of labels on bottles.
- The confusability of bottles due to similar shapes and colors.
- Noise levels in working environments that may cause messages to be misunderstood.
- The position of displays and controls on equipment that may make it difficult for the health-care provider to read the former while operating the latter.

- The height of beds.
- The position of call buttons for the use of patients.

Anthropometrics, by which the dimensions of equipment are matched to the anatomical dimensions of the users, should also be included. Particularly in an ethnically heterogeneous society such as the United States, anthropometric considerations can be critical. Differing heights and arm lengths can mean a difference of as much as 30 cm in positions that can be reached by individuals sitting at a given console. Similar problems are the spacing of controls and the size of keyboards and buttons. There are also more subtle problems. It is common to make use of color coding in displays and controls, and increasingly, as computer graphics displays are built into automated or semi-automated equipment, colored displays are being used. But the naive use of color is not necessarily helpful, and a substantial proportion of people, particularly males, are color-blind to a greater or lesser degree. When color coding is used in a display, it is essential that color-blindness be considered and some other dimension, such as shape or brightness, be used to support the encoding of information in displays. It is left as an exercise to the reader to think of other such problems in ergonomics.

Almost every problem at this level of design can be solved with existing data. Books such as Grandjean (1980), Sanders and McCormick (1993), and Boff and Lincoln (1988); tools such as MANNEQUIN, available from Humancad (Melville, NY); databases such as ERGOBASE, available from Biomechanics Corporation of America (Deer Park, NY); or general texts such as that of Norman (1988) all provide suitable sources of information for the designer. It should be noted that the increasingly international nature of science and medicine can give rise to error prone systems. For example, the United States is one of a relatively small number of countries in which the position of a switch is coded so that UP = ON. The majority of countries that the writer has visited use DOWN = ON. Hence, equipment purchased from one country may have control and display conventions that violate the expectations of a user in another country—a potent source of error. Such a violation of expectation is usually easy to change by changing the physical orientation of a switch.

More difficult to change, and potentially more dangerous, is the fact that the color coding of electrical wiring differs widely between countries. It is not unknown for a technician unfamiliar with the coding conventions in one country to make a mistaken assumption when wiring a plug to carry main electrical supplies. (American readers may like to ask themselves how they would wire a cable that has one brown wire, one white wire and one yellow-and-green wire, a recent European standard.) But knowledge exists to avoid the vast majority of problems at the level of physical ergonomics of equipment design and workplace layout. There is no excuse for designers failing to make use of this knowledge. The appropriate use of existing data can solve most problems of local scope.

Individual Behavior

With the growing attention to human error in the last 20 years, there has been increasing study of the way in which errors arise at the level of individual behavior. General treatments of error in the context of the psychology of the individual will be found in Reason (1990), Norman (1981, 1988), Senders and Moray (1991), and Rasmussen and Batstone (1989). These are descriptions of the psychology of the individual, of single operators processing information and making decisions on their own.

An important distinction in the errors of individuals is between slips, mistakes, and violations (Norman, 1981; Reason, 1990). *Slips* are defined as errors where the person correctly assesses what needs to be done, and acts accordingly, but slips in carrying out the intention. A common example is typing errors, but the classical slip of a surgeon's knife is of course another, as is a slip of the pen that leads to the writing of an incorrect drug name or dose when the physician has formed a correct intention but fails to carry it out.

A *mistake*, by contrast, is an error in which the person fails to form the correct intention as to what act to perform, and so performs an action which, judged objectively, is incorrect. Note that the word *objectively* may be misleading. People may fail to decide on the *correct* action because the information they receive is inadequate or unclear. A decision based on the best information available may seem to be in error if the person taking it lacks training or experience as to what action is required. There may be a failure of memory as to what to do given the situation that has arisen. Although there may be a better solution available to someone who has perfect information and perfect training, it does not follow that the decision actually made is in error. That decision may be the best possible under the circumstances. Insofar as results fall short of what was required, it is a fault of the system for failing to support adequately the person required to make the decision.

A *violation* is usually defined as a deliberate choice to behave in a nonstandard way, such as when violating normal procedures. Violations may occur because a person believes that a nonstandard procedure provides a better chance of success than the standard procedure, or can in special cases be an example of deliberate malice. Violations can, however, also occur unconsciously if a strong pattern of behavior has been formed by habitual conscious violation so that the behavior ultimately becomes virtually automatic.

Violations and their place in the natural history of error are further discussed later in this chapter. For now, attention is focused on what slips and mistakes tell us about the origins of error. Note that even if all the design problems at the level of physical ergonomics were to be solved, slips and mistakes could still occur. Even with the best possible equipment, situations will occur where the information available is insufficient to lead to an unambiguous diagnosis of the situation, or where memory fails to guide the decision maker to choose the correct action.

Two major psychological factors affect the probability of error. The first factor is that people tend to avoid reasoning their way to solutions, and prefer to pattern match. When pattern matching, people decide that a present situation is identical to one that has occurred before and that it more or less resembles. The second mechanism is that given uncertainty as to what action to take, people will choose one that has worked before, and the more often they have successfully used a particular action, the more likely they are to choose it. These two mechanisms have been called *similarity matching* and *frequency gambling* (Reason, 1990). They appear to be pervasive, and are very strong causes of errors.

The psychology of similarity matching and frequency gambling is fairly well understood, and the implications are obvious. It is rather difficult for systems designers to find practical advice on how to counteract them, although again the standard texts mentioned earlier often contain some discussion. Appropriate counterstrategies can be based on an understanding of the psychological mechanisms involved. For example, we can think of an observer receiving information as being faced with the decision matrix shown in Fig. 5.2.

The target that is hit or missed may be an alarm, a shadow of a tumor, the name on a label, or in general anything, either perceptual or in memory, where a decision is required as to whether an event has really occurred. Usually misses and false alarms are thought of as errors, and there is a strong theory, the theory of signal detection, that describes what factors cause these errors. For example, a poor signal-to-noise ratio obviously limits detectability, reducing hits and true negatives. If an observer receives a signal that is ambiguous, whether a spoken message, a reading on an instrument, or an X-ray of a possible tumor, the ambiguity is resolved by the brain by taking into account the a priori probability of the event and the costs and payoffs associated with alternative outcomes.

All else being equal, an ambiguous event will be perceived as being the event with the highest subjective probability, and ambiguity will tend to be resolved in favor of an outcome that has the best expected payoff in terms of

STATE OF THE WORLD

event actually occurred   event did not occur

JUDGEMENT BY OBSERVER

event occurred

HIT TRUE POSITIVE CORRECT IDENTIFICATION	FALSE ALARM FALSE POSITIVE
MISS	TRUE NEGATIVE CORRECT REJECTION

event did not occur

FIG. 5.2.   Decision matrix for signal detection theory of decision making.

the costs and values involved. If a person expects many targets to occur, most of their responses will be in the upper row of Fig. 5.2, whereas if targets are expected to be rare, most of the responses will be in the lower row. If false alarms have unimportant consequences, and detections are valuable, then again, most of the responses will be in the upper row, and in the lower row for the opposite situation. Note that which row the observer favors is dependent on the subjective expectations and subjective estimation of payoff, not merely on the detectability of the signal. It must be emphasized that it is not a matter of the observer failing to take enough trouble over the decision. The content of perception is itself governed by probabilities and payoffs, not just the conscious decision as to what action to take (Swets & Pickett, 1982).

There is an intimate relation between correct and erroneous behavior in situations to which the theory of signal detection applies. For a given physical situation, the percentages of hits (correct diagnosis of the state of events) can only be increased if accompanied by an increase in false alarms, whereas false alarms can only be reduced by reducing hits unless the properties of the displayed information are altered. Thus, the probability of correct behavior can only be increased at the cost of errors. Suppose we are faced with a situation in which the observer is missing too many signals. The theory of signal detection states that the probability of hits can be increased by changing the observer's expectation or by changing the payoffs associated with each of the cells in the matrix. But if hits increase for a given signal-to-noise ratio, the false alarm ratio will also *inevitably* increase. (See Swets & Pickett, 1982, for a discussion in a medical context.) If a low false alarm rate together with a high hit rate is required, then the only way to reduce misses is to redesign the system so as to increase the signal-to-noise ratio. The lighting may be altered, a new instrument that is more sensitive may be introduced, the legibility of a display or a document may be increased, and so forth. Admonition, training, and the like will not alone bring about the desired ratio of hits to false alarms when the signal is weak.

Some errors are due not to decision making but to the properties of memory. If the nature of the errors is such as to suggest that fallible memory is the cause of mistakes, then steps will have to be taken in light of what is known about the psychology of memory. This may involve an increase in the quality and frequency of training, provision of some form of decision aid whether computerized or traditional hard copy, or providing the person with access to the assistance of others. (Note that this involves a move from the psychology of the individual to the psychology of teams and groups.)

Many aspects of attention and decision making are well understood. For example, a basic fact about visual attention is that people need to look directly at a display to read highly detailed information. The area of maximum acuity and color vision is only about 1.5° diameter, and eye movements do not occur more frequently than about 2 per second in most tasks outside the laboratory (Moray, 1984, 1986). Hence, the rate at which an observer can monitor a complex visual display is extremely limited. It is also known that it is very difficult for observers to remain alert for periods of more than 10 to 20 min when keeping watch for rare events. The interaction of attention with fatigue is known to produce a kind of tunnel vision.

Among the most important aspects of decision making, particularly in situations where an observer is trying to decide what has happened in an unusual situation, is confirmation bias, a tendency to select a likely hypothesis quickly and thereafter accept only evidence supporting that hypothesis. This applies not only to medical diagnosis, but to the diagnosis of almost all situations, from equipment malfunction to understanding unexpected changes in a balance sheet or unexpected changes in the pattern of work by a team or a group of individuals. (For important discussions of decision making, see Sage, 1981, and Klein, Orasanu, Calderwood, & Zsambok, 1993).

Errors are made by individuals. The properties of a system defined by PHYSICAL DEVICE and PHYSICAL ERGONOMICS in Fig. 5.1 determine such things as the number of eye movements required to carry out a task, the signal-to-noise ratio of a display, how well memory is supported, and so on. That is, there is a clear causal path outward from the core of Fig. 5.1 toward INDIVIDUAL BEHAVIOR. We shall now see that there are equally clear causal paths inward from the outer layers of the system.

Team and Group Behavior

Although errors are made by individuals, most work situations are such that a person is a member of a group or team either directly or indirectly. The distinction between teams and groups is a matter of the degree of formal allocation of roles. Where several people carry out a single task and each has a specific role, the collection of people is a team. Where the grouping is transitory and informal, it is a group. Thus passengers in a train or aircraft are a group, whereas the crew form a team. In the medical context, the professionals in an operating theater are a team, each with well-defined roles, whereas patients in a waiting room constitute a group.

A common problem and cause of error seems to be the way in which the structure of a team dissolves into an informal group under certain circumstances, undermining the formal patterns of authority and responsibility and losing the cohesion and mutual support that are characteristic of a well-integrated team. On the other hand, situations sometimes arise where individuals are thrown together unexpectedly to operate as a team. In such cases, the advantages of a coherent team are not likely to be found, and indeed, the constraint of a team setting may itself give rise to errors if those involved are unused to working together.

Human reliability is the obverse of the tendency of humans to make errors. Research on reliability has drawn attention to several problems in teams and groups. In order to make nonhuman hardware systems more reliable, it is common to make use of redundancy. For example, if the failure rate of a component in a hardware system is higher than can be tolerated, the system can be made more reliable by incorporating several of the same components in parallel. If it is known that the probability that a particular motor will malfunction is 0.01 per hr, then by having three of them in parallel, the probability that all three will fail at the same time is 0.000001. (This assumes, of course, that the causes of malfunction are random events.) Attempts to use multiple humans in this way will, in general, fail.

The social dynamics of a group of people performing the same task are unpredictable. If people believe that another will check on whether they have performed a task correctly, the accuracy of performance may increase or decrease. It may increase if the performer believes that the second person will check to see if the work has been done poorly and will penalize the performer for bad work or reward him or her for good work. The quality of work may well decrease if the relation between two people is such that each thinks that the other will catch any imperfections in the work without there being any penalty involved. Which result is seen will depend on the nature of the hierarchy of authority in the team. In general, it is extremely difficult to predict whether having two or more people check one another's work will improve performance, degrade it, or leave it unchanged.

In strongly hierarchical teams where a piece of work must be checked by several people, there is no guarantee that accuracy will improve. Frequently, such hierarchical checking becomes merely a formality, and signing off is little more than a signature to certify that each person has seen the prior person's signature. In a case discovered by the writer, a set of operating procedures had been signed off by more than four levels of authority, and no one had discovered that the original documents were completely flawed. This is likely to happen where the work to be checked is extensive, complicated, and highly technical. The higher levels of authority are often remote from the technical expertise needed to check the details of the system, and signatures are merely used to satisfy bureaucratic guidelines. This is particularly true where management is seen as a profession in itself devoid of technical expertise in the particular domain being managed, a common philosophy today where it is assumed that someone who can manage one organization can therefore manage any other without the need for technical knowledge.

In a strongly hierarchical organization, those higher up cannot have time to check in detail work performed by those low in the hierarchy, because the quantity of information far exceeds the ability of those high in a hierarchy to process it. Generally, it overwhelms even those low in the hierarchy. Hence the technique of signing off on quality checks is often largely a superstitious behavior performed mainly to provide an audit trail of apparent authority should legal action question the system's efficiency.

It is clear that the structure of authority in teams and groups is critically important in reducing error. If a team has a very strong hierarchy, then it will be difficult for juniors to question decisions made by those at a higher level of authority even when the latter make errors. Furthermore, there will be tendency for those low in the hierarchy to be afraid to show initiative. On the other hand, if the hierarchy fails due to the absence or other unavailability of a person in authority, then the team may well dissolve into a group in which no one has clear authority to make decisions or to take action. Where intelligent or automated hardware is incorporated as part of a team (a situation becoming increasingly common with the advent of expert systems and other computerized aids), there is a similar problem. How can people decide whether and how to query the decisions of the computerized aids? Are they to be regarded as authorities of last resort? If not, when should their advice be accepted and when rejected?

It seems reasonable to expect that in a team, the collective knowledge, skill, and wisdom will be greater than that available to any individual in the team. On the other hand, social dynamics often prevent such collective expertise from becoming available. The phenomenon of groupthink (Janis, 1972) is well established. In situations of uncertainty, members of a team or group will often tend to reinforce each other's assumptions. Where one might expect that the knowledge or skill needed to solve a problem must be more plentiful where there are more people, the ability to explore alternatives is actually reduced by the tendency of the group to come to a premature agreement that then insulates its members from further alternatives. During the accident at Three Mile Island, the team in the control room became unable to think of any explanation other than their initial hypothesis about the nature of the fault in the plant. At the end of the shift, a person who had not been involved in their discussions took a fresh look at the data and offered a completely different explanation, which was in fact the correct one (President's Commission Report, 1979). If collective behavior is to reduce error, ways must be found to reduce the tendency to groupthink and to free people from the tendency of hierarchical authority to bias the choice of action.

As discussed earlier, individuals tend to perceive and make judgments that are strongly determined by expectations about relative probabilities of events, costs and payoffs, similarities between current situations and past situations, and the success of certain actions in previous situations. All of these tendencies can be reinforced by the presence of other members of a team. There are interactions between personality styles and the degree to which people will accept authority or reject it. Social dynamics play a major role in the success with which a team can deal with unexpected events. Where the social structure of a team is well designed, where there is a fluent and free flow of information among its members, and where authoritarian hierarchical control is correctly designed (not necessarily minimized), exceptionally good and error-free performance from the group as a whole can be found. Where the social dynamics are not correctly designed, errors can go undetected, uncorrected, unobserved, and unreported. This can lead to catastrophic consequences and may hinder collective learning or improvement of performance over time. Particularly interesting is the work of people like Klein et al. (1993) and Rochlin, LaPorte, and Roberts (1987) on what makes teams effective.

The structure of authority in a team or group becomes critical when an abnormal situation occurs. This is particularly true when a beyond-design-basis event occurs whose possibility has not been foreseen. Such events frequently require a new pattern of organization and responsibility, but if the team is very tightly structured, no one will feel that they have the authority or responsibility to make radical changes in the way things are done. Teams need a balance between authoritarian coordination and flexible self-organization. It is only the latter that permits the evolution of new ways of behaving in situations for which no rules exist. On the other hand, flexibility, although supporting innovation, tends to lead to unpredictability, and the latter is often unacceptable to higher levels in an organization.

The behavior of the individual can thus be altered extensively by membership of a team or group. In some teams, individual ability may be repressed and constrained. In others, it may be nurtured and used in efficient cooperation where the strengths of one member make up for the weaknesses of another. What is clear is that how individuals will behave cannot be predicted without knowing the dynamics of the social setting within which they are working.

Organization and Management

The effects of organizational and managerial behavior affect the probability of error in yet more global ways. It is at these higher levels that policy decisions are made that indirectly but powerfully act downward to constrain the degrees of freedom in the behavior of teams and of individuals. Managerial influences can also act upward from the core of Fig. 5.1 by policies that determine the choice of equipment and the design of physical facilities.

A good example of organizational and managerial behavior that can have a major effect on the probability of error at the level of the individual is that of setting policy for shiftwork and hours of work. Just as there is a wealth of knowledge available to support the ergonomic design of equipment, so is there abundant evidence for the effect of shift work patterns, shift length, and circadian rhythms on error. It is known that errors greatly increase in the small hours of the morning, and often in the early hours of the afternoon. Errors in human information processing begin to increase significantly for shifts longer than 12 hours, and in physically demanding jobs, the errors often begin to increase at shorter intervals. Where shift work is used, certain patterns of shift rotation are less likely to cause errors than others (Tepas & Monk, 1987), and it is a managerial responsibility to decide what pattern of shift work will be implemented and how long working hours will be.

An obvious example of how managerial policies may increase or decrease error is to be seen in the long-standing problem of the hours worked by junior doctors in hospitals. The decision that junior physicians must work shifts that are frequently up to 24 hours in length is clearly a managerial policy. It is worth noting that because of the danger of increasing errors as time on shift increases, there is no other setting in hazardous industrial or military settings where people are permitted let alone required regularly to work the hours that are commonplace in hospitals.

Managerial decisions can have profound indirect effects through the choice of equipment. Decisions by management to invest in certain kinds of equipment can clearly cause major changes in the behavior of people who use the equipment. Insofar as managerial level members of an organization seldom have technical expertise or human-factors expertise, the control that they exercise over the quality of equipment, of operating instructions, of training in completely new technology, and so on is often at best remote. The opportunities for errors depend on the tasks given to members of an organization and the equipment with which they are provided. Those decisions are usually the province of management, if only because of the central role of financial constraints on what can be purchased. Although they have a direct effect by deciding what equipment to purchase, what shift schedules to implement, and so forth, they have little direct face-to- face communication with the health-care deliverers, and in that sense their effect is indirect and remote.

Just as one might expect that the collective knowledge and wisdom of a team should be greater than that of its individual members, so one might expect that the wisdom of an organization should be even greater. This will be true only if the organization is designed in such a way that it can, as whole, benefit from experience. Corporate memory seems to be extremely volatile unless very particular care is taken to enhance it. It is common to find that organizations are extremely rigid and unable to learn from past experience. Records of past errors and accidents are not kept, nor are they used to discover how to change the system. Frequently there is little significant change in practice following an error or accident, although those most closely involved are blamed and may be treated severely.

The ability of an organization to learn from the past errors of its members depends to a large extent on the attitudes and the managerial culture that are developed in the organization. If an organization is to learn, that is, to change its behavior as a result of past errors in a way that is reflected in the behavior of its members, then it must acknowledge errors that occur. If people report errors, those reports must be taken as information on the basis of which the organization can make constructive change. They must not be suppressed as undesirable "whistle blowing." The organization may need to accept the risks of making public the occurrence of errors. Errors should be seen, as they are in control theory, as signals for a needed change in practice. If, as commonly occurs, they are at all costs concealed, no learning will take place, and they will occur again. (An example of a system that permits "no fault" reporting of errors is the Aviation Safety Reporting System [ASRS] operated by the Federal Aviation Administration [FAA]. The ASRS allows pilots to report their errors anonymously so that the incidents can be recorded and analyzed without the pilot being blamed. These reports are fed back to the airline community through a publication that is available on a monthly basis, serving as a reminder of the kind of errors that even the best pilots may make.)

The whole notion of an organizational culture is linked to leadership at the managerial level. A special case is the notion of a *safety culture*. The attitude of members of an organization to safety, to corner cutting, or to violations is largely determined by managerial behavior. An open and flexible style of management can promote organizational learning so that individual violations of rules, which may occur because someone has found an objectively better way of performing some task, can become incorporated into the officially permitted or encouraged patterns of behavior. Likewise, an extreme rigidity of behavior can be imposed by authoritarian organizational structures, which will almost certainly render an organization unable to cope well with unforeseen events. Rigid rules do not necessarily make for a safer practice, especially in systems subject to many dynamic disturbances.

Legal and Societal Pressures

The outermost layers of Fig. 5.1 are causes of error that are usually remote from individuals but are still powerful. Behavioral options available to those working in a system may be tightly constrained by regulatory rules. Only certain drugs may be administered, only certain procedures undertaken. Violations of regulatory rules may have a heavy financial penalty attached. In the United States, the pattern of regulation in most industries has until recently been strongly prescriptive. Regulations prescribe what should be done under as many foreseeable situations as possible.

Except for the simplest systems, such an approach is doomed to failure. As systems and organizations become ever more complex, the number of possible events far exceeds the number of rules that can be thought up by regulators, and even where rules exist there are many problems. First, as the number of rules increases, it becomes increasingly difficult for people to learn, recall, and obey them all. Second, an individual or organization is often left in a very poor situation to deal with events unforeseen by the regulations—what is permissible in a situation for which there is no explicit regulation? Third, there is a grave danger that satisfying the regulations becomes an end in itself. The fact that the purpose of the rules is to regulate behavior with respect to some problem that affects the system is forgotten. People begin to feel that it is necessary and sufficient to carry out the rules as written, even though there may be far better ways to perform some task, and even when tasks exist for which no procedure has been specified. Treating rules as necessary and sufficient is the ultimate protection for the actor. No one can be blamed for following the rules, even when doing so causes undesirable consequences in a particular set of circumstances.

In addition to formal regulation, there is, particularly in the United

States, enormous pressure on behavior caused by the threat of legal action. In a highly litigious society, there is a fear of leaving undone what should be done and, equally, of doing things that are unnecessary. The legal system puts great pressure on an organization to make rules so that the organization will be protected by the rules. Rules allow an organization to pass blame down the hierarchy to individuals who break the rules, whether by accident or design. This in turn constrains behavior and can cause errors both of omission and commission.

Regulation and litigation are themselves driven by larger issues of social, cultural, and political pressures. Both decision making and overt behavior can be distorted by the requirements of society. A decision as to what treatment to use may be altered by knowledge of the economic situation of a patient. Pressures by shareholders for greater profits may affect decisions about work practices, manning levels, or the purchase of equipment. Cultural beliefs may render certain treatments unacceptable. At an individual level, the fears and hopes of a patient may exert strong pressure on a physician as to the choice of treatment. Union practices may predispose people to behave in certain ways, and antagonism to unions by management may be of equal influence.

ERRORS AND CONSTRAINTS

Consider again the statement that errors are made by individuals. This is often easy to see, as when people pick up a bottle next to the one that they wish to pick up, or misread a thermometer or blood pressure measuring instrument. At other levels, it is more difficult to attribute error to individuals, as when a board of management makes a collective decision about policy. But here, even if it makes sense to think of the collective decision as being an error of collective management, the decision is supported by the discussions, votes, and other activities of individuals, which in turn lead to the collective decision. The preceding sections of this chapter have discussed some ways in which influences from different levels of a system can impinge on an individual to cause or reduce error. These influences can be collectively described as the *constraints* that act on an individual to cause errors.

Constraints from Equipment

There are constraints on individuals' behavior from the equipment with which they work, acting from the center of Fig. 5.1 outwards. The equipment with which a person works is a filter that limits the information available for decisions and a filter that limits what actions can be taken. The accuracy of diagnosis, for example, depends very greatly on what measurements can be made. Computerized Axial Tomography (CAT) or Magnetic Resonance Imaging (MRI) provide data that are simply unavailable in the absence of the relevant equipment. The existence of computer databases and electronic communication alters the way in which decisions are made at all levels compared with an organization that uses only paper records.

Constraints from Physical Ergonomics

The size, shape, and legibility of displays, the positions and shapes of controls, the quality of alarms and communication subsystems, to name but a few physical properties of equipment, all constrain the way in which a person can acquire and use information and exercise control. These constraints force a person into certain patterns of thought and behavior, and those in turn can lead to a greater or lower probability of error. Even at the organizational level this is true. The quality of communication systems, the response time of computer databases, the time it takes to arrange meetings, and even the quality of air conditioning in a meeting room can have an effect on the quality of decisions made.

Constraints Within the Individual

The accuracy with which individuals acquire information and use it for decision and control is subject to the basic limitations of human psychology. These include the limits on the accuracy of perception, the volatility and accessibility of memory, the dynamics of attention, and the precision of motor skills. Other limitations arise from the complex interactions of emotion, motivation, judgment, and decision making. The human as an information processor is limited in rate and accuracy, and is biased by expectations and value systems. These limits impose powerful constraints on behavior.

Constraints in Teams and Groups

Constraints on the behavior of individuals arise from social dynamics within teams and groups. They include social pressure to conform by other members of the team, hierarchical patterns of authority and responsibility, habits of work that have grown up in a group, folklore about better ways of doing things that is passed from older members to new members, and tendencies toward groupthink. Other constraints are the quality of communication between members and whether a team is always composed of the same people or called together in an ad hoc way. Recently these factors have been emphasized by research into what is called *situation awareness*, the degree to which members of a team know from moment to moment what one another are doing and the extent to which they share a common understanding of what is happening.

Constraints from Organization and Management

Management exercises constraints in a great variety of ways. These include particularly the molding of organizational culture, which is deliberately or inadvertently developed with an emphasis on a particular combination of safety, profit and service, and the standards to which the members of the organization are expected to adhere. Patterns of shift work, available budgets, manning levels, response to the reporting of errors by oneself or others, openness to suggestions and innovation, even levels of pay and the pattern of bonuses for good work or sanctions for bad work all affect the way in which the individual members of the organization work, as well as the quality of their judgment and action.

Legal and Regulatory Rules

Legal and regulatory constraints play a very important role in determining behavior. The degrees of freedom available to an individual may be very tightly constrained by fear of litigation and by prescriptive regulatory rules. The first of these will lead to inherently conservative behavior, in which risky choices with a potential for better outcomes are rejected in favor of options that are believed to be fail-safe, not in the sense that they are necessarily best for the patient involved, but best in the sense that the person making the decision is safe from legal recourse even if things turn out badly. The second tends to promote rigid behavior, which cannot deal well with events unforeseen by the regulation and inhibits innovation.

Societal and Cultural Constraints

Society exerts pressures by its general expectations and even by pressure on individuals to adopt a particular philosophy of life. These in turn affect the options that will be considered by the individual, the choices that will be made, and the amount of risk that can be tolerated. For the health-care provider, such pressures may be in the form of demands from people outside the profession, or constraints from within the profession such as a limit on the numbers allowed to enter it, attitudes to the political structure or financial structure of the career, and so on. Questions of triage and rationing of care are intimately connected with social philosophy, whereas religious beliefs may prohibit the use of certain techniques or drugs.

DECISIONS, OUTCOMES, AND ERROR MANAGEMENT

One approach to error is to study the psychology of error. One can try to identify psychological mechanisms that cause error, but the list of constraints derived from Fig. 5.1 suggests that such an approach to error reduction is not likely to be successful.

What causes a person to record incorrectly the temperature of a patient? If the cause can be identified, then we can hope to make changes to the system so as to reduce the probability of such errors in the future. The cause of a misrecording may be due to a slip, in that the person writes down incorrectly a value that was correctly read, or it may be an error of perception when reading the thermometer, followed by correctly writing down the misperceived value. However, the cause of that misreading or miswriting may not be the perceptual or motor mechanisms of the individual. It may have arisen because the person is under great pressure to complete the task rapidly, having been urgently asked for a reading from a member of the health-care team. The person may have recently come from a meeting in which the statistical distribution of temperature in people with a certain illness was discussed, and the knowledge of the expected value of the reading from this kind of patient may predispose the observer to make the error. The display on the thermometer may have been degraded, making it difficult to read. If the data are to be recorded using a keyboard, there is a finite probability of typing errors, probably in the region of 10^{-3} to 10^{-4} per keystroke for average typists. Although the error emerges as the behavior of the individual, the cause may lie in one of the constraints listed here or from a variety of other causes. Hence, the most appropriate way to try to reduce the probability of error may not be to look at the psychology of the individual.

More generally, it is important to realize that all the constraints listed here interact. Even if the cause of *this* particular error were correctly identified and steps taken to prevent it, we can never be sure that no error will arise on the next occasion that the opportunity arises. *This* error may not recur, but in removing it, the system may have been changed in a way that makes another error more likely. (An obvious example is when computers are introduced: Errors of handwriting obviously decrease, whereas typing errors increase. Which has the higher probability, and under what circumstances?) A fundamental principle is that when we make any change in a system, that change propagates effects *throughout* the system and may cause many changes that were not foreseen. These may be at a different level in Fig. 5.1 from where the change was made.

The impact throughout the system of a change at a single level is most readily seen when changes occur at the outer levels of Fig. 5.1. Consider a change in regulatory philosophy from one where the regulations specify exact behavior to one where no specific behavior is specified, and the regulated organization is allowed to perform a task in any way it chooses, subject to satisfying the regulatory authority that the way chosen is safe. (This difference is the difference in regulatory philosophy, for example, between the United States' nuclear industry and that of Canada or the United Kingdom. The former spells out in great detail what must be done by all utilities under as many situations as possible. The latter allow utilities to do anything they can convince the regulators is safe.) The relaxation of the requirement to follow a specified pattern of behavior may have a great effect on how individuals perform their tasks, and hence one may expect a large change in the probability of error. That change may increase error or reduce it: It may increase in some parts of a system and decrease in another. The net effect on those who are served by the system may be difficult to predict. Similarly, if management changes the number of hours that a person must work, that change will have widespread effects throughout the system. These may include the level of fatigue experienced by individuals or the members of the group or team with whom they interact, the way in which handover at shift changes may alter, and changes in morale that may have important effects.

Even a change in the physical ergonomics of a piece of equipment may have more than a local effect. If a new kind of display is installed, the new symbology may be easier or harder to read. This will change the time required to carry out tasks, and may change the pattern of verbal communication between the person using the equipment and other members of the team. This in turn may change the length of time for which the patient is under the anesthetic and hence the number of operations performed each day, and so on. It is important to understand that when *any* element of a system is changed, the result is better described as a new system, not just as the old system with a change. Particularly in large complex systems, it is often very difficult, if not impossible, to foresee the results of quite minor changes because of the tendency for changes to propagate their effects throughout the system with varying time delays and unforeseen interactions (Ashby, 1956).

A lesson from this line of thought is that if one wants to reduce error and its consequences, the most productive approach may not be to ask for the particular psychological mechanism that caused the error. It may be more productive to try to change the system as a whole in such a way that the undesirable behavior or the undesirable outcome does not happen, *however it was caused.* When an error manifests itself, it will do so in the context of one of the constraints described earlier. It may seem to have been an error caused by a poorly designed piece of equipment, an inappropriate pattern of attention, a failure to communicate within a group, overlong hours of work, or an overly tight constraining regulatory rule. But it does not follow that the level of the constraint that seems to have caused the error is the same level at which one should try to prevent the recurrence of the error.

It is also important to distinguish clearly between errors and undesirable outcomes. The fact that a person's actions lead to an undesirable outcome does not mean that the person made a mistake. As we have already seen, there are many situations where, if a person follows a well-defined rule, the outcome may not be what is expected because there is some factor present that was not foreseen by the rule makers. Similarly, it can often be the case that a slip, mistake, or violation leads to a better outcome than would have the "correct" behavior as defined by a rule. In particular, violations are often ways to explore new and better solutions, and errors may also, albeit accidentally, lead to similar discoveries. Both in understanding how errors arise and in designing systems for their management, it is most important to keep the distinction between error and outcome distinct.

The systems approach to error management requires that possible solutions be considered at many levels of constraint. Suppose one were to find that infection is sometimes spread within a hospital because syringes are occasionally inadequately sterilized. One solution would be to change the rules that must be followed by personnel to ensure that sterilization is properly performed. More efficient training could be introduced. Better displays and alarms might be designed to ensure that the syringes were autoclaved for the requisite period of time. Personnel might be dismissed or fined for inadequate performance.

Those solutions are at the level of the individual, and assume that by acting to make perception, attention, memory, and skill more efficient, the errors could be reduced or eliminated. But a completely different solution is to change to disposable syringes, so that there is no need to perform sterilization at all. The first solution involves ergonomics and the psychology of individual behavior, and invokes constraints at the two innermost levels of Fig. 5.1. The change to disposable syringes is a solution at the level of organizational and managerial levels that acts indirectly through the innermost level by changing equipment. No blame or requirement to try harder is imposed on the personnel. (But note that with a change in practice the opportunity for *different* errors arises.)

Either of the suggested approaches could solve the problem. The choice between them might be constrained by economics rather than by an understanding of psychology. The important point is that each solution has its impact on the system in quite different ways. If the solution adopted still places responsibility on individual people to remember to perform tasks in a particular way, similar individual errors will probably recur in the future, even if team constraints in the form of supervisory checking were to be implemented. If the chosen solution alters the system so that there is no opportunity for the particular error to recur, then even though other errors may occur, that particular one will not.

Global or managerial solutions will not always best reduce error. There will be many cases where redesigning equipment or retraining personnel is the correct solution, and such redesign may be carried out by the user rather than the manufacturer. The point is that *no change has only local effects.* All changes propagate throughout the system. When trying to eliminate a particular error, the proposed solution and the consequences of that solution must themselves be examined in the context of the entire system. What constraints will be made stronger? What constraints will become irrelevant?

The problem of error management can be thought of as a search through the system for ways to constrain the possibility of particular errors without relaxing the constraints on others. Given an emphasis on the desirability of constraining outcomes, a detailed analysis of the psychological causes of error may not be required. Moreover, the search for solutions in the psychology of the individual may lead to solutions that will be extremely difficult to implement. How can one ensure that someone will always be alert when checking the work of another person? Almost always, it will be better to find constraints on behavior that remove the need for such checking. If the organizational culture favors safety and service rather than profit, how can that philosophy best be transmitted to newcomers? Rather than emphasizing an understanding of the deep psychological causes of error, it will be better to find ways to make safety look after itself. This does not imply that the psychological causes of error can be ignored, but that a complete systems approach to designing a health-care system is required in which the approaches from many disciplines are integrated.

CONCLUSION

Errors are made by individuals, but individuals work within systems. Systems are composed of people, things, information, and the relations and interaction among them. Systems can be analyzed and described in many ways, and the different ways emphasize different sources of constraint on the individuals who work in them. For any error that must be prevented, there will be many places in the system at which to intervene and many different ways of intervening. To use a classical metaphor, to concentrate too closely on the details of a particular error often leads to locking the stable door after the horse is stolen. To do so ensures that if there is another similar horse and a similar thief, we will be able to prevent the loss of the second horse. But in fact, in large, complex systems such as health-care delivery, there is an infinite number of horses and an infinite number of thieves. When any horse goes missing we should consider not merely locking doors, but rebuilding stables, retraining personnel—or even keeping animals other than horses.

The systems point of view emphasizes versatility in searching for solutions and shows that concentrating on ever-tighter local constraints will simply leave the system increasingly vulnerable to unforeseen events. It also emphasizes that the way to reduce error is to examine systems at all levels of constraint description. Although the process may be more time-consuming and difficult than looking for simple solutions at the level where the error is detected, it offers the hope of a greater measure of controlled adaptability, innovation, and versatility. Human error is not only a property of humans—it is a property of systems that include humans. In the end, the best way to prevent errors in giving medication may be as simple as changing the font of the print used in labels, or as complex as changing the purchasing strategies of the entire hospital. By using a systems approach with its potential for rich solutions, the number of errors can be reduced and their consequences mitigated.

Errors will always occur, and it is perhaps as well to reflect, particularly in a litigious society, that although the way in which death comes to each of us may be due to an error, death itself is not an error, but a result of life.

REFERENCES

Ashby, W. R. (1956). *An introduction to cybernetics.* London: Chapman and Hall.

Boff, K., & Lincoln, J. (1988). *Engineering data compendium.* New York: Wiley.

Gaines, B. R., & Moray, N. (1985.) *Development of performance measures for computer-based man-machine interfaces.* DCIEM-PER-FIN:JUL85, Downsview, Ontario.

Grandjean, E. (1980). *Fitting the task to the man.* London: Taylor and Francis.

Janis, I. L. (1972). *Victims of groupthink.* Boston: Houghton Mifflin.

Klein, G. A., Orasanu, J., Calderwood, R., & Zsambok, C. E. (1993). *Decision making in action: Models and methods.* Norwood, NJ: Ablex.

Moray, N. (1984). Attention to dynamic visual displays. In R. Parasuraman (Ed.), *Varieties of attention* (pp. 485–512). New York: Academic Press.

Moray, N. (1986). Monitoring behavior and supervisory control. In K. Boff, L. Kaufmann, & J. Beatty (Eds.), *Handbook of perception and human performance* (pp. 40-1–40-51). New York: Wiley.

Norman, D. A. (1981). Categorization of action slips. *Psychological Review, 88,* 1–55.

Norman, D. A. (1988). *The psychology of everyday things.* New York: Basic Books.

President's Commission Report on the Accident at Three Mile Island Unit 2. (1979). New York: Pergamon.

Rasmussen, J., Duncan, K., & Leplat, J. (1987). *New technology and human error.* Chichester, England: Wiley.

Rasmussen, J., & Batstone, R. (1989). Why do complex organizational systems fail? *Summary proceedings of a Cross-Disciplinary Workshop in "Safety Control and Risk Management."* Washington, DC: World Bank.

Reason, J. (1990). *Human error.* Cambridge, England: Cambridge University Press.

Rochlin, G. I., LaPorte, T. R., & Roberts, K. H. (1987). The self designing high-reliability organization: Aircraft carrier flight operations at sea. *Naval War College Review,* 76–90.

Sage, A. P. (1981). Behavioral and organizational considerations in the design of information systems and processes for planning and decision support. *IEEE Transactions on Systems, Man, and Cybernetics, 11,* 640–678.

Sanders, M., & McCormick, E. J. (1993). *Human factors in engineering and design* (7th ed.). New York: McGraw-Hill.

Senders, J. W., & Moray, N. (1991). *Human error: Cause, prediction and reduction.* Hillsdale, NJ: Lawrence Erlbaum Associates.

Swets, J. A., & Pickett, R. (1982). *Evaluation of diagnostic systems.* New York: Academic Press.

Tepas, D. I., & Monk, T. H. (1987). Work schedules. In G. Salvendy (Ed.), *Handbook of human factors* (pp. 819–843). New York: Wiley.

6 Misinterpreting Cognitive Decline in the Elderly: Blaming the Patient

Georgine Vroman
Ilene Cohen
Nancy Volkman
Geriatric Mental Health Clinic, Bellevue Hospital Center, New York

This chapter is based on experience and material derived from the Computer Memory (Cognitive Rehabilitation) Program of the Bellevue Hospital Center's Geriatric Mental Health Clinic, New York City. The program has been in existence since 1984.

The chapter focuses on the communication process between patients (and their support persons) on the one hand and their physicians (and other health-care personnel) on the other. The role of cognition, that system of interrelated mental processes on which this exchange of information depends, is analyzed. A detailed discussion follows of some common misperceptions and misinterpretations that can occur during this exchange. There is a listing of some of the underlying causes of these misunderstandings. Throughout, the chapter stresses the possibility that these human errors could seriously harm the patients' treatment and the doctor–patient relationship.

The central point of the chapter is that many of these potentially harmful misunderstandings can be prevented. This requires that all parties involved become actively engaged in the process of prevention. Not only the doctors and the rest of the health-care personnel have to make this commitment, but also the patients themselves and those who support them.

Although the material to be discussed is relevant to the treatment of all patients, the focus is on the elderly. This growing segment of the population in the United States is particularly vulnerable to misunderstandings that can occur between doctors and their patients.

HEALTH CARE AND THE ELDERLY

The encounter between patients and their doctors can be perceived as a cognitive event. It is a form of communication between patients and their clinicians during which an exchange of information takes place. This information can be verbal or nonverbal, and it may occur on a conscious or an unconscious level. This means that the information is not conveyed by words alone; it also means that the participants may not even be aware of certain exchanged bits of information that nevertheless become part of their interpretation of what has occurred.

On the one hand, the unfamiliar environment of the examining room may provide intimidating and unsettling clues to the patients. The same may be true for the impersonal rituals of the medical examination itself. On the other hand, doctors are likely to interpret the appearance and behavior of their patients on the basis of previous medical and personal experience, which may not necessarily apply in the individual case. For instance, doctors may unwittingly share certain prejudices and stereotypical ways of thinking that are held by the general population. The result could be an error in judgment about their patients' personal characteristics. Even the doctor's diagnosis and thus the patient's treatment may then be influenced.

The two sides of the doctor–patient exchange are unequal in status. This particular form of communication is always asymmetrical. There are several reasons for this asymmetry: Basically, the two participants perform opposite roles, that of the authority figure and of the supplicant, respectively. Further, they usually do not share the same socioeconomic and educational background and may not use the same vocabulary. In addition, they may not belong to the same ethnic, cultural, or religious groups, and each side could hold a different set of beliefs and values. A gender or age difference could further complicate matters.

The existence of these discrepancies may be insufficiently realized by either side. And even where this is not the case, the interpretation of the significance of these differences may not be the same. A mismatch in communication could occur in which either side misperceives or misinterprets the information provided by the other. In the relationship between doctor and patient, such misunderstandings could have serious consequences.

For many reasons, the chance of this type of error occurring is greater in the treatment of elderly patients than for younger ones. First, there is the inequality in status already mentioned, which may be more pronounced when complicated by a generation gap between doctor and patient. Then, elderly individuals are further at a disadvantage because they frequently suffer from sensory losses, specifically poor vision or hearing. Finally, there is the likelihood that their cognitive functions, such as attention, concentration, and memory have declined.

Physicians may misinterpret an elderly patient's temporary confusion in the unfamiliar and stressful surroundings of clinic or hospital ward as a sign of developing senility. Elderly patients, who may be hard of hearing, can become overwhelmed by the many simultaneous demands on their concentration and memory. Chances are that some elderly patients will not be able to understand their doctor's instructions clearly, let alone accurately remember and follow them.

In both instances, a human error has occurred that could have disastrous consequences: Doctors may misdiagnose their patients' cognitive state and consider them to be demented. Patients may not comply with their doctor's orders and therefore could not receive the treatment they need.

The problems caused by this unequal exchange are not limited to the primary therapeutic relationship between doctors and patients. Trouble may also occur in the contact of patients with health-care administrators and supporting personnel at many levels of authority. Members of the patients' families and of their other support groups, who traditionally mediate between patients and health-care givers, could get caught in an ambivalent position.

Some aspects of these asymmetric exchanges may well be hard to avoid and could be typical of the discourse between inherently unequal partners. This situation may even have some benefits, because it can define the therapeutic, or caregiving, relationship more clearly and make it less ambiguous. Other inequalities, however, can become counterproductive, even harmful, and most could be avoided.

Recent developments in the education and practical training of physicians and other health-care workers have helped foster a greater awareness of possible misunderstandings between themselves and their patients. Better skills to prevent such problems have also been promoted.

Also, all hospitals in the United States are now legally required to post copies in several languages of the *Patient's Bill of Rights* in public areas. Patients receive one copy and must sign it to indicate that its contents have been explained. Another copy can be retained for their own information. The patient's family or other support group also receives this information if they ask for it. Most hospitals demand that staff doctors are familiar with the contents of this document. Furthermore, most hospitals now routinely employ *patient advocates* who will assist patients who get into a conflict with their health-care providers or the hospital administration.

The Patient's Bill of Rights spells out the patients' rights to information about their diagnosis and treatment and about the personnel who treat them. Patients must give consent for any treatment or procedure and may refuse to accept these. If treatment is refused, the patients must be told about the medical consequences of such a refusal and be informed of alternative treatments. Further, patients can decide to participate in research or refuse to do so. The privacy and confidentiality of the patients and their treatment is guaranteed.

Finally, this Bill of Rights gives information on how to complain when patients are dissatisfied with their treatment—for instance, via the patient advocate. Procedures to appeal any decisions in matters of conflict are also spelled out.

The establishment of these rights has certainly helped in promoting a measure of balance in the relationship of patients and their health-care providers. It also provides a means to resolve any conflicts that might develop. However, not all patients may fully understand this document, even after an explanation, and some may not know how to use it properly. As a means of last resort for resolving patients' grievances against their health-care providers, there is, of course, the controversial malpractice suit.

When discussing patient treatment, it is important to distinguish between the terms *illness* and *disease*. Although the terms are commonly used interchangeably, disease should properly be reserved for a specific pathological condition, whereas illness defines the way the disease manifests itself in the individual patient. It is, in other words, the disease that causes the illness of a particular person.

These days, most physicians accept the fact that their patients' illnesses cannot be considered separately from those individuals' situations in society. Many patients themselves have become much better informed about disease and illness. They know more about available treatments and are not only more knowledgeable consumers of health care, but capable of taking an active part in their treatment.

This improved knowledge and the preparedness to participate actively in one's own treatment may be more the exception than the rule for elderly patients. This situation is linked with the generally more limited educational background and the greater social isolation that are typical for this age group. Elderly patients are also at greater risk to be suffering from cognitive decline. Further, misconceptions about cognitive impairments are quite common among elderly patients themselves and their caretakers, and even their health-care providers may share these opinions. There are, after all, still considerable gaps in the understanding of the normal aging processes.

Finally, the persistence of prejudice and stereotypical thinking about the elderly can turn suspicions into paranoia when old people are confronted with seemingly insensitive treatment. Some elderly people may react with depression or rebellion when faced with social or emotional losses and a loss of independence.

Clearly then, the care for the health and well-being of elderly patients represents a challenge that demands special attention.

THE NATURE OF COGNITION

This chapter's focus on the cognitive aspects of patient treatment requires a more detailed discussion of the nature of cognition itself.

Cognition can be conceived of as an information processing system consisting of three stages: INPUT (intake of information, including sensory perception) → PROCESSING (integration and storage of information in accessible form, including memory and planning) → OUTPUT (language, expression and action).

Although these three stages do succeed each other, there is no simple unilinear sequence. Each stage influences the two others, and feedback loops of information connect all three. There exists an interrelatedness between these stages that results in the whole system becoming affected when any of the stages is damaged.

Cognitive functions can be defined as all those processes that allow persons to obtain, integrate, and retain information about themselves, other people, and the environment and that enable them to share this information and act on it. In Fig. 6.1, the INPUT stage contains the cognitive function of perception, mediated by the sense organs. The PROCESSING stage contains the cognitive functions of memory, attention and concentration, logical thinking, problem solving, and planning. The OUTPUT stage contains language functions, expression, and action. Only the most important cognitive functions are mentioned.

Figure 6.1 further illustrates the interrelatedness and interdependence of the cognitive system's three stages by indicating the *feedback loops* that exist between them. It also shows that the exchange of information between the individual and the environment (and the latter includes both the *internal environment* of the human body and the *external, or social, environment*) takes place in both the INPUT and the OUTPUT stages. Wood (1990) stressed the importance of attention in integrating and mediating the exchange of information between the internal and external environment.

The three stages of the cognitive system interact as follows: The INPUT is determined by the quality and quantity of the incoming information. Thus, this stage depends on the status of the sense organs, the sensory perception, which permits the person to become aware of the available information; that is, the incoming information must be received. Without adequate attention and concentration, however, the incoming information may not be noticed. Further, the information must be recognized.

Attention and concentration are cognitive functions belonging to the PROCESSING stage, as is memory, a cognitive function (influenced by previous experience and education) that makes recognition possible. Rec-

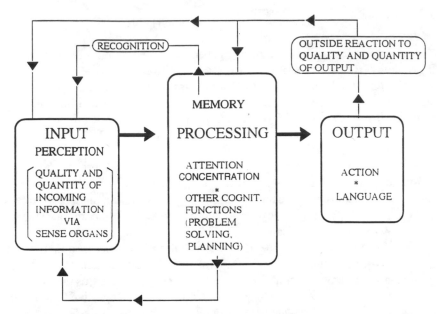

FIG. 6.1.   Diagram of the three stages of the cognitive process (heavy arrows) and the principal functions for each stage. Each stage influences the two others. The thin arrows indicate the feedback loops of information that demonstrate how the stages are interrelated.

ognition is needed in order to make sense of the incoming information. However, during this interpretation of incoming information, misconceptions can occur.

As De Bono (1970) argued so convincingly, the process of making sense of what is being observed is often ruled by the partly unconscious search for patterns and the tendency to favor the obvious and the familiar. On this basis, a person forms a set of premises that lead to further interpretations, and that will become the starting point of the analytical, step-by-step process of problem solving that follows. De Bono called this the "vertical thinking process" and acknowledged its essential, analytical nature. However, he warned that the results can only be as good as the premises from which this process derives. He advocated a retraining of the thought processes, especially perception, in "lateral thinking."

In lateral thinking, alternative ways of looking at available information are systematically tried out, and the more obvious and familiar explanations for what is being observed are routinely challenged. This way the premises themselves are being questioned, unproductive routines can be discarded, and creative approaches to the problems at hand might suggest themselves. Addressing problems as an interdisciplinary team should encourage lateral thinking, because each participating discipline has a chance to challenge the thinking patterns of the other disciplines.

The PROCESSING stage influences not only the INPUT stage but also the OUTPUT stage. Its planning and problem solving components lead to action, and its language-generating functions make verbal communication possible. Thus, the individual can interact with the external environment.

The OUTPUT stage itself feeds back information to both the PROCESSING stage and the INPUT stage as follows: The response from the outside world to the quality and quantity of a person's OUTPUT reflects the ways in which this person is being perceived by others.

Such a person will notice and interpret this reaction, which may then determine how future contact with the external environment will be handled. For instance, people may avoid as too burdensome any conversation with individuals who have a serious speech impediment, but whose language is otherwise intact. The poor speakers notice this avoidance and may interpret it as being judged mentally deficient and unworthy of the effort at communication. They may then withdraw as well, and more social isolation and a further deterioration of speech could be the result.

In summary, cognition is communication, determined by the quantitative and qualitative nature of one's cognitive status. This communication takes place at the interface of the individual and the physical and social environment. However, the capability to function well cognitively must go hand in hand with opportunity. Cognitive functions, as all functions, must be exercised and practiced in order to develop and remain at their optimum. Being able to interact socially must go together with a chance to practice this skill, or it may be lost. Therefore, any circumstance that isolates a person can harm that individual's cognitive functioning. Among these circumstances could be living alone, being physically incapacitated, or having trouble hearing or seeing. Lacking the means or the opportunity to meet other people, and this includes adequate transportation, falls within this category.

The next part of this chapter specifically addresses cognition in the elderly.

COGNITIVE FUNCTIONING IN THE ELDERLY

There are many reasons why older persons are particularly vulnerable to cognitive problems. Some have little connection with cognitive capabilities as such, and could be better described as conditions or situations that are typical for an aging population and that influence cognitive functioning indirectly.

For instance, physical or social isolation and a lack of mental or sensory stimulation can reduce both the quantity and quality of incoming information and the opportunity to exercise one's cognitive abilities. Older individuals are also more sensitive to medication, the side effects of which could include impaired cognitive functioning. There are circumstances that cause *cognitive overload,* where cognitive capabilities become overburdened and can no longer function as they would under normal circumstances. Further, dehydration and hypothermia, and also certain infections and metabolic abnormalities, can cause cognitive functions to deteriorate, as can a lower tolerance for alcohol with age. In many cases, these various conditions are amenable to intervention, and the effects, including the impaired cognitive functioning, can be reversed.

Still, no matter what the underlying cause, many elderly people themselves commonly assume that even minor cognitive lapses, routinely disregarded in younger persons, signify the onset of dementia, particularly Alzheimer's disease. Health-care personnel, caretakers, friends, and family who deal with the elderly frequently share these assumptions. What elderly people and others perceive as a failing memory can actually be a difficulty with attention and concentration and may require a different kind of treatment. (See also Wood & Fussey, 1990, and Sohlberg & Mateer, 1989, on the crucial importance of attention in cognitive functioning.) The significance of a patient's minor acts and omissions that are the result of diminished cognitive functioning are often exaggerated by others. Genuine cognitive impairment, on the other hand, is often perceived as deliberate.

To what extent cognitive decline is an inevitable part of normal aging is still not definitely and systematically determined. Most neuropsychological tests for cognitive functions have statistically established normal scores ranged by age group, usually in decades. But not all tests cover the same range: The much-used Wechsler test (Wechsler, 1981) gives normal scores up to age 74, whereas the Randt test does so up to age 90 (Randt, Brown, & Osborne, 1980). Such a difference in range renders some tests less suitable when working with the very old.

In the cognitive rehabilitation program at the Bellevue Hospital Geriatric Clinic, marked differences in performance were found for people in their 60s, 70s, 80s and 90s. For patients with normal test scores for their age group we found that, with increasing age, attention and concentration problems can become more pronounced and that there is more of a chance of cognitive overload. Also, as a person grows older, memory for recent information may suffer. Further, the ability to perform multiple-step tasks diminishes and the reaction time tends to get longer with increasing age. (See Willis, 1985, for a more detailed discussion.)

For many elderly persons, being able to drive allows them to maintain their independence. However, cognitive losses, some of them ignored or not fully realized, could affect driving ability. As early as after the age of 55, reduced reaction time and the need for better lighting for all activities may begin to influence driving skills. For most drivers of 75 and over, these skills will definitely have deteriorated. Medications, fatigue, arthritis, and poor vision and hearing may be at fault, but prolonged reaction time and a certain inflexibility in making appropriate, rapid decisions contribute to the problem (see also Malfetti & Winter, 1986, 1991).

Doctors who treat elderly patients should ask about driving habits and caution their patients about the risks associated with diminished cognitive functioning. These professionals must be held responsible if they fail to do so and accidents occur (Bedway, 1992). Several tests exist for assessing driving fitness. Gianutsos, Campbell, Beattie, and Mandriota (1992) gave an overview of commonly used driving fitness tests. These authors also described their own computer-augmented test. The Transportation Research Board has established a Task Force on the Older Driver, which studies issues relevant to this age group. As many authors point out, there are significant individual differences in capability within each advanced age group. Training, the use of technical aids, and experience can all make a significant difference.

INTERPRETING COGNITIVE DECLINE
OF ELDERLY PATIENTS IN A CLINICAL SETTING

Patients cannot be effectively treated for *any* ailment unless they are cognitively competent. They must be able to understand their treatment and to cooperate in its implementation. Both tasks require adequate cognitive functioning. If patients are not capable of performing these tasks, harmful errors may occur. In that case, the help of the patients' relatives, friends, or caretakers could be enlisted.

When elderly persons become ill, they should have a comprehensive medical examination. Because the elderly are prone to cognitive decline, it is essential to evaluate their cognitive status before implementing their treatment. Also, their cognitive status should be reassessed periodically during the course of treatment. If cognitive decline is present, its effect on a particular patient at a particular time has to be evaluated, including any changes in daily life functioning. Possible remedies for this decline must be decided on and implemented. All this is part of an ongoing process that has to keep pace with any changes in the patients' medical or cognitive status or in their living conditions.

In order to assess cognitive functions, patients are usually given a series of neuropsychological tests. This procedure has two basic purposes: (a) The determination of the status of a patient's cognitive functions. (b) A contribution to the correct diagnosis of any cognitive deficit present and its underlying cause, whether this deficit be a dementia or not.

These two goals complement each other. The latter, the contribution to diagnosis, is essentially the concern of the physician, usually a psychiatrist or neurologist. Knowing the correct diagnosis is important for predicting the future course of the underlying disease and makes a more effective treatment possible. Such is the case when the cognitive impairment turns out to be a secondary symptom of a different primary condition, such as depression or anxiety.

Testing and studying the patients' cognitive capabilities is usually done by a psychologist. These procedures give a more detailed insight into the specific symptoms for each form of dementia or other cognitive impairment and their variations among different patients. Clarifying areas of relative strength and weakness helps develop a more effective treatment and can address the effect of the cognitive losses on the patients and their families.

The term *dementia,* already introduced, needs a further definition. Cohen (1986) does so as follows: "an overall decline in intellectual functioning characterized by impairments in memory, judgment, and abstraction, accompanied by aphasia, agnosia or apraxia, any or all of which are sufficient to interfere with work and social functioning" (p. 84). This definition is based on that of the American Psychiatric Association's *DSM-III* (1980). It has been criticized by psychologists as being too medical, and as not acknowledging the significant contribution that neuropsychological observation and theory could make to differential diagnosis and treatment.

Among the primary dementias, we can distinguish Alzheimer's disease, multi-infarct dementia (caused by multiple small blockages of cerebral blood vessels), and other forms due to vascular causes, such as stroke. Other dementias are part of neurological diseases such as Parkinson's disease and Huntington's chorea or some encephalitis (including that of AIDS and Lyme disease). Finally, there are dementias caused by toxic substances (such as alcohol abuse) or lack of oxygen. In addition, there are the many cases in which dementia is a secondary, and possibly reversible, symptom of another condition. An important distinction is the "pseudodementia" caused by depression. Therefore, whenever dementia is found, a diagnosis should always be based on a thorough medical examination.

The ideal course of action would be for all geriatric patients to undergo neuropsychological testing. This may not always be feasible due to lack of time or qualified personnel. Therefore, most health-care providers resort to any of a number of comparatively simple cognitive screening tests that will at least indicate which of their elderly patients need a more detailed investigation.

Some of the more popular of these screening tests should be mentioned.

The scope of this chapter does not allow a detailed discussion of their comparative merits. (See Poon, 1986, for such a discussion of the most frequently used neuropsychological tests for a geriatric population.) Three of the most widely used screening tests are the Mini-Mental-Status Examination (MMS; Folstein, Folstein, & McHugh, 1975), the Global Deterioration Scale (GDS), and the related Brief Cognitive Rating Scale (BCRS); both by Reisberg (1983).

The most frequently used MMS is easy to administer after some training. It tests a patient on simple cognitive tasks related to time, place, and memory of generally known facts. The MMS also tests attention and calculation, delayed recall, the naming of common objects, the ability to follow multiple part instructions, writing a sentence, and copying a geometric figure. Points are given for correct performance on each part of the test, and the maximum overall score is 30. A score of 23 is generally considered as the cutoff point for the existence of mild dementia.

Reisberg and colleagues did comparative studies on the MMS and the GDS and found a good correlation; however, they found the MMS to be useful in cases of mild dementia but not in severe cases or for longitudinal studies where the continued deterioration of a patient has to be followed (Reisberg et al., 1986). One problem with the MMS is that the reported total numerical score creates the impression of one absolute quantitative measure. Knowing *how* this score was arrived at, by looking at the different parts of the test, provides some qualitative indications.

The GDS and BCRS both rate the quality of a patient's cognitive functioning on a variety of daily life tasks and express the degree of dementia in numbered stages, from 1 through 7, where 1 is normal and 3 forms the borderline between essentially benign age-related cognitive deficits and mild Alzheimer's disease (see Reisberg et al., 1986, for a detailed discussion). The problem with these tests is that they depend on the observations of an experienced clinician *and* the cooperation of someone who knows the patient well.

Psychometric test batteries, such as the well-known Wechsler Adult Intelligence Scale (WAIS; Wechsler, 1958; revised, 1981) and the Randt Memory Test , also called the New York University Memory Test (Randt, Brown, & Osborne, 1980), test patients on a wide array of cognitive functions. The administration of these tests requires considerable training; they are time consuming; and they have their proponents as well as their critics. The administration of the Randt test can be facilitated by a computer-assisted version (Glosser, 1984).

The Randt test (which has several equivalent versions to reduce the effect of specific learning when repeated) has been used in the Bellevue Computer Memory Program. It was found to be reliable (i.e., little variation of the results by individual patient, between test versions, and between testers).

The Randt, however, is not sensitive enough to confirm changes that are less than major found over time in our patients' performance on their cognitive exercises. Also, on retesting, some patients showed almost no change on the Randt, although there was considerable deterioration in daily life. We interpret the latter finding as a demonstration that our patients do better in the focused and undisturbed environment of the testing room than in conditions of possible cognitive overload in daily life.

Recently, considerable interest has developed in the so-called *ecological validity* of both tests and treatment. Ecological validity demands that treatment be directed toward improving a patient's handling of the demands and tasks of daily life and that tests be designed to measure such improvements in functioning (Hart & Hayden, 1986).

Erickson and Howieson (1986) reviewed studies where test scores were compared with actual daily life functioning, before and after treatment (often called "intervention") that had addressed both cognitive and practical aspects of the tasks of daily life. The authors found that sometimes the functioning had improved, although the test scores had not changed. The reverse was also found, that is, that the person's level of functioning had not improved, but the test scores had.

The authors concluded that finding the right test instrument to measure changes in daily performance remains a challenge that must be met before clinical research can become credible. They also wrote: "Improved psychosocial functioning and consumer satisfaction may count for more than improved scores on learning and memory tests" (Erickson & Howieson, 1986, p. 75). The implication here is that test scores are no reliable indication of how patients (the consumers of the treatment) actually function in daily life and of how satisfied they are with the results of their training.

The Bellevue Hospital's Geriatric Mental Health Clinic

At the Bellevue Hospital's Geriatric Mental Health Clinic, patients are offered a *comprehensive treatment plan with a team approach* in which all health-care workers participate as an interdisciplinary team. The services of the clinic are discussed briefly as one example (among many possible ones) of a serious effort to meet elderly persons' health needs in an effective and respectful manner.

Although patients can be self-referred, usually this is done by an outside physician or some institution. The main sources of reimbursement to the clinic are Medicare and Medicaid, and patients should be eligible for one or both of these entitlements. Because Bellevue is a municipal public hospital, sometimes exceptions can be made for patients who need the clinic's services but are not eligible for Medicare or Medicaid.

All patients in the clinic are routinely seen by a *physician* experienced in geriatric medicine, who also makes the necessary medical referrals. It is important to be aware of medical conditions that can cause secondary cognitive deficits. The patients also see a *psychologist* who supervises any testing needed and offers psychotherapy if the patient is a likely candidate. The psychologist collaborates with the *psychiatrist* on the diagnosis of possible mental illness and on the need for psychopharmacological treatment, including that for insomnia. A *neurologist* will be consulted if a dementia or other neurological condition is suspected. A *social worker* discusses with the patients what their needs and resources are and offers specific services. A *nurse-practioner* gets involved in routine medical procedures and discusses the intended medical treatment with the patient. The mental health division of the geriatric clinic works in close cooperation with its medical division.

Although the multidisciplinary approach contributes greatly to effective and humane treatment of the clinic's clients, misunderstandings still occur when a patient shows symptoms of cognitive decline. The Bellevue Geriatric Mental Health Clinic also offers *cognitive rehabilitation,* a program for patients who either complain of cognitive impairment (usually memory problems) or who are found to be suffering from them. The cognitive rehabilitation specialist at this clinic usually holds a PhD in clinical psychology and has had special training in cognitive rehabilitation. See Gianutsos (1991) for a discussion of the most appropriate educational or professional backgrounds for the practitioners in this as yet unlicensed rehabilitation specialty.

In general, cognitive rehabilitation programs offer their clients a variety of cognitive exercises as well as counseling and training to improve the management of cognitive deficits in daily life. Elliott (1990), a patient with severe memory problems, reported on her own successful regime, regularly supervised by a psychologist, that combined the use of memory aids and the introduction of behavioral changes.

Neuropsychological testing, usually done by a psychologist, assesses cognitive functioning, using a battery of established tests chosen from the many that are available (see also Poon, 1986, for a further discussion of such tests). Some cognitive rehabilitation programs, although not the one at the Bellevue clinic, include tests of their own devising or modifications of established tests. Most of the tests used at Bellevue, such as the Randt, have been discussed. Testing helps in the diagnosis and provides guidelines for the kind of exercises the individual patient needs as well as the level of difficulty at which these should start.

The Bellevue Geriatric Clinic's cognitive rehabilitation program is built around a core of specially designed computer exercises that address a range of cognitive functions, such as different forms of memory, concentration, and problem solving (Gianutsos, Blouin, & Cochran, 1984). These exercises can be presented at different levels of complexity, and the results can be tabulated. There is also a wide choice of educational programs and puzzles available. The course of treatment is tailored to the patient's individual needs. The exercises can be administered by trained volunteers working under expert supervision.

Patients are encouraged to develop *strategies* that help to improve their performance on the exercises. If necessary, effective strategies are demonstrated. The computer exercises are intended to strengthen cognitive functions but do not directly address tasks of daily life. However, patients are taught how such learned strategies can be used for such tasks, as learning to look for visual patterns in a visual memory task can help in finding one's way in unfamiliar surroundings. Developing personally meaningful associations when trying to memorize unrelated items may help in recalling dates or events. Many patients report some success in generalizing such strategies to situations in their daily lives.

The development of *metacognitive skills* is another important goal. Cicerone and Tupper gave a good description of metacognition, drawing on work by A. L. Brown. Quoting this author, they wrote that the concept can be best summarized as "'thinking about thinking' or 'knowing about knowing.'" Further paraphrasing Brown the authors distinguished "two major categories—(1) activities which are concerned with *conscious reflections* on one's cognitive abilities, and (2) activities which are *self-regulatory* mechanisms during an ongoing attempt to learn or solve problems" (Cicerone & Tupper, 1986, p. 75). Insight and self-knowledge are essential components of metacognitive skills and goals of metacognitive training.

The Bellevue cognitive rehabilitation program specifically addresses patients' self-perceptions about their cognitive performance. Patients often think that they are functioning worse than they are in reality and become very depressed. However, they can be helped to evaluate their cognitive functioning more objectively. They can also learn to develop effective adaptive strategies that make use of their strengths and help them cope with their cognitive losses. More realistic self-perception leads to higher self-esteem, especially when added to the sense of success they experience when learning to do the exercises, even if done at a simple level and with some help. Assistance in finding solutions for specific cognitive problems in daily life provides a further sense of empowerment, even to those with deteriorating cognitive skills.

People who are losing their cognitive capabilities are suffering a devastating loss of identity. They often feel helpless when trying to cope with these fundamental changes in themselves. Helping such individuals to develop metacognitive skills can go a long way toward reducing the disturbing effects of these changes. These skills also enable patients to participate more actively in their own treatment and in decisions that involve their lives.

The patients in the Bellevue program are allowed to experience success while learning to meet challenges. Even moderately demented patients were found capable of mastering some exercises at a simple level and improving their scores over time. We have also documented that gains in performance tend to persist even if the treatment is interrupted for periods of several months, as does the ability to use earlier learned strategies (Vroman, Kellar, & Cohen, 1989; see also Squire, 1986).

The cognitive rehabilitation therapist often mediates between patients and staff, such as a nurse or social worker, by explaining to these workers the degree of the patients' cognitive decline and its effect on the individual patient. Specific ways of dealing with cognitive problems may then be suggested to relevant staff or members of a patient's support group. The cognitive rehabilitation therapist may cooperate with staff in deciding on appropriate solutions and may get involved in their execution, including work with support groups.

For a review of cognitive rehabilitation as a neuropsychological specialty, see Gianutsos (1991) and Wood and Fussey (1990).

MISINTERPRETING COGNITIVE DECLINE
IN THE ELDERLY

It is our position that, in general, misinterpretations occur when there is a mismatch in communication between the parties involved. Either available information is misunderstood, ignored, or not even noticed, or information is withheld for some reason or not communicated effectively. Usually the blame must be shared by the partners in this communication process.

However, in the case of the doctor–patient exchange, there is always an inherent asymmetry that puts most, but not all, of the responsibility for change on the health-care provider. The process of change must begin with the awareness that such a mismatch of communication could develop and that there are ways to prevent this. Failure to make such a change could lead to harmful errors in judgment that may influence the patients' treatment. This mismatch in communication is more likely to develop when the patients are elderly, and the resulting misinterpretations are often blamed on these patients' deteriorating cognitive abilities.

The focus of this chapter is not on errors in interpretation that are the result of professional incompetence on the part of the health-care giver. Here, symptoms of significant medical conditions could be missed, resulting in the wrong diagnosis or treatment. Also, specialists might neglect to look beyond their own specialty—for instance, when an ophthalmologist does not think of diabetes when examining a patient with glaucoma or cataracts. These are important examples of human error on the part of the health-care provider that can harm the patient.

Rather, this chapter's concern is with the scope of the medical perception when it does not go beyond the medical condition and excludes the person of the patients themselves. Personal factors such as age, gender, ethnic background, living conditions, and the availability of financial and social resources may be viewed in a perfunctory or stereotypical fashion. The proper attention and the necessary time may be lacking for the way patients voice their own complaints, and thus, important symptoms may be overlooked. Possibly, no serious effort is made to assess the impact of symptoms such as arthritis on a patient's lifestyle. Elderly patients, on their part, may interpret a necessary striving for efficiency by health-care providers as the age prejudice they encounter in other contexts. The practice of no longer routinely making house calls has robbed physicians of the chance to learn, in an informal way, about their patients' daily surroundings and lifestyle.

Even when there is a team in place that cooperates in assessment and treatment, elderly patients do not necessarily benefit from such a multidisciplinary approach for several reasons: (a) There could be a lack of communication between the team members themselves—for instance, when there is no open and egalitarian discussion or when members are too defensive about their own territories. (b) The team members might share conscious or unconscious misconceptions about the special needs and circumstances of a geriatric population and there is no dissenting voice to challenge these potentially harmful errors in interpretation.

At the very least, misconceptions between elderly persons and health-care providers can cause mistrust and frustration. In the worst case, the physician–patient relationship can become seriously damaged. The physician may then no longer take enough time for the patient in question and could base diagnosis and treatment on incomplete information. Patients, feeling neglected, may refuse to follow doctor's orders, look for answers to their needs elsewhere, or simply stop coming.

Some patients may attempt to regain control over their own condition by seeking out alternative forms of healing, which could conflict with that of their regular doctors and result in harmful conditions. A recent survey in the New England Journal of Medicine (Eisenberg et al., 1993) found this trend to be a major problem. It estimated that in 1990, one in three Americans used one or more of the 16 listed forms of unconventional medicine (some of these, such as biofeedback and acupuncture are considered to be effective for some conditions) at a cost of $13.7 billion. Of this amount, $10.3 billion was out of pocket, which is equal to all out of pocket hospital costs for all hospitalizations in the United States over that year.

Of patients using unconventional medicine, 47% did so without the knowledge of their regular doctors. Chronic conditions (e.g., back pain, anxiety, depression, and allergies), rather than acute ones, were involved. The most frequent users were White, well-educated, and in the higher income brackets. The authors attributed this extensive use of alternative medical practices to a lack of communication and advocated that doctors not only become more familiar with unconventional medicine, but *always* ask about its use when taking a medical history. Only then could this possibly harmful form of noncompliance be detected. Others, commenting on this study (e.g., Angier, 1993), related the patients' own dissatisfaction with standard medical practice to their perception that it is "less caring" and "too invasive"; in other words, patients blame their doctors' attitudes.

Situations Leading to Misinterpretations of Cognitive Decline in the Elderly

The following discussion of the different types of misinterpretations that can occur between elderly patients and their doctors will also bring together some earlier material in a more systematic way. The text must be read with the thought in mind that these misunderstandings can all, in their own way, lead to potentially harmful human errors in judgment: The doctors may not have all the necessary information for proper diagnosis and treatment; the patients, on their part, may not comply with doctors' orders because they either do not understand or do not remember them. Some patients may also be unable or unwilling to follow their doctor's instructions for a variety of reasons. In any case, because of these largely preventable human errors on both sides, the treatment will not be effective, the patients' needs will not be met, and harmful conditions may result. And if detected, these misunderstandings are likely to be blamed on the elderly patients' perceived cognitive impairment, and not on failures in communication.

For the sake of clarity, the following section will list *by number* some of the more common examples of misinterpretations of cognitive decline in the elderly. Suggestions for prevention will be part of the discussion.

1. Health-Care Giver and Patient Do Not Share the Same Experience and Background and May Use a Different Vocabulary. This results in different interpretations of medical terms and concepts, although each side may assume that their meaning is shared with the other side. This discrepancy may even extend to the names of body parts, such as when patients use euphemisms like "back" for "buttocks." Medical anthropologists like Kirmayer (1992) have reported on the often misunderstood attempts of some patients to make sense of their frustrating medical condition on their own terms. In this study, patients were found to impose their own definitions, values, and belief systems on their chronic ailments and the required invasive therapies. This way they were able to accept, for instance, conditions like kidney failure and the necessary hemodialysis.

To remedy this situation, the *physician* should be alert to the possibility of disparate vocabularies and guard against it. Specifically, health-care workers should not automatically assume that adequate communication has taken place, or blame the patient's cognitive state when misunderstandings become obvious. It is essential to take a careful and methodical medical history that covers all bodily functions. When in doubt, the patients should be asked what they mean.

Older patients may list all their complaints before coming to the point and the doctor needs to find a way to cut this process short. The possibility of self - diagnosis must also be kept in mind: Some patients may have already concluded that they are becoming senile or are dying of cancer. Being alert to this possibility and taking the time to provide convincing reassurance may prevent much needless suffering. Even a diagnosis of cancer should never be given to patients without simultaneously discussing treatment options. Otherwise the patients may conclude that their case is hopeless and that the doctor has abandoned them.

The *nurse* or *nurse-practioner* could obtain much of the needed information before the doctor sees the patient. All patients should be asked for a list of their medications or bring their medicine containers to their appointment. Special attention should be paid to routinely taken, over-the-counter drugs (e.g., aspirin, cough medicine, vitamins, and certain sleeping pills). Patients tend not to report those even if prescribed by a doctor. Information on habits like smoking and alcohol consumption could be obtained. Patients should also be asked for permission to obtain transcripts from physicians they see outside the clinic. The nurse-practioner could also instruct the patients on how to present their most important complaints most efficiently to the doctor.

After the doctor has seen the patient, the nurse-practioner is the logical person to make sure that the doctor's orders are understood and will be followed. As a rule, instructions are best given in writing, broken down in steps, and using simple terms to avoid cognitive overload. Patients often do not grasp or retain more than part of the instructions. Educational material could be useful in explaining the nature of the disease and the treatment, but it should be discussed in addition to being given out. At each visit it may be necessary to go over the instructions again. Telephone calls may be needed in the interim to make sure that patients follow the orders. This job requires much patience and perseverance, but it will be worth it, because it has been shown that even patients with memory problems and poor attention can learn and retain some simple facts (Squire, 1986).

Diets should also be discussed, and the services of a *nutritionist* may be necessary. Many patients do not count snacks as part of their food intake. Diabetics, for instance, often think that fruit or cheese can be eaten in unlimited amounts. Even otherwise well-educated diabetics may not under-

stand how insulin works and inject their regular dose when skipping break-
fast and, misinterpreting or ignoring the warning signs, risk hypoglycemic
shock.

Patients should realize that they have to take an active role in their own
treatment. This includes learning to present their complaints effectively and
cooperating with health-care personnel in asking questions until instructions, and the rationale behind them, are thoroughly understood and written
down. Terms should be explained and patients should insist that their
concerns receive proper attention .

*2. There Are Insufficiently Realized Differences in Ethnic, Cultural, or
Religious Backgrounds.* The scope of this chapter allows only a brief
discussion of this important and complex source of cognitive misunderstandings. Some examples follow: Strictly orthodox Jews or Muslims, whose
religions forbid close contact between men and women, may not permit the
medical examination of a patient by a doctor of the opposite sex. People from
India or descendants from a Hispanic culture may make a distinction between
foods with "hot" and "cold" qualities and consider some dietary combinations incompatible and harmful. These same people may also adhere to a
medical system, or a concept of physical and psychological health, that does
not coincide with the Western, biology-based medicine of their doctors. For
instance, these patients may believe that illness is the result of a disturbed
balance in their bodies. Reversing this imbalance would require measures
derived from non-Western medicine or intervention by another traditional
medical practitioner from the same cultural background.

Ingrained value systems or religious rules prescribing what is clean or
unclean or what is beneficial or detrimental to one's health may interfere
with a patient's compliance with doctor's orders. Doctors may be totally
unaware of such a discrepancy in beliefs and values. The patients may not
realize that their belief systems about health and illness are not shared by or
known to their doctors. Patients might also decide to follow their own
convictions and disregard their doctor's orders. In any case, the resulting
noncompliance could be harmful to their health.

The best way to prevent such a harmful situation from occurring is to be
forewarned. In a big city hospital like Bellevue, serving an extraordinarily
diverse population with many immigrants, the chance of such a situation
developing is very real. In Bellevue a special questionnaire is used to explore
such differences in background at the time of a patient's admission. Workshops for staff are also held, drawing on the special expertise of members of
its multi-ethnic personnel.

Patients should be encouraged to voice any reservations they might have
when their treatment is in conflict with their value systems.

3. Certain Concurrently Existing Medical Conditions May Influence Cognitive Functioning.   Both health-care personnel and patients should keep in mind that many elderly patients typically suffer from several medical conditions simultaneously. It is quite common for an elderly patient to have high blood pressure, arthritis, diabetes, insomnia, and poor vision due to cataracts simultaneously. Not only do these conditions exercise a cumulative effect on the patient, but they also require that several medications be taken, often according to a complicated schedule, and not necessarily compatible with each other. Many medications are incompatible with alcohol.

Some medications can influence cognitive functioning. They may also interact unfavorably in ways not predictable from separate use. In addition, elderly people are generally much more sensitive to medications than younger individuals, possibly because they have trouble eliminating these substances that may then accumulate within the body. Finding the proper dosage for an elderly person can be quite difficult. The health-care provider should watch for any adverse effect of medication on cognitive functions, whether this be sluggishness or excitation and insomnia. Infections, even minor ones like a cold or a toothache, can also impair cognition, as can dehydration, hypothermia, or hyperthermia and unreported or unsuspected alcohol abuse.

A special class of co-occurring medical conditions negatively affecting cognition should be discussed separately. This is the *loss of sensory powers that patients are not aware of.* Such losses can have serious consequences. Patients with insidiously progressive hearing losses may start blaming others for "mumbling"; the others, in turn, reproach these patients for "never listening." This may result in a confrontational situation. Patients who do not notice losses in visual acuity (e.g., due to cataracts or glaucoma) may not realize that their clothes are dirty and stained, conditions that may be misinterpreted as the tell-tale signs of a beginning dementia.

Even less widely appreciated are the permanent visual field deficits that can occur after a stroke or other brain injury. In the case of *hemianopia*, patients do not see objects in the affected half of both visual fields, opposite to the affected half of the brain. Because these field deficits are not perceived as black or empty areas, patients often do not even know that they exist. Under these conditions, the brain compensates by supplying information that could have been there, but is not. For instance, depending on which side of the visual fields is affected, a 3 can be read as an 8, a C as an O. These patients typically misread letters, numbers, and words and skip parts of sentences, often complaining that the text "does not make sense." They "see" their complete dinner tray, but not the specifics and may ask why there is no butter, although it is there on the tray.

Hospital personnel and family members usually become impatient and think that the patient is confused or demented. The patient becomes disoriented and paranoid. This condition is often called *visual neglect* . It has been pointed out (see, e.g., Gianutsos, Glosser, Elbaum, & Vroman, 1983) that this term puts the blame on the patient and that *hemi-imperception* would be more accurate. The term *hemi-inattention* indicates a limited awareness of part of one's visual fields. This condition, which usually accompanies hemianopia, appears to exist occasionally with intact visual fields.

To remedy this situation, *doctors and other health-care personnel* should be alert to the effect of concurrent medical conditions and of medications on the cognitive functions of elderly patients. Such an awareness is the best safeguard against faulty conclusions and their consequences. Possible losses of sensory functions should be tested for, specifically hearing and vision, but also the sense of touch and smell.

As far as hemianopia is concerned, this condition often goes unnoticed by health-care givers as well. Any elderly patient who has suffered a stroke, a multi-infarct, or a head injury, or who is suspected of having done so, should be tested for possible visual field deficits. The consequences of such a condition are serious. In any situation where an unexpected obstacle could occur within the affected area of the visual field, there is the danger of an accident to the patient or to others.

Many patients recover sufficiently to want to start driving again. However, their ability to do so safely should be assessed by an expert. The stroke or other brain damage may have compromised the cognitive processing needed for driving. Also, possible visual field deficits rarely recover, and the ability to compensate for this loss may not be adequate. There is no way that these people can be prepared for the unexpected, and the results may be disastrous. A physician who treats such a patient bears a great responsibility and should not simply assume that this person will know when to stop driving.

The *elderly patients* must also be alerted to the possibility that certain conditions will make them less cognitively intact and learn to handle such a situation accordingly. They, or people close to them, should watch for changes in behavior when a new medication is introduced. Patients should postpone important decisions when they feel ill or distracted. Elderly persons, especially, should drink enough fluids to prevent dehydration. They may have to be warned that tolerance for alcohol decreases with age.

If hemianopia exists, this should be demonstrated to the patient and its consequences discussed. Patients can be trained always to scan both visual fields and to watch for common mistakes such as the misreading of price tags. What cannot be taught is to expect the unpredictable, such as a child suddenly running out of a driveway. These patients must be prepared to give up driving.

Patients, and those close to them, should be as alert as the doctors to the possibility of unsuspected sensory losses and be made aware of the consequences. Cognitive rehabilitation can make an important contribution in raising patients' awareness of the various, often unsuspected, conditions that can affect their cognitive functioning.

4. Patients Suffer from Cognitive Overload.   When the quantity or the complexity of the input received by someone becomes so great that it cannot be processed readily, that person can no longer comprehend what is happening. Similarly, circumstances may demand that someone execute several tasks simultaneously. This demand may overtax the available cognitive processing ability, and the person could become unable to act at all.

Elderly patients are particularly prone to such effects and have difficulty functioning in unfamiliar surroundings. Specifically, visiting a doctor's office or being admitted to a hospital can be a frightening experience. Moreover, it subjects these patients to more impressions and instructions than they can deal with, especially when they are already apprehensive, and may have impaired hearing or vision. Disorientation, even panic reactions may occur, and accidents can happen. Hospitalized elderly patients can become so disoriented when having to go to the toilet at night that they may try to climb over restraining guard rails, fall out of bed, and break their leg or hip.

A recent study by Pashler (1993) casts doubt on the ability of *any* person to carry out two tasks at once. That author, drawing an analogy with computer processes, hypothesizes that at the neuronal level, parallel processing does take place. At the level of decision making or memory retrieval, however, the process is more like time sharing, in which activities are actually done one at a time. One could speculate that for elderly people, such a time-sharing process, if it exists, would be compromised by impaired recent memory and difficulty in shifting attentional focus. This would make it hard to remember an extra task and difficult to shift one's attention to it.

This situation can be remedied when *health-care personnel and others* dealing with elderly patients become aware of situations where cognitive overload could occur and try to reduce its impact. A family member or friend might be encouraged to come along to the doctor's office or be available to intervene on the patient's behalf on the hospital ward. A *nurse* or *nurse-practioner* can get much initial information from the patient in a more neutral and less complicated setting.

After the patient has seen the doctor, the nurse-practioner may also take time to explain the received instructions and review how the medication has to be taken and how to implement the recommended changes in lifestyle. A family member or friend could be present at these discussions as well. Furthermore, it may be desirable to familiarize the patient ahead of time with the procedures, the ward, and the personnel involved in scheduled surgery.

The *patients* can also do a great deal to reduce the chance of cognitive overload. When scheduled for an office visit, they should have a list ready of topics to be discussed and questions to be asked. If necessary, they should arrange for someone to be there who could write down instructions. Even being brought to the appointment or taken home will relieve a stressful situation. Hasselkus (1992) analyzed what happens when an elderly patient's caregiver is present during the doctor–patient encounter. Briefly stated, the caregiver shifts, as needed, between the following roles: (a) being a go-between who explains to either party what the other means, (b) making certain that important topics are covered, and (c) supplying some information directly on behalf of the patient.

Elderly patients should respect the demands on their doctor's time but insist on clear answers to clearly stated and well-rehearsed questions. Further, they should insist on legible, complete instructions on their medicine containers, which should be easy to open. The *pharmacist* carries an important responsibility in these matters.

5. Patient Noncompliance.   Usually this means that patients do not keep appointments or do not follow doctor's instructions. These two behavior patterns can be exasperating for those who are trying to set up an effective system of check-ups and treatment. Such patients are often considered irresponsible, not motivated, or even demented. In reality, they may simply not remember the appointment (often made several weeks or months in advance) or do not understand the instructions. When this occurs repeatedly, patients may be dropped from the clinic roll, without necessarily being warned that this has happened. Sometimes the doctor adjusts by not prescribing any but the most innocuous medications for fear of accidental overdosage. The harmful consequences of noncompliance are obvious, but health-care personnel that fail to explore its underlying causes must share the responsibility with the patient for this serious human error.

This problem might be remedied if *health-care personnel* understood that elderly people, especially when they live alone, may require extraordinary measures to help them comply with their treatment plan. Telephone contact after a missed appointment may help, but a call to remind the patient on the day before the appointment is more likely to solve the attendance problem. Weekly appointments might be scheduled at the same time of the same day of the week. We found that even moderately demented patients eventually learned to come for their weekly appointments without reminders. In a clinic such as Bellevue's, which is set up with the expressed goal to make it easier for elderly patients to remain at home and still receive comprehensive health care, such measures should be routine.

Instructions should be written out clearly, in simple, nontechnical terms and in letters large enough for an elderly person to read. The patients should be clear about the rationale behind the instructions. Patients could be asked to repeat the instructions to the nurse, thus ensuring that the information is correct and complete.

Finding out whether patients have a *support group* that could help get them to appointments or supervise the taking of medications is another important way to prevent noncompliance. The Bellevue Geriatric Mental Health Clinic's team, usually through its social worker, routinely enquires about such matters and tries to get family members or other concerned individuals involved. The cognitive rehabilitation therapist has been successful in working with friends and neighbors of some moderately demented patients in establishing more unconventional support groups that help patients make decisions needed for their care.

Senior centers or groups with a religious affiliation may provide both companionship and escort services. Felton and Shinn (1992) and Felton and Berry (1992) found that elderly patients receive important support, as measured in self-esteem and a sense of belonging, from groups more than from individuals. These findings are underreported because traditional interviews and questionnaires tend to concentrate on person-to-person relationships.

The *nurse or nurse-practioner* should check regularly to make sure that wrong habits or misconceptions have not re-established themselves between visits. The label on medicine containers should contain all necessary information and be clearly legible. Medicine dispensers that indicate when the medicine has to be taken are a helpful aid. A one-time effort at instruction is not likely to be effective. Repetition and exploration of different ways to get the necessary points across are essential, such as role playing, posters, pamphlets, workshops, and videotapes.

The *patient* should try to keep track of appointments and instructions by, for instance, using a pocket calendar and always keeping appointment slips and clinic cards together. The proper use of memory aids like calendars, memos to themselves, and medicine dispensers can be taught by the nurse or nurse-practitioner. The cognitive rehabilitation therapist also could be consulted or be directly involved in this effort. Establishing routines can be helpful.

Patients should cooperate in providing all necessary information about their medications and the physicians they see outside the clinic. The clinic should post, or have available in printed form, the rules of the clinic, the names of clinic personnel who are relevant to the individual patient, and instructions on how they can be reached by phone. Hospital admission services routinely provide information booklets to newly admitted patients that usually contain a copy of the Patient's Bill of Rights. The Bellevue Geriatric Mental Health Clinic also provides its patients with a copy of the Patient's Bill of Rights, adapted to its specific population.

It is important for elderly patients in particular to realize that it will take effort on their part to become effectively involved in their own care, but it will be worth it. With such involvement they will become more self-reliant and less open to criticism and scolding.

6. Misunderstandings Due to Unfamiliarity with the Actual Needs and Resources of Individual Patients.   This problem might be remedied if *health-care personnel* investigate these patients' needs and resources carefully. This information should be documented in a way accessible to all members of the team, who should routinely consult it. The administrative mechanisms for this sharing of information are generally in place; however, it may require conscious efforts and supervision to ensure consistent compliance with yet another bureaucratic duty. A team approach to each patient's care is one of the better guarantees that such sharing will take place in an environment where different responsibilities are necessarily assigned to different people.

The patients' own priorities should be known and respected whenever possible. These should not automatically take second place to greater effectiveness in treatment or to the patients' "best interests" as perceived by health-care personnel. For example, a fiercely independent, elderly patient whose multiple disabilities included deafness and legal blindness, refused to accept placement in a nursing home or even household help, although she was aware that her living conditions were becoming hazardous and unsanitary. Mentally alert, although deteriorating, she remained in her beloved apartment, coping somehow, until she died there. It may become necessary to intervene in similar situations, but such a decision should never be taken lightly and not unilaterally.

Finding out about unconventional support groups, which may only need someone to get organized, is a possible solution. Such a group may require much time and diplomacy to get in place and to become effective. The health-care team, or one of its members, could then try to work out a solution with this support group. This group might include friends, neighbors, relatives, home attendants, members of a congregation, neighborhood caseworkers, and even a court-appointed financial guardian, as well as the patients themselves. Every attempt should be made to preserve the patients' sense of identity and their self-reliance. A failure to do so may lead to discord and rebellion. Caretakers have been known to physically abuse or scream at a troublesome patient. Some patients have expressed their frustration by hitting or cursing those responsible for them.

The *patients* may have to learn to be explicit about their own situation and priorities, and should understand the implications of their choices. They should also be helped in exploring all options as realistically as possible and in accepting the best available compromise. They should then be given support to make it work.

This process took more than a few years for one patient, a divorced, retired, successful businesswoman, who had deteriorated to the point that she routinely lost her way, no longer took care of her personal needs, and had put on a great deal of weight. With the help of a good friend, a succession of home attendants, her financial guardian (who had earlier been appointed with her consent and cooperation), and the cognitive rehabilitation therapist, she finally accepted the fact that she needed a full-time home attendant. This may not be the end of her story, but for now it is a workable solution. Hers is an example of the step-by-step process of finding solutions for patients that keep pace with a progressive dementia. She still comes for her weekly cognitive rehabilitation session, in which her performance is stable if at a very simple level. These sessions give her a sense of belonging and of accomplishment and, thus, empower her. The regular sessions also provide the clinic with a way of monitoring her health status and of getting her to come in for her periodic check-ups.

DISCUSSION

The interaction between patients and health-care providers has been analyzed as an exchange of information between two essentially unequal partners. Although they share a common goal, namely to relieve the patients of their complaints, this goal is perceived differently by each partner in the exchange, a difference that neither party fully realizes. Misunderstandings may lead to serious errors.

Medical perception, as chronicled by Foucault (1963, 1975) has changed focus over the last 2 centuries. In the early 18th century, each disease was studied and treated as a distinct species, although the patient's body was recognized as the seat of the disease. By the end of the 18th century, illness came to be considered as an undesirable social phenomenon that would eventually disappear—and consequently so would the need for hospitals— once society was made just and healthy. This point of view led to the establishment of public health. In the beginning of the 19th century, the renewed interest in autopsies encouraged a pathological anatomical perception of disease that went beyond the organs to the tissues.

Subsequently, developing technology deepened this form of perception: The contributions of listening and touching were added to the faculties of seeing and observing that were already in use. Once again, doctors learned to perceive their patients and their illnesses in a different way. In this century, elements of these earlier forms of medical perception were developed further, and technical advances such as microscopy, biochemistry, and radiological and electronic means of visualization have become indispensable tools. Although the earlier forms of medical perception are still recognizable, they have coalesced with these advances and now also include a better understanding of the ways the patients themselves, as individuals and as social beings, contribute to their own illness.

There is still no end to the changes in store. The privileged position of the medical profession itself is now being challenged. Among the issues being contended are medical accountability, the right of access to medical knowledge and medical records for patients and their agents, and their right to participate in medical decisions, including the right to die. These are concepts that need to be further defined, and the rights and responsibilities involved will have to be spelled out for doctors and patients alike. Although some progress has been made, the debate is still far from completed. The roles of hospitals and departments of public health are also being questioned. Finally, the matter of how to provide affordable, equal access to good medical care for all is among the most pressing dilemmas that our society must solve.

Some of the problems in the doctor–patient interchange have been discussed as mismatches in information that can seriously compromise the relationship and the patients' treatment. Scrutiny of these bilateral misinterpretations as cognitive events revealed several underlying causes. Ways were also suggested of preventing the potentially harmful human errors that could result from these misinterpretations.

The position was taken that some comparatively simple remedies could greatly increase the effectiveness of treating all patients, especially the elderly. The most important goal is preventing misconceptions from jeopardizing the patients' treatment and consequently their health. But also important is increasing the mutual trust and satisfaction of doctor and patient.

Some basic necessities are understanding the elderly person's cognitive processes, knowing about the patient's individual situation, and competence in helping the patient deal with the cognitive demands involved when needing medical care. Common sense, an efficient division of labor between members of the health-care team, and the creative exploration of better means to communicate are all essential. The patients themselves should make a determined effort to become better consumers of health care by becoming better informed. Patients should also be able to express their own needs effectively and insist on getting the information they need. Above all, they need solutions to their problems that respect their own wishes and decisions and that allow them to retain their identity and independence as much and as long as possible.

Much of the material discussed in this chapter could be useful in dealing with any patient. It is most applicable to elderly patients who do not suffer from more than mild cognitive losses (a score of 3 or less on Reisberg's Global Deterioration Scale). Still, many patients in the later stages of dementia can learn to do some simple tasks as well as continue to make certain individual decisions. It will remain possible and beneficial to treat the emotional needs and the remaining abilities of those patients with consideration and respect.

ACKNOWLEDGMENTS

The Bellevue Hospital Geriatric Clinic's Computer Memory Program was started in 1984 with a seed grant from the United Hospital Fund of New York and later received a partial grant from the Metropolitan Jewish Geriatric Center. The Auxiliary to the Bellevue Hospital Center has been closely involved in the Program since its inception and has funded it since June 1988. We gratefully acknowledge the Auxiliary's continued encouragement and support and that of Dr. Michael Freedman, Director of Geriatric Medicine. Dr. Rosamond Gianutsos of Cognitive Rehabilitation Services, Sunnyside, New York, has always provided help and advice. We thank her for this and for her comments on an earlier version of this chapter. We also thank Drs. Lucia Kellar and Candace Martin of the Geriatric Clinic for their suggestions. We gratefully acknowledge the cooperation of the Clinic's team members. And finally, we thank the patients whose participation in the Memory Program continues to inspire us and the many student volunteers whose work with the patients was indispensable. The opinions in this chapter are solely those of the authors and do not necessarily represent those of the Bellevue Hospital Center and its agents.

REFERENCES

American Psychiatric Association. (1980). *Diagnostic and statistical manual* (3rd ed.) Washington, DC: Author.

Angier, N. (1993, January 28). Patients rushing to alternatives. *The New York Times*, p. 12L.

Bedway, B. (1992, November 16). What's your responsibility for older patients who drive? *Medical Economics*, pp. 114–123.

Bellevue Hospital Center. (1989). *Patients bill of rights*. New York: Health and Hospital Corporation.

Cicerone, K. D., & Tupper, D. E. (1986). Cognitive assessment in the neuropsychological rehabilitation of head-injured adults. In B. P. Uzzell & Y. Gross (Eds.), *Clinical neuropsychology of intervention* (pp. 59–83). Boston: Martinus Nijhoff.

Cohen, D. (1986). Psychopathological perspectives: Differential diagnosis of Alzheimer's disease and related disorders. In L. W. Poon (Ed.), *Handbook for clinical memory assessment in older adults* (pp. 81–88). Washington, DC: American Psychological Association.

De Bono, E. (1970). *Lateral thinking*. London: Penguin.

Elliott, M. (1990, September-October). From the patient's point of view. *Cognitive Rehabilitation*, pp. 8–10.

Erickson, R. C., & Howieson, D. (1986). The clinician's perspective: Measuring change and treatment effectiveness. In L. W. Poon (Ed.), *Handbook for clinical memory assessment in older adults* (pp. 69–80). Washington, DC: American Psychological Association.

Eisenberg, D. M., Kessler, R. C., Foster, C., Norlock, F., Calkins, D. R., & Delbanco, T. L. (1993). Unconventional medicine in the United States. Prevalence, costs and patterns of use. *New England Journal of Medicine, 328*(4), 246–253.

Felton, B. J., & Berry, C. (1992). Groups as social network members: overlooked sources of social support. *American Journal of Community Psychology, 20*(2), 253–261.

Felton, B. J., & Shinn, M. (1992). Social integration and social support: Moving "social support" beyond the individual level. *Journal of Community Psychology, 20*, 103–115.

Folstein, M. F., Folstein, S. E., & McHugh, P. R. (1975). Mini-Mental-State: A practical method of grading the cognitive state of patients for the clinician. *Journal of Psychiatric Research, 12*, 189–198.

Foucault, M. (1975). *The birth of the clinic. An archaeology of medical perception* (A. M. Sheridan Smith, Trans.). New York: Vintage Books. (Original work published 1963)

Gianutsos, R. (1991). Cognitive rehabilitation: A neuropsychological specialty comes of age. *Brain Injury, 5*(4), 335–368.

Gianutsos, R., Blouin, M., & Cochran, E. L. (1984). *Manual and software, Cogrehab (Vol. 3)*. Bayport, NY: Life Science Associates.

Gianutsos, R., Campbell, A., Beattie, A., & Mandriota, F. (1992). The Driving Advisement System: A computer-augmented, quasi-simulation of the cognitive prerequisites of driving after brain-injury. *Assistive Technology, 4*(2), 70–86.

Gianutsos, R., Glosser, D., Elbaum, J. & Vroman, G. M. (1983). Visual imperception in brain-injured adults: Multifaceted measures. *Archives of Physical Medicine, 64*, 456–461.

Glosser, D. (1984). *Randt Memory Test. Computer scoring program*. Software and instructions for Apple and IBM. Bayport, NY: Life Science Associates.

Hart, T., & Hayden, E. (1986). The ecological validity of neuropsychological assessment and remediation. In B. P. Uzzell & Y. Gross (Eds.), *Clinical neuropsychology of intervention* (pp. 21–50). Boston: Martinus Nijhoff.

Hasselkus, B. R. (1992). The family caregiver as interpreter in the geriatric medical interview. *Medical Anthropological Quarterly* (new series) *6*(3), 288–304.

Kirmayer, L. J. (1992). The body's insistence on meaning: Metaphor as presentation and representation in illness experience. *Medical Anthropological Quarterly, new series, 6*(4), 323–346.

Malfetti, J. L., & Winter, D. J. (1986). *Drivers 55 plus: Test your own performance.* Falls Church, VA: AAA Foundation for Traffic Safety, and New York: Safety Research and Education Project, Teachers College, Columbia University.

Malfetti, J. L., & Winter, D. J. (1991). *Concerned about an older driver? A guide for families and friends.* Washington, DC: AAA Foundation for Traffic Safety, and New York: Safety Research and Education Project, Teachers College, Columbia University.

Pashler, H. (1993). Doing two things at the same time. *American Scientist, 81* (1), 48–55.

Poon, L. W. (Ed.). (1986). *Handbook for clinical memory assessment of older adults.* Washington, DC: American Psychological Association.

Randt, C. T., Brown, E. R., & Osborne, P. P., Jr. (1980). A memory test for longitudinal measurement of mild to moderate deficits. *Clinical Neuropsychology, 2,* 184–194.

Reisberg, B. (1983). The Brief Cognitive Rating Scale and Global Deterioration Scale. In T. Crook, S. Ferris, & R. T. Bartus (Eds.), *Assessment in geriatric psychopharmacology* (pp. 19–35). New Canaan, CT: Mark Powley.

Reisberg, B., Ferris, S. H., Borenstein, E. S., Sinaiko, E., de Leon, M. J., & Buttinger, C. (1986). Assessment of presenting symptoms. In L. W. Poon (Ed.), *Handbook for clinical memory assessment of older adults* (pp. 108–128), Washington, DC: American Psychological Association.

Sohlberg, McK. M., & Mateer, C. A. (1989). *Introduction to cognitive rehabilitation: Theory and practice.* New York: Guilford.

Squire, L. R. (1986). Mechanisms of memory. *Science, 232,* 1612–1619.

Vroman, G., Kellar, L., & Cohen, I. (1989). Cognitive rehabilitation in the elderly: A computer-based memory training program. In E. Perecman (Ed.), *Integrating theory and practice in clinical neuropsychology* (pp. 395–415). Hillsdale, NJ: Lawrence Erlbaum Associates.

Wechsler, D. (1981). *Manual to WAIS, rev. 1981.* New York: Harcourt, Brace, Jovanovich.

Willis, S. L. (1985). Toward an educational psychology of the adult learner: Cognitive and intellectual basis. In J. E. Birren & K. W. Schaie (Eds.), *Handbook of the psychology of aging* (2nd ed.) (pp. 818–844). New York: van Nostrand Reinhold.

Wood, R. Ll. (1990). Toward a model of cognitive rehabilitation. In R. Ll. Wood & I. Fussey (Eds.), *Cognitive rehabilitation in perspective* (pp. 3–25). London: Taylor & Francis.

Wood, R. Ll. & Fussey, I. (Eds.). (1990). *Cognitive rehabilitation in perspective.* London: Taylor & Francis.

7 Using Statistics in Clinical Practice: A Gap Between Training and Application

Roberta L. Klatzky
Carnegie Mellon University

James Geiwitz
Advanced Scientific Concepts

Susan C. Fischer
Anacapa Sciences, Inc.

A woman who had undergone a mastectomy for breast cancer was concerned about being at increased risk for an independent occurrence of cancer in the remaining breast. She asked her doctor what the chances were of such an occurrence, and he responded, "about 1 in 12." "Isn't that better than normal?" she asked in surprise, to which he replied, "Well, the chance of having cancer in both breasts is well below that of having cancer in one."

The doctor is confusing a priori (beforehand, without knowledge of a previous cancer) and a posteriori (afterwards, knowing that cancer has occurred in one breast) probabilities of having cancer in two breasts. This confusion is hardly unexpected. Medical training does not make a person immune to the kinds of statistical reasoning fallacies that have been widely documented in other populations. Indeed, medical school does not provide a substantial background in statistics. Yet, statistical knowledge is increasingly in demand by patients and medical practitioners, and without such knowledge, errors can occur. This chapter describes a gap between the need for statistical knowledge and the degree of expertise characteristic of physicians in medical practice. It uses, as evidence for this gap, results of a study called the MD-STAT project. Finally, the chapter offers suggestions for reducing the gap between statistical knowledge and its need in medical practice.

WHAT IS STATISTICAL KNOWLEDGE?

The term *statistical knowledge* can be used in several senses. It can refer to knowledge of data regarding a particular disease—for example, the magnitude of risk factors and the nature of survival given various courses of treatment. Statistical knowledge can also refer to an understanding of statistical concepts themselves—for example, knowledge of relative risk or significance tests or survival functions. Statistical knowledge is also knowledge that underlies the process of drawing inferences from data—sometimes called procedural knowledge. This might include, for example, knowledge of how to perform a test for statistical significance. In addition, statistical knowledge might also be considered an attitude about statistics. A positive attitude toward statistics is certainly not independent of knowledge about statistical concepts; the two are in fact mutually supporting. Even experts might agree that it is possible to lie with statistics, but it is generally harder to lie to the knowledgeable than the ignorant, who mistrust statistics accordingly.

The accessibility of the various kinds of statistical knowledge to the physician varies considerably. Most physicians can readily acquire the first type of statistical knowledge, that of statistical data. The availability of data about many diseases, even to lay persons, is obvious in today's society. Newspapers and popular magazines trumpet the latest findings about risk factors. The relation between cholesterol and heart disease, silicone breast implants and immune disorders, and body shape (apple vs. pear) and breast cancer are examples of widely publicized findings.

Large-scale databases have increasingly made it possible to provide physicians with data from epidemiological studies of risk and prognostic factors. For example, the Surveillance, Epidemiology, and End Results (SEER) program of the National Cancer Institute (NCI) provides data on the incidence of a large number of cancers. These data have been used to address issues such as dietary etiology of cancer (Schatzkin, Baranovsky, & Kessler, 1988). The Breast Cancer Detection Demonstration Project (BCDDP), with over a quarter of a million participants, was used to build a predictive model that assigns an individual's probability of developing breast cancer over a particular interval (Gail et al., 1989). A database of 75,000 women undergoing systemic adjuvant therapy for breast cancer was used to assess the effects of various therapies in a worldwide collaborative study (Early breast cancer trialists' collaborative group, 1992).

Practicing physicians can access data in a variety of ways. They can read journals received in the office or at a library; they can attend information sessions at conventions; or they can access computerized data services, such as the MEDLINE system of the National Library of Medicine (NLM), an online bibliography of vast scope. Searches of MEDLINE can be facilitated by Grateful Med, an interface program. The NCI has recently instituted an advisory system, CancerNet, which is available on an electronic network. This provides summary information and NCI recommendations along with references to related articles.

This is not to say that access to data is not problematic. For a busy physician, the requirement to visit a library or learn to use a computer search interface may render a data source unusable. Availability and ease of access are very different. But access to data is only part of a potential bottleneck between the burgeoning research literature and the physician's (and ultimately, the patient's) understanding. The other impediment may lie in knowledge that is essential to interpreting the reported results.

As has been noted, data constitute just one type of statistical knowledge; another concerns statistical concepts per se. One important concept, for example, is that of conditional probability, which underlies measures of risk. Concern with a risk factor focuses on the probability of incurring a disease on the condition that the factor is present. For example, the probability of a heart attack among people who smoke is a conditional probability. Comparative measures of risk are necessary to determine whether a potential risk factor actually has an effect. This is accomplished by comparing the risk within a population for which the factor is present to that of a control population, for which the factor is absent. When evaluating the comparative effect of some risk factor, physicians may encounter such terms as relative risk, odds ratio, or attributable risk. Risks may be combined (e.g., the smoker may also be obese), which leads to questions about the independence of multiple risk factors. It may be of interest to measure probabilities over time. For example, the survival function indicates the probability of being alive at each point in time after some life event has occurred, such as incurring cancer. Often, these functions have well-defined mathematical forms (e.g., exponential).

Even elementary statistical concepts are often subject to misunderstanding. A common confusion with regard to conditional probabilities (Dawes, 1988) is in the direction of prediction: $Pr(A|B)$[probability of Event A, given Event B] is treated as $Pr(B|A)$ [probability of Event B, given Event A]. Consider that the probability of a person being from Spain (Characteristic A) if she speaks Spanish (Characteristic B) is not the same as the probability of a person speaking Spanish (B) if he is from Spain (A). In order to know the relationship between these two measures, one must know what are called the base rates, $Pr(A)$ and $Pr(B)$, or their ratio. By the laws of probability:

$$Pr(A|B) = Pr(B|A) \times \{Pr(A)/Pr(B)\}. \tag{1}$$

In terms of the example, the term within the brackets is the probability of being from Spain, divided by the probability of speaking Spanish—a relatively small number.

For a medical example, Miller (1987) reported data associating obesity with breast cancer: "71% of those with breast cancer were obese, compared with 54% of the controls" (p. 95). This provides $Pr(obese|cancer)$ but not $Pr(cancer|obese)$. A physician, estimating the probability of an obese patient's developing breast cancer, might misread these statistics and assert that the probability is .71. Another common error is to subtract the figure for nonobese women from that for obese women, producing .71 − .54 = .17. In fact, these calculations cannot be performed correctly without knowing the probability of breast cancer for women in general. The controls provide an estimate of the obesity rate in the general population. If we assume the base rate for breast cancer to be .05 and use Equation 1, the probability of breast cancer in obese women would be:

$$Pr(obese|cancer) \times \{Pr(cancer)/Pr(obese)\} = .71 \times \{.05/.54) = .06 \quad (2)$$

The comparable figure for the nonobese is .03. The treatments one would prescribe for someone with a risk of 6% are quite different from those indicated for someone with a risk of 71%.

Knowledge about how to make statistical inferences is another type of statistical competency noted previously. A statistic is generally described as a quantity computed from data in a sample. *Statistical inference* has been defined as "drawing conclusions about populations from samples" (Hoel, 1960, p.1; cited in Dawes, 1988, p. 317). In the classical statistical approach, the inference is made by first generating what is called a *null hypothesis* about the population, then testing whether the statistical data are sufficient to reject that hypothesis. The null hypothesis might be, for example, that there is no difference between two groups of individuals with respect to the average on some measure. Another approach to statistical inference is called Bayesian, after an 18th-century theorist, the Reverend Thomas Bayes. In this approach, the sample statistic is treated as one source of data. That statistic is then combined with prior beliefs to determine a new belief about the population.

Rather than using formal rules for statistical inference, people often make inferences by using quick-and-dirty heuristics, or rules of thumb. For example, they may estimate the proportion of the population having some characteristic by the ease of retrieving individuals with that characteristic from memory. The problem with such heuristics is that they can introduce considerable error. Having a friend with a rare disease would be likely to lead someone to overestimate the probability of that disease in the general population, for example.

DETERMINING THE GAP: THE MD-STAT PROJECT

Determination of the gap in physicians' statistical knowledge is based in part on data from a project designed to investigate how statistical knowledge might aid physicians who are making medical decisions. That project is referred to as the MD-STAT project. There were three components of the project: One was a survey of physicians engaged in clinical practice involving patients with concerns about the breast; the second was a survey of medical schools; and the third was an intensive task analysis of the practice of a single physician. The surveys of physicians and medical schools were designed to gather information on the average physician's training in statistics, access to statistical data on breast cancer, and use of computer technology.

The survey of physicians was administered to a sample selected to vary in size of population center and geographical region. Physicians were selected who had practices involving breast care. In metropolitan and urban regions, the physician sample largely consisted of breast specialists, but in rural regions it included general practitioners and gynecologists. The survey of medical schools was conducted with the goal of determining the amount of statistical training in their formal curricula. The sample consisted of 40 medical schools, with at least one school from each state having a medical college for which a listing of statistical courses was available.

The task analysis of the single physician's practice was designed to illuminate the medical decision-making process as it is applied to patients with breast cancer and to determine the need for statistical information at various points in this process. An experienced physician with a large number of breast cancer patients was used as the subject matter expert (SME) for the task analysis. He was asked to go through a typical decision-making sequence from initial diagnosis of breast cancer through the prescription of treatment to the final follow-up examinations. Basic knowledge-acquisition techniques, supplemented by analytic probes, were used to identify the loci and types of decisions and the statistical data on which they were based.

The ultimate goal of the MD-STAT project is to facilitate the acquisition of statistical knowledge by practicing physicians. Of the four types of statistical knowledge we have mentioned, the most critical in this context are data pertaining to a disease of interest and an understanding of concepts that are used in presenting and interpreting the data. It is unlikely that most practicing physicians would be required to have the procedural knowledge necessary to conduct formal inference. The goal of MD-STAT is to promote understanding in a practical sense and thus help physicians take advantage of available data. This means helping physicians understand basic statistical concepts that are likely to appear in formal presentations of data and trying to "inoculate" them against common errors in interpretation and reasoning. Of course, promoting a positive attitude is also important.

STATISTICAL NEEDS AS REPORTED BY THE SUBJECT MATTER EXPERT

The MD-STAT project was intended in part to determine whether practicing physicians perceived a need for statistical data in the clinical setting, and if so, what types of statistical concepts they needed to understand in the research literature. Recall that the sampled physicians were involved with breast care. The need for statistical data in this context can be better understood by reviewing a typical sequence of events that occur when a physician treats a breast-care patient. For this purpose, a run-through of the task analysis provided by the SME is summarized in the following section.

There are three major tasks in the management of breast cancer patients: diagnosis, prescription of treatment, and scheduling of follow-up events. The decision-making process within each of these three tasks is described by first indicating the subtasks and the cognitive steps in the task and then focusing on the steps in which statistical reasoning is (or should be) involved.

Diagnosis

A breast cancer case begins when a symptom or indication of breast cancer is presented to the physician. There are four typical sources of presenting symptoms: self-examination, mammogram, clinical examination, and patient complaint (other than a lump). These four sources are listed in order of relative frequency. A lump detected by self-examination is by far the most common initial indication, accounting for around 80% of the SME's cases (see Kinne & Kopans, 1987). A suspicious mammogram is the second most frequent source and is becoming more frequent as mammograms become part of many women's annual physical examination. A mammogram picks up smaller tumors than self-examination does, and it also points a finger of suspicion at lesions that are not palpable. Relatively few cancers are first brought to the physician's attention by clinical examination or the patient's complaint of breast pain or nipple discharge.

Following the presentation of the initial indication, the physician gathers more data, first by clinical examination and a mammogram (if they are not the source of the initial indication). Next more data are gathered by a fine-needle aspiration. The aspiration will draw fluid from a cyst, causing it to collapse; this outcome is an indicator of a benign breast condition. If the lesion does not collapse or if the fluid contains blood, the fluid is analyzed cytologically for cancer cells. This analysis produces virtually no false positives, but a potentially substantial proportion of false negatives—that is, determinations of no cancer when in fact cancer exists. Given this level of test error, a negative fine-needle biopsy is often followed up by a core-needle biopsy (which examines tissue rather than cells) or an excisional biopsy, in which the lesion is removed and examined.

The first major decision in breast cancer diagnosis is whether or not to perform an excisional biopsy. Data from the clinical examination and the mammogram have, by this time, indicated a certain probability of carcinoma, and data from the fine-needle aspiration add their own estimate of the probability of cancer. At this point, the patient's risk factors are considered. If the patient has higher risk than average, for example, because breast cancer has occurred in her mother or sister, this too adds to the physician's estimate of the likelihood of cancer. On the basis of this information, the physician decides whether or not to perform an excisional biopsy.

If the probability of cancer is very high, the physician may perform a lumpectomy, then biopsy the tumor. A lumpectomy removes more tissue than an ordinary excisional biopsy. It is more likely to be preferred to the ordinary biopsy if certain other conditions apply, for example, if the cancer appears to be localized, indicating that lumpectomy rather than mastectomy might be the ultimate treatment. The decision to perform a lumpectomy is often based on what physicians call the triple test: If the clinical evidence indicates cancer, and the mammogram concurs, and the fine-needle biopsy concurs as well, then the probability of a false positive approaches zero, and the more extensive procedure is performed.

Statistical information is relevant to almost every step in the diagnostic task. For each of the general sources of initial indication of disease, there is a certain probability assigned to each symptom. For example, if a mammogram indicates a lesion with spiculated, ill-defined margins, the probability of carcinoma was estimated by the SME as .99. Indeed, even the type of tumor is strongly indicated by such a mammogram pattern—it is an invasive ductal carcinoma. Similarly, lesions in mammograms with certain other characteristics, such as asymmetrical tissue density or clustered microcalcifications, have associated probabilities of being malignant. If two or more of these secondary characteristics are present, the probability of cancer rises. A lesion investigated with a fine-needle biopsy and found to be negative for cancer cells still has some probability of being malignant.

The excisional biopsy has an associated probability of cancer: The chance of a palpable mass that is examined by biopsy being cancerous was estimated by the SME as between .20 and .30. This latter probability is affected by nonmedical considerations, such as the threat of malpractice suits. The SME indicated that in litigious areas of the country, the probability of a biopsy turning up a carcinoma is less than .10, whereas in countries such as Sweden, with socialized medicine and no malpractice suits, the probability approaches .40. Presumably, the reason is that physicians with a strong prospect of being sued are more likely to order biopsies when there is relatively little probability that cancer is present.

It should be noted that the radiologist reading the mammogram and the pathologist performing the biopsy are subject to the same uncertainties that the physician faces in the diagnostic process. There are very few radiologists who specialize in mammograms. Radiologists make errors at a certain rate, and many of their errors are false negatives. A false negative is a dangerous kind of error because it leaves a growing cancer undetected. Pathologists also make errors; in particular, they are likely to make errors with fine-needle biopsies because they are generally more skilled in histologic analyses (tissue) than they are in the cytologic analyses (cells) required for such biopsies.

Prescription of Treatment

Once the breast cancer is diagnosed, the next major decision-making task is to prescribe a treatment for the cancer. The treatment options include various kinds of surgery (lumpectomy, quadrantectomy, and the mastectomies: simple, radical, and modified radical). Other options that may or may not accompany breast surgery are chemotherapies (tamoxifen, cytotoxic chemotherapy, immunotherapy), radiation therapy, and ovarian ablation, which may be chemical, surgical, or radiological. Prophylactic therapies (for healthy breasts under risk) include breast surgery and tamoxifen chemotherapy. There are reconstructive therapies as well, such as implants or surgical transfer of tissue from other parts of the body to the breast area.

To prescribe one or a combination of these breast cancer therapies, the physician needs to know primarily the tumor type (ductal or lobular? invasive or noninvasive? estrogen receptive?), the number of tumors (is it multifocal?), and the stage of tumor development (has it metastasized?). The task of prescribing a treatment, then, is much like the task of diagnosis. It depends on conditional probabilities of this form: If the tumor is X, then treatment Y will be effective with probability P. Similarly, the side effects of treatment Y will appear with probability P'. Also, reconstructive therapies such as prosthetic implants have a certain probability of causing problems. These probabilities must be estimated by available statistics. The prognosis after a treatment, for example, is inferred from statistics about its effectiveness with past recipients. In sum, there are similar demands for statistical inference among the tasks of diagnosing breast cancer, prescribing a treatment, and selecting a reconstructive process.

Scheduling of Follow-Up

After treatment, the physician continues to monitor the health status of the patient. In the case of breast cancer, this may mean frequent mammograms and clinical examinations. Medical norms may recommend examination at regular inspection periods, but these recommendations may be based on relatively pragmatic factors. Construction of an optimal screening program is a very difficult problem (Eddy, 1980). Mammograms present a certain risk of causing cancer while they are detecting cancer. Medical examinations of all types cost time and money. What is the probability that a heretofore undetected breast tumor will grow to a dangerous size before the next examination? Will a noninvasive tumor become invasive before the next examination? Generally, the probability that a noninvasive tumor will ever become the more dangerous invasive type is unknown. Nevertheless, the medical decisions made during the follow-up period should be based on the same kind of probability estimates that inform the decisions in the diagnosis and prescription phases of the process.

STATISTICAL NEEDS AS REPORTED BY THE SURVEYED PHYSICIANS

The task analysis conducted with the SME reveals extensive need for the understanding and evaluation of statistical data. This conclusion was augmented by data from the survey of breast-care practitioners. Their statistical needs were assessed with several types of questions. One question asked whether the physician encountered specified contexts in which a use for statistical data might arise and if so, whether data were actually used in that context. A second question asked the physician about the types of issues patients raised that might call for consideration of statistical data. A third question asked the physician to rate a set of statistical concepts with respect to their importance in understanding medical data. Finally, physician respondents were asked to indicate their agreement with statements like "Statistical data are more and more in demand by patients."

As is consistent with the SME's evaluation, the surveyed physicians' most commonly cited use of statistical data was for evaluating the patient's prognosis and communicating it to the patient. Almost every respondent indicated having been asked a question about prognosis of the general form, "What are my chances of survival?" A second widely cited use for data was for diagnosing breast cancer and communicating the diagnosis to the patient.

Physicians' responses also indicated that patients frequently ask questions about the effects of risk factors, such as: "My mother had this disease; how likely am I to get it?" or "Does alcohol really promote breast disease?" Answering such questions requires knowledge about risks and how to compare them. A need for statistical knowledge was also evidenced in questions about diagnostic test results; nearly half the physicians reported hearing questions like, "Are my test results similar to the average?"

When asked to indicate the importance of various statistical concepts, respondents from the physicians' survey rated most of those listed as at least moderately important. The highest importance rating went to the concept of a clinical trial, a medical research design in which patients are randomly assigned to groups that receive distinct treatments and are then followed for some period of time. The least importance was assigned to the z-score, the value of an individual on some measure relative to the mean value on that measure, in standard-deviation units. Terms relating to measures of risk and risk comparison received moderate to high ratings.

A final indication of the importance of statistics in a clinical practice comes from the survey respondents' agreement with certain statements. Very strong agreement was found on the statement, "Statistical data are useful not only to researchers but also to the practicing physician."

PHYSICIANS' STATISTICAL TRAINING

The MD-STAT project clearly documents the extensive need for statistical knowledge on the part of practicing physicians. Whether a physician recommends a program of diet and exercise or bypass surgery to a heart patient, the recommendation is presumably based (at least in part) on the effects of such interventions in the past. Knowledge about such past interventions constitutes, in some sense at least, a sample statistic. What is called for is inference from that statistic. The project assumes that knowledge of relevant data and statistical concepts will better inform the physician making the recommendation and will help in dealing with patients.

As more data related to risk and prognostic factors are amassed, patients as well as physicians express more interest in what is known. The practicing physician is the primary link between the patient and the research literature. The efficacy of that link depends, however, on the physician's ability to synthesize and interpret the data. The statistical training undergone by the typical practicing physician is discussed next.

Part of the MD-STAT project was a survey of medical schools to determine the nature of their statistical offerings and requirements. Of the 40 medical schools that were sampled, two thirds required some type of training in statistics, and almost all offered statistics courses as electives. Of those that did require course work in statistics, the average number of courses required was slightly greater than one. Most of the schools offered only those courses that they required plus one or two additional courses. Several of the schools offered multiple statistics courses as electives through their epidemiology, biostatistics, or medical informatics departments. These courses covered a wide range of topics (e.g., probability, multivariate statistics, epidemiological methods, computers and decision making, and research methods).

The medical school survey indicates that the typical physician is required to take at most a single statistics course. Responses from the physicians' survey agree with that estimate. The respondents' statistical training was examined by considering the number of courses they had taken at the undergraduate and medical school level, as well as through continuing education. Overall, the average number of courses reported having been taken was 1.35. The time since the most recent course averaged 18 years.

Not surprisingly, with the reported level of statistical training, the physician respondents did not feel terribly knowledgeable about statistics. When rating their statistical knowledge on a 1–10 scale, 85% used only the bottom half of the scale, and almost one third marked themselves at the very bottom. The respondents also tended to disagree with the statement, "Medical schools currently give about the right number of courses in statistics for those who will go on to have a clinical practice."

A gap between training and application clearly emerges from these data. On the one hand, physicians see increasing need for using statistical data in their practices; on the other hand, they have very limited training and little confidence in their statistical knowledge.

EFFECTS OF THE GAP:
ERRORS IN STATISTICAL APPLICATION

A general model of the application of statistics in clinical practice might be formulated as follows: The clinician begins with a set of observations; for example, characteristics of a patient's mammogram. This information is fed, along with required parameters, into a reasoning algorithm, that is, a well-specified procedure. The observations and parameters are passed through the algorithm to produce a "correct" inference.

For example, the clinician might like to determine the probability that cancer is present, given an observation from a mammogram. Someone who wishes to compute that probability usually begins with knowledge about the reverse conditional probability—the probability that the particular observation would be present, given cancer (i.e., how characteristic of cancer patients the observation is). This is a situation in which Bayesian reasoning can be applied, in the form of Equation 1. Recall that the equation states that $Pr(A|B) = Pr(B|A) \times \{Pr(A)/Pr(B)\}$. It indicates that if reasoners know the base rates, that is, the probabilities of the disease and the observation in the general population (or their ratio), they can use a formula (the algorithm) to come up with an estimate of the desired information.

Statistical reasoning errors can arise in this model because clinicians have an incorrect (or missing) algorithm or because they have incorrect knowledge about required parameters. An appropriate algorithm may be applied properly but with the wrong parameters; the algorithm may be invalid in some way (with right or wrong parameters); or there may not be an algorithm to apply at all. In the latter case, the clinician presumably defaults to some more qualitative, heuristic approach.

Indeed, it is unlikely that a practicing physician would be an adept

Bayesian reasoner. The Bayesian approach as a description of human reasoning in general has been criticized on several bases, including the inability
of humans to do the appropriate manipulations mentally and the unavailability of relevant data in many cases (Wolf, Gruppen, & Billi, 1985). What
is more likely is that the physician will take a heuristic approach to determining the desired information. Heuristic approaches, as noted earlier, are
subject to producing erroneous results.

The Representativeness Heuristic

A number of heuristics and their consequences have been documented. One
is the *representativeness heuristic:* In decision-making tasks in which people
must generate their own estimates of the probability of a condition or an
event, they often base estimates of probability on the degree to which
situational characteristics are representative of the event. In doing so, they
ignore its true probability (the base rate). Thus, if students are described as
nerds, most subjects will assign a high probability to their being computer
science majors, even though such majors are relatively uncommon (Tversky
& Kahneman, 1973).

   Physicians, who are subject to the same representativeness pitfall, might
estimate the probability of a woman's having breast cancer, given a positive
mammogram, as very high. Presumably they do so because a positive
mammogram is representative of women with breast cancer. In fact, they are
quite mistaken (Eddy, 1984). Assume that the woman in question has a
personal and medical history that would suggest that her cancer risk is .05,
that is, the probability of her not having cancer is .95. Assume that the
mammogram has a hit (correct detection) rate of 80% and a false alarm
(false positive) rate of 20%. If 100 women of her type were subjected to
mammography, 5 would be expected to have cancer, and the following
groups would be expected:

   A. Cancer: No; Mammogram: Positive
      $95 \times .20 = 19$
   B. Cancer: No; Mammogram: Negative
      $95 \times .80 = 76$
   C. Cancer: Yes; Mammogram: Positive
      $5 \times .80 = 4$
   D. Cancer: Yes; Mammogram: Negative
      $5 \times .20 = 1$

Given that the patient has a positive mammogram, she must be in either

Group A or C. The probability that she is in Group C (that she has cancer), given the mammogram, is the number of cases with cancer and a positive mammogram, divided by the total number of cases with a positive mammogram, or $4 / (4 + 19) = .17$. Most subjects guess the probability to be much higher, closer to the probability of a positive mammogram given that cancer is present (.8, in this example).

Do physicians really make this kind of mistake? Dawes (1988) reported on a surgeon who was recommending prophylactic removal of undiseased breasts in women classified as high risk. The surgeon so advised his patients because (as he read the research) such high-risk women have greater than a 50% chance of developing breast cancer. Using research data to compute the probability presents another picture: Women in the high-risk group were said to constitute 57% of the population but have 92% of the cancers. The overall incidence rate in the population was .075. From these data, the probability of developing breast cancer, given that the woman is in the high-risk group, can be calculated by Equation 1. The ratio of the base rates (probability of breast cancer, divided by probability of being high risk) is $.075/.57 = .13$. The reverse of the desired probability (that of being high risk, given cancer) is known to be .92. The product of these probabilities is .12. That is considerably less than the "better than 50% chance" on which the surgeon was basing his advice. The surgeon boasted that he had removed both breasts from 90 women in the last 2 years for prophylactic reasons.

In a carefully designed study, nine clinical physicians estimated the probability that each of 1,531 patients had pneumonia, which was confirmed or disconfirmed through X-rays (Christensen-Szalanski & Bushyhead, 1981). Although the physicians were sensitive to the presence and absence of clinical phenomena, they consistently overestimated the probability of pneumonia. The researchers concluded that the physicians were misled by the representativeness heuristic, tending to believe that a person entering a clinic is sick until shown to be well.

The Availability Heuristic

Another common heuristic has been termed the *availability heuristic* (Tversky & Kahneman, 1973). Physicians estimating the probability of a woman's experiencing a relapse of breast cancer have to avoid the tendency to use their own patients as a basis for that judgment. Those patients represent a small and biased sample, but they are available to the physician's memory. When asked to estimate the probability, the physician finds it difficult to discount personal experience. Availability of reports in medical journals also fuels this type of error. For example, rare and exotic diseases are overrepresented in medical journals, leading to a significant overestimation of their probability (Dowie & Elstein, 1988).

Other Reasoning Pitfalls

Reasoning errors can arise not only from particular heuristic approaches, but from misunderstanding basic concepts. One common error has been called the *conjunction fallacy*. This consists of estimating the probability of two events occurring together (in conjunction) as higher than the probability of the less probable event alone (with or without the other event). In one study, internists were asked about the probability of certain symptoms, given that a 55-year-old woman had a pulmonary embolism (Tversky & Kahneman, 1983). A sizable majority of the internists (91%) mistakenly believed that the probability of two symptoms, dyspnea and hemiparesis, occurring together was greater than the probability of hemiparesis alone (but see Wolford, Taylor, & Beck, 1990).

Another reasoning pitfall is the *framing effect:* Suppose a physician sees 48 breast cancer patients per year. Two treatments are possible, with the following outcomes predicted: If treatment A is prescribed, 12 patients will survive. If treatment B is prescribed, there is a .25 probability that 48 patients will survive and a .75 probability that no patients will be saved. Which treatment would you prescribe if you were the physician? Although the expected outcome is identical, most people given such a choice choose treatment A, the sure thing, over B, the calculated risk. Now consider the following: If treatment C is prescribed, 36 patients will die. If treatment D is prescribed, there is a .25 probability that no one will die and a .75 probability that all 48 will die. Given the choice between C and D, most subjects now choose D, the risk, over C, the sure thing (Tversky & Kahneman, 1981). Yet treatment A is equivalent to treatment C; in both cases 12 people live and 36 die.

Why should C be shunned and A be sought, given that they are equivalent and the alternatives have equal expected outcomes? The only real difference is the way in which the treatment outcomes are phrased, but these differences in phrasing elicit radically different preferences. This outcome is called a framing effect. A rational decision maker would be indifferent to these options, because they are logically identical, but humans are rarely completely rational in their choices. They show an irrational preference for the sure thing when outcomes are framed in terms of gain (people will survive) and a similarly irrational preference for risky (probabilistic) treatments when the outcomes are framed in terms of loss (people will die).

Physicians and cancer patients are subject to the same pitfalls. In one study of cancer patients, the patients chose riskier treatment options when the possible outcomes were framed as loss, in particular, in terms of possible death (Llewellyn-Thomas et al., 1982). One must take great care to ensure that the framing of information on outcomes of various treatments does not influence treatment choice.

Even statistical experts make errors in statistical reasoning. A recent publication of the National Research Council (Committee on Risk Perception and Communication, 1989) indicated several types of experts' errors that have been documented. These errors include drawing conclusions from too limited a number of samples, inferring regular patterns from random variations, interpreting ambiguous data as in line with predictions, and being overconfident about predictions. Nonexperts could be expected to make these mistakes to an even greater degree.

CLOSING THE GAP: HOW CAN WE IMPROVE STATISTICAL REASONING?

One approach to improving physicians' reasoning and decision making is through computer programs variously known as knowledge-based systems, decision aids, and expert systems. The concept of automated decision-aiding systems emerged at least 3 decades ago (Ledley & Lusted, 1961). These systems generally incorporate a knowledge base of facts and procedures for using them. Expert systems are so called because the underlying knowledge is derived from experts. When applied to a specific problem, the decision aid requires that the user enter data about the problem's characteristics. These characteristics are processed by rules in conjunction with available knowledge. The decision aid goes beyond a conventional database retrieval system in that its outcome casts light on the user's decision. The term *decision aid* itself suggests a program that provides some of the information that is useful in decision making, often quantified or structured data, but leaves other elements to the user. Although an expert system may provide a recommended answer, in practice it generally functions more like a decision aid in that it contributes to decision making rather than governing it.

Undoubtedly the best known medical expert system is MYCIN (Shortliffe, 1976), which is used to identify the source of a bacterial infection and to recommend treatment. Recently the NLM developed AI/RHEUM, a diagnostic system for rheumatologic diseases (Kingsland, Lindberg, & Sharp, 1986). The NLM has also developed an expert system for automated indexing of the medical literature (Humphrey, 1989). Interpretation of medical images has also been aided with knowledge-based systems. Swets et al. (1991) described a two-part interactive decision aid that first prompts the user to rate a given image with respect to each of a series of scales, then uses the obtained scale values to estimate the probability of malignancy, which is then communicated to the user. An analysis of breast-cancer diagnoses with and without the decision aid revealed a clear advantage for the aided condition. That advantage increased with the difficulty of the diagnosis.

An extension of the initial MD-STAT project is envisioned in the form of a computerized decision aid that would enable physicians to retrieve quickly statistical information about risk and prognostic factors related to breast cancer and help them to interpret it. The system would incorporate an extensive base of information about risk and prognostic factors in breast cancer. Equally important is an online help system that would provide help in dealing with statistical concepts. For example, users who encounter the term *relative risk* would be able to retrieve its definition. They could go beyond the definition to an entry that discusses various measures of relative risk and their usefulness. The help system would also include warning flags about common reasoning errors such as the reversal of conditional probabilities.

It is acknowledged, however, that computerized systems have hardly been met with widespread use in the medical community. For example, the NCI recently conducted a formal evaluation of its Physician Data Query (PDQ) system, which disseminates information about clinical trials to physicians (Czaja, Manfredi, Shaw, & Nyden, 1989). Despite intensive publicity about the system and efforts to make it maximally available to health-care workers, fewer than 50% of physicians in cancer specialties were aware of its existence, and the level of reported use was low. The system had very little use by community physicians who were not cancer specialists. In fact, the greatest use was by employees of the Cancer Information Service who provide information to the public.

Obviously, constructing an automated decision aid for physicians dealing with statistical issues is a challenging prospect. Nonetheless, findings of the MD-STAT project indicate interest in such a system and a firm belief in the importance of statistical data by a substantial subset of practicing physicians.

ACKNOWLEDGMENT

This work was supported by the National Science Foundation, Division of Industrial Innovation Interface, as Project 9160456, "Feasibility of Aiding Decisions through Automated Data Retrieval from the Perspective of Practicing Physicians."

REFERENCES

Christensen-Szalanski, J. J. J., & Bushyhead, J. B. (1981). Physicians' use of probabilistic information in a real clinical setting. *Journal of Experimental Psychology: Human Perception and Performance, 7,* 928–935.

Committee on Risk Perception and Communication. (1989). *Improving risk communication.* Washington, DC: National Academy Press.

Czaja, R., Manfredi, C., Shaw, D., & Nyden, G. (1989). *Evaluation of the PDQ system: Overall executive summary.* (Report to the National Cancer Institute under Contract No. N01-CN-55459). Champaign: University of Illinois, Survey Research Laboratory.

Dawes, R. M. (1988). *Rational choice in an uncertain world.* Orlando, FL: Harcourt Brace Jovanovich.

Dowie, J., & Elstein, A. (Eds.). (1988). *Professional judgment.* Cambridge, England: Cambridge University Press.

Early breast cancer trialists' collaborative group. (1992). Systemic treatment of early breast cancer by hormonal, cytotoxic, or immune therapy. *The Lancet, 339,* 1–15.

Eddy, D. M. (1980). *Screening for cancer: Theory, analysis, and design.* Englewood Cliffs, NJ: Prentice-Hall.

Eddy, D. M. (1984). Variations in physician practice: The role of uncertainty. *Health Affairs, 3,* 74–89.

Gail, M. H., Brinton, L. A., Byar, D. P., Corle, D. K., Green, S. B., Schairer, C., & Mulvihill, J. J. (1989). Projecting individualized probabilities of developing breast cancer for white females who are being examined annually. *Journal of the National Cancer Institute, 81,* 1879–1886.

Hoel, P. (1960). *Elementary statistics* (2nd ed.). New York: Wiley.

Humphrey, S. M. (1989). MedIndEx system: Medical indexing expert system. *Information Processing & Management, 25,* 73–88.

Kingsland, L. C., III, Lindberg, D. A. B., & Sharp, G. C. (1986). Anatomy of a knowledge-based system. *MD Computing, 3,* 18–26.

Kinne, D. W., & Kopans, D. B. (1987). Physical examination and mammography in the diagnosis of breast disease. In J. R. Harris, S. Hellman, I. C. Henderson, & D. W. Kinne (Eds.), *Breast diseases* (pp. 54–86). Philadelphia: Lippincott.

Ledley, R. S., & Lusted, L. B. (1961). Medical diagnosis and modern decision making. *Proceedings of Symposia in Applied Mathematics, 14,* 117–157.

Llewellyn-Thomas, H., Sutherland, H. J., Tibshirani, R., Ciampi, A., Till, J. E., & Boyd, N. F. (1982). The measurement of patients' values in medicine. *Medical Decision Making, 2,* 449–462.

Miller, A. B. (1987). Early detection of breast cancer. In J. R. Harris, S. Hellman, I. C. Henderson, & D. W. Kinne (Eds.), *Breast diseases* (pp. 122–134). Philadelphia: Lippincott.

Schatzkin, A., Baranovsky, A., & Kessler, L. G. (1988). Diet and cancer. Evidence from associations of multiple primary cancers in the SEER program. *Cancer, 62,* 1451–1457.

Shortliffe, E. H. (1976). *Computer-based medical consultations: MYCIN.* New York: Elsevier.

Swets, J. A., Getty, D. J., Pickett, R. M., D'Orsi, C. J., Seltzer, S. E., & McNeil, B. J. (1991). Enhancing and evaluating diagnostic accuracy. *Medical Decision Making, 11,* 9–18.

Tversky, A., & Kahneman, D. (1973). Availability: A heuristic for judging frequency and probability. *Cognitive Psychology, 5,* 207–232.

Tversky, A., & Kahneman, D. (1981). The framing of decisions and the psychology of choice. *Science, 211,* 453–458.

Tversky, A., & Kahneman, D. (1983). Extensional versus intuitive reasoning: The conjunction fallacy in probability judgment. *Psychological Review, 90,* 293–315.

Wolf, F. M., Gruppen, L. D., & Billi, J. E. (1985). Differential diagnosis and the competing-hypotheses heuristic: A practical approach to judgment under uncertainty and Bayesian probability. *Journal of the American Medical Association, 253,* 2858–2862.

Wolford, G., Taylor, H. A., & Beck, J. R. (1990). The conjunction fallacy? *Memory & Cognition, 18,* 47–53.

8 People Versus Computers in Medicine

Thomas B. Sheridan
Massachusetts Institute of Technology

James M. Thompson
Harvard Medical School

The art and science of medicine has evolved significantly over the past century, as evidenced by, among other things, the increase in average longevity, the reduction in infant mortality, and improvements in the management of pain. Nevertheless, and in spite of improved technology and better knowledge, medicine remains vulnerable to human error in many forms, and in some ways this vulnerability is exacerbated by the greater complexity of medical systems. Further, the evolution in medicine has come at a dollar cost, which has been increasing significantly in recent years. One cannot read a newspaper or watch TV these days without becoming aware of the criticisms of the U.S. health-care system: that our health care administrative costs are double those of our neighbor Canada, that a large fraction of our population is not getting any health care, and, by doctors' own testimony, that a certain fraction of health-care effort need not exist at all.

Into this chaos, over the last decade, came the computer, one of the very few things that is getting cheaper for the equivalent service it renders. Or, to put it another way, whether it is bits, bytes, mips, or megaflops, the computer power per dollar has been doubling every several years, and promises to continue to do so. Therefore, it is natural to ask whether new forms of computer-based technology can compensate for human fallibility in medicine, as in other areas, and play a significant role in solving the health-care economic crisis. Unhappily, the answer is often "no" on both counts.

There is a tendency to look for simple "silver bullet" solutions to problems. Although technology has often been proffered as such a silver bullet for many of the world's ills, the public has come to know better. In this vein, the relation between humans and computers has come to be viewed as a kind of shoot-out—humans versus computers. Whereas some enthusiastic technologists see the new super-sophisticated computers as the ultimate victor over the ever more challenged but flawed human, many other people, with some justification, see computers as unable to make judgments in proper context, and, as humans themselves, deep down they hope that humans will "win."

This chapter looks briefly at the dollars, but more specifically at the performance enhancement and error reduction aspects of computers in medicine. Dollars is the denominator, and performance (error reduction) the numerator of the ratio that we all wish to maximize. This treatise asserts that neither the human nor the computer is best, that a close collaboration of the two is the best and will produce a "win–win" situation.

PEOPLE OR COMPUTERS, WHICH WILL IT BE?

New computer-based medical technology is expensive, at least at the developmental stage. Under the present crisis in health-care economics, there is a natural tendency to say that high tech is not necessary, it is breaking the bank, and to use a harsh phrase for health care, it is overkill. A typical allegation is that there are too many magnetic resonance imaging (MRI) units around. Every hospital wants one not because it is needed, but because it is high tech and presents a good image of progress and delivery of the best available medical care. The common cry is for hands-on primary care physicians, not overpriced specialists and their high-tech armamentarium. Caring humans are needed, not computers or robots.

Such an attitude is understandable and contains much truth. However, let us not throw out the baby with the bath water. If one were to remove systematically all reasonably high technology from all hospitals, it would be disastrous. The point is not people versus computers anyway. That is an entirely wrong way to view the problem. The problem is really how to make the best use (in terms of both economics and quality of health care) of technology and humans in combination.

Interestingly, the American space program has been suffering from a similar imagined dilemma between human and computer. There has been a feud between the "humans-in-space" advocates and the proponents of more computers and robots. Though simple versions of the latter have already proven their utility in exciting deep space probes, the human-in-space proponents seem anxious about robots for the more complex and varied tasks of near space, or give them little credibility. Probably the computer advocates have promised too much too fast. Far from working to enhance the collaboration of astronauts and computers, such feuding has generated political antibodies that have inhibited the desired collaboration. Meanwhile, both Japan and the European Community have been moving past the United States in applying computers and robotics in space. May the same not happen with respect to human–computer collaboration in medicine?

SOME IMPORTANT DEFINITIONS

In this section we give brief definitions of some important concepts that are discussed in later sections of this chapter. For those interested in a more detailed treatise, a list of suggested readings is presented at the end of this chapter.

Error. A usually acceptable definition of a human error is an action that fails to meet some implicit or explicit standard of the actor or of an observer. Agreeing on exactly what that "standard" is has been a topic of considerable debate. This is addressed in more detail subsequently.

Expert System. An expert system is a computer-based system, with its associated knowledge base and algorithms, that can draw conclusions and give advice on a particular subject. Its design is not limited to a specific architecture and can be based on such diverse systems as neural networks and fuzzy logic systems. It is important that an expert system be able to explain the rationale behind its decision, as this is a necessary condition for acceptance by its user.

Integrated Display. Display integration means combining representations of multiple variables in a meaningful relationship. For example, a patient in the intensive care unit (ICU) will have a multitude of monitors (e.g., pulse oximeter, electrocardiogram [EKG], noninvasive and invasive blood pressure recordings, ventilator settings, and airway parameters). Each monitor has its own set of displays and alarms that add to the complexity of the patient care in the ICU setting. An integrated display would combine the outputs of these independent monitors into a single coherent display. If designed properly, the integrated display would ease the user workload, decrease error, and reduce the response time to critical situations.

Artificial Intelligence (AI). AI is not easy to define. Workers who associate themselves with the term develop computer programs that seem intended to perform in ways that are traditionally regarded as intelligent (e.g., to play chess, recognize faces, or give advice). Some AI workers are said to seek better understanding of the human mind by computer simulation of mental functions. Others are said to seek improved computational capability—without reference to the mind. Perhaps in between are those who seek improved cooperation between mind and machine. Hollnagel (1989) asserted that the primary role of AI in human–computer interaction (HCI) is not in understanding the humans but in helping them by increasing their functional capability. AI techniques can augment the human in functionality by visual pattern recognition and discrimination, by monitoring for a particular variety of signals or patterns, by checking for insufficiency or redundancy or consistency, by offering expert advice, and by providing defaults when necessary (recommended decisions when input data regarding the circumstances are insufficient for "best advice").

Neural Networks.   A neural network is a self-organizing group of interconnected computational units that are based on artificial neurons. They are artificial neurons because they are presented as models of biological neurons. Each individual computational unit accepts sensory inputs from the environment (or other units), processes them to produce an output signal, and passes on the resulting signal as an output of the network (or to other units). They are considered self-organizing because weights or interconnections between neurons are increased or decreased by the computer program as a function of whether incremental changes make the performance better or worse. The pattern of interconnections between units and the path of propagation of data differs among various network architectures. Because of their ability to learn from example, neural nets can be trained to perform a well-defined function or task, such as data analysis or medical diagnosis. However, as problem complexity increases, the required training time (and the number of test cases required) increases exponentially.

Fuzzy Logic.   In everyday experience, a human operator's knowledge about variables or relationships between variables can be called "fuzzy." That is, the observations and thoughts of most people most of the time may be said to be mentally modeled and/or communicated to other persons in terms of fuzzy sets of natural language words that have overlapping meanings (such as *cold*, *warm*, *hot*, *very hot*). Computers need to be given more precise data. Thus, whereas a human would likely make a fuzzy statement such as "I feel very hot" when talking to a physician, a computer could only understand an exact statement like "my temperature is 99.8 degrees." Fuzzy logic will put such statements as "I feel very hot" into a form usable by the computer (by translating the human term *very hot* to corresponding levels of *relative truth* or *membership*).

Fuzzy rules and logic have been applied in many and varied kinds of human–machine interactions. For example, Buharali and Sheridan (1982) demonstrated that a computer could be taught to "drive a car" by a person specifying rules, as they came to mind, relating when to accelerate and by how much, when to brake and how much, and when to steer. At any point in the teaching process, the computer "knew what it didn't know" with regard to various combinations or conditions of rules, and then could ask the supervisor-teacher for additional rules to cover its "domains of ignorance."

Human–Computer Interaction.   Literally this phrase means any interaction between human and computer, including the human programming the computer as well as scanning and reading from it. However, the term is most often applied to the human use of computers in situations where the computer and its database are the end objects and are not mediators in the control of a physical dynamic system such as a human patient. For well-defined situations, the computer is now capable of interpreting and "understanding" information from various physiological measurements, building up databases of "knowledge," associating what is known from the past with what is currently happening, recommending decisions about what actions to take, and explaining how it arrived at those recommendations. Human decision making can improve by considering such recommendations and reasons, provided there is sufficient feedback to the human about whether following the recommendations succeeded or failed in the diagnosis or therapy, and the human in turn feeds this back to the computer.

For more information, refer to the suggested reading list at the end of this chapter.

Telediagnosis.   The term telediagnosis refers to diagnosis of a patient, such as inspection, auscultation, or palpation, by remote means, for example, by using a combined audio-video communication link and a teleoperator.

Teleoperation, Telesurgery.   A teleoperator is a robotic machine that extends a person's manipulating capability to a location remote from that person (Sheridan, 1992). A teleoperator includes artificial sensors of the environment, an instrument for moving these in the remote environment, channels of communication to and from the human operator, and devices to apply forces and perform mechanical work in the remote environment. In the medical context, this can mean operating, through natural or surgically imposed openings in the skin, on internal organs or tissues as in laparoscopy, colonoscopy, or arthroscopy. Or it can mean a physician in one city inspects or operates on a patient in another city by remote control of the remote video camera position and manipulator hand.

CURRENT APPLICATIONS OF COMPUTERS
IN MEDICINE

Familiar hospital computer systems include intelligent monitors, computerized tomography (CT) scanners, and MRI. There are a number of emerging technologies with applications in medicine that utilize the increased computational power of today's personal computers and workstations. Some of these systems are briefly discussed subsequently.

Neural Networks

Neural network systems have been proposed and tested in a variety of clinical situations (Baxt, 1991; Miller, Blott, & Hames, 1992; Weinstein et. al., 1993). They are becoming increasingly accurate, and thus have become successful in providing a valuable second opinion in clinical situations. In a blinded, prospective study, Baxt (1991) compared the accuracy of the physicians and an artificial neural network in the diagnosis of acute myocardial infarction (heart attack) in 331 patients presenting to an emergency room. The physicians had a diagnostic sensitivity of 77.7% and a diagnostic specificity of 84.7%. The neural network had a sensitivity of 97.2% and a specificity of 96.2%. Sensitivity is the probability of a positive test result for a patient who has the diagnosis under consideration (also known as true-positive). Specificity is the probability of a negative test result in a patient who *does not* have the diagnosis under consideration (also known as true-negative).

The error-reduction capabilities of this described system are substantial. This neural-based system would be useful in the setting of a busy emergency room, where time pressures contribute to diagnostic errors. The same system would also be useful to a family practitioner in a rural setting who has a difficult job being a jack-of-all-trades and whose training and cardiac experience might be suboptimal when compared to that of a big-city specialist. In addition, there are many areas of the country that unfortunately do not have a health-care facility within easy access. In such a situation, lay people, working with the help of the neural network system, could be the first line in the diagnosis and treatment of a cardiac patient. This would definitely be preferable to no care at all.

Fuzzy Logic Systems

Fuzzy logic systems, which originated in the United States, have been extensively developed in Japan and in other nations. Here in the United States, fuzzy logic has only recently been applied to medical systems. Ying, McEachern, Eddleman, and Sheppard (1992) demonstrated the use of a fuzzy expert system to regulate the mean arterial blood pressure in postoperative cardiac intensive care unit patients. Sodium nitroprusside (SNP) is a potent vasodilator with a quick onset and short duration that is delivered by continuous infusion. It is often used in an intensive care setting to control a patient's blood pressure, but its use requires frequent monitoring. Because the medical care provider is often engaged in other activities, and the patient's condition can change rapidly, a previously appropriate infusion rate could quickly result in dangerously low blood pressures. To decrease the occurrence of this type of drug error and to improve the quality of patient care, automatic control delivery systems have been proposed and tested (Martin, Schneider, & Smith, 1987; Packer, Mason, Cade, & McKinley, 1987; Meline, Westenskow, Pace, & Bodily, 1985). The article by Ying et al. (1992) described the results that were achieved by closed loop regulation of an SNP drip by a fuzzy controller. A fuzzy controller is defined in this clinical situation as an expert system that uses fuzzy logic for its reasoning. A discussion of closed-loop control theory with some medical applications was presented by Westenskow (1986).

Expert Systems

Over the last decade, as computers have become more powerful and inexpensive, we have seen an improvement in the sophistication of the algorithms used in expert systems. The size of the knowledge base used in such systems has increased as well. Expert systems have been proposed and implemented in a wide variety of clinical situations. This number will probably increase dramatically in the future. The Expert Systems Program at the Computer Science Branch (CSB) of the Lester Hill National Center for Biomedical Communications (LHNCBC), established in 1984, has a number of expert systems projects under development (program description from the CSB LHNCBC). Among these are the AI/RHEUM system in rheumatology, the medical expert systems evaluation project, and the AI/COAG hemostasis consultant system.

As expert systems become more acceptable, the difficult human-factors question concerning how much advice is too much probably will remain. That is, at what point, as an advice-giving system becomes more reliable and more user-friendly, does the operator accede to its advice perfunctorily, in effect abandoning responsibility for critical independent thought. This leads quite nicely to the legal question of liability, which is discussed later in the chapter.

In addition to providing a valuable second opinion to the health-care provider, expert systems can also be used for the continuing education of such providers. There has been an explosion of information in medicine, with a concomitant increase in the number and complexity of diagnostic and therapeutic modalities. From the feedback provided by the expert system, the health-care provider will be exposed to the latest knowledge base, diagnostic techniques, and therapeutic developments pertaining to a wide variety of disease processes. As any medical student and resident can tell you, the best place to learn is from your patient. The expert system will assist us in this learning process.

Medical Record Keeping and Database Systems

The maintenance of accurate patient records has always been an important part of total patient care. This becomes even more important in the current medical legal climate. Medical record computer systems have been designed to do more than simple word processing. Such systems are available with templates that insure consistently complete documentation of the patient's visit and can be instructed to do a quality assurance analysis based on reminders designed by the physician (McDonald et al., 1984). In addition to insuring a complete medical record, the system will also guarantee that an important piece of information is not overlooked.

There are other advantages to a computer-based medical record system. Poor penmanship, which is a major problem with handwritten notes, no longer would be an issue, and errors would be avoided. There are additional methods of controlling errors in automated medical records systems. Schwartz, Weiss, and Buchanan (1985) described a number of checks that can be used for medical data. An example of an error check based on validity is a range check. A range check will question the entry of a value that is outside of a predetermined range (for example, an age of 188). A name field check might have a pattern check that will question input if it does not match a predefined pattern (for example, a telephone number would have three digits, a hyphen, three digits, another hyphen, and four digits). A dictionary check would verify that the entry is an acceptable word (an example is a spelling check seen in some word-processing programs). A consistency check, which is an error check based on redundancy, will detect errors by comparing an entry to previously entered data (for example, an entry of an ovarian cyst for a 26-year-old male).

Improvements have also been made with computer-based hospital information systems. Information related to various aspects of patient care, such as lab data, X-rays, and test results, is entered into a central system that then allows immediate retrieval of information concerning the patient. Work is currently being done on standards so that such records could be made available nationwide. The storage and retrieval of patients' old records would be easier, especially between hospitals and physician offices. Quality assurance issues would become centralized. Diagnostic and therapeutic decisions could be analyzed and then reviewed nationally to define a range of patient care that could be used to define a more realistic national standard of care.

A computer-based hospital system would also decrease the incident of forced errors. A *forced* error is one in which task demands exceed physical capabilities or resources available. For example, an emergency room physician is sometimes forced to make a decision on the basis of inadequate patient information. Instant information about the patient's medical history, allergies, and medications would decrease the incidence of these forced errors.

POTENTIAL SYSTEMS

Teleoperators to Assist with Patient Care in Isolation Units

There are numerous instances where physical contact between the health-care provider and the patient is not possible or is undesirable. From the health-care provider's viewpoint, such a situation may exist when the patient is in space or has a highly contagious disease, or if there is a substantial risk of assault associated with the evaluation and treatment of a person, such as a schizophrenic during an acute psychotic episode. From a patient's standpoint, this situation would exist if the patient is immunosupressed, whether from chemotherapy or as a result of drugs used for the suppression of tissue rejection in a postoperative organ transplant. In either case, close physical contact between the patient and the health-care worker would be minimized if the isolation unit were equipped with a teleoperator system. As an example, a recent heart transplant patient (who would be immunosupressed to prevent rejection of the donor heart) could have vital signs, EKG, and relevant pressures monitored remotely. Ventilator settings and intravenous fluid rates, as well as the infusions of vasoactive agents, antibiotics, and other agents could all be adjusted by the teleoperator. Minimizing the patient's contact to potentially infectious agents would definitely be in this patient's best interest.

Use of Online Dynamic Models for Prediction and Control in Therapy

The expert system traditionally is a static database of production rules and a network of computer frames. However, in delivering anesthesia or planning surgery or other interventions, dynamic models can be repetitively set with initial conditions (based on current measurements plus hypothetical sequences of actions) and run in fasttime to determine best predictions of what will happen (in a few minutes or a few days). This technique has already been proven in the control of supertankers, aircraft, nuclear power plants, and other complex dynamic systems.

Generation of Virtual Environments for Visualization or Training

One of the "hottest" (and therefore potentially oversold) fields is that of virtual reality. In virtual reality, computer graphics and human interface devices such as head-mounted displays and touch stimulus-generating gloves are combined to give the user a sense of being present in an artificial, actually computer-generated, world. The user can move around in the artificial space, change viewpoint, and actively modify the environment. Perhaps most important, variables that are not normally observed visually (e.g., chemical concentration) can be displayed in relation to space and time, as can magnitude scales not normally experienced as well as in juxtaposition to what is normally observed, namely the human body. This also can be used for training surgeons (and other health-care workers) where the surgeon trainee "operates" on a computer-modeled patient (or part of one) that is continually observed on a screen using an "instrument" that conveys the proper tactile sensations, all without endangering any real patient. The rationale is the same as for using flight simulators in pilot training.

Robotic Systems to Assist with Patient Care

Mechanical robots have been under gradual development in industry and aerospace for 2 or more decades. Such robots may be turning a corner of safety, cost, and practicality for such uses in hospitals and nursing homes. The Japanese have had a national program called "Silver Robotics" and the French one called "Sparticus" to develop such robots. Neither of those programs has yielded usable products, but the trend is there, and provided we are patient and do not expect instant magic (which is what has come to be expected from robots), there will surely be important benefits.

Telediagnosis and Acute Patient Care

In the late 1960s, a telediagnosis link was established between the teaching hospitals of Boston and Logan Airport following an aircraft accident in which many persons died for lack of medical attention because the ambulances were blocked by rush-hour traffic jams in the tunnel connecting the airport and downtown Boston. Trials with this telediagnosis communications link showed that with a nurse physically present with the patient, adequate diagnosis of minor trauma could be rendered by a remote physician, and the nurse could then provide whatever therapy was immediately necessary.

Technology has changed significantly in the 25 years since the Boston experience. Many have complained about the lack of health-care services in rural settings. Could the current 911 system be redesigned to act as a centralized telediagnostic referral center? If so, then persons in remote locations could be linked to triage health-care workers, who could evaluate, treat, or triage the caller to the appropriate specialist. The telecommunications network already is in place. Information highways, with audiovideo links through the phone systems and cable TV networks, are already being tested in the United States. Perhaps a telediagnostic 911 system is just a matter of time.

OTHER ISSUES RELATED TO THE USE OF COMPUTERS IN MEDICINE

Health-Care Reforms

Consider a situation that uses expert rules that has resulted in an increase in medical error. As part of a health-care reform plan, in 1990 the Maine legislature approved a 5-year program that proposed specific practice guidelines (expert system rules if you will) for various clinical situations. Participation by health-care workers was optional. The program works as follows: If the health-care providers follow the algorithm (associated guidelines), then they are given protection by the state from any malpractice lawsuits associated with the diagnosis and treatment of that condition. The state also agreed that the decision by participating health-care providers not to use these guidelines could not be used by plaintiffs as evidence of substandard care. An article in the Wall Street Journal (Felsenthatl, 1993) reviews this and presents a case in which a boy "wasn't given a neck X-ray when he was taken to an emergency room after a mountain-bike accident shortly after the experiment began. Four days later he nearly collapsed, . . . prompting an X-ray that showed he had broken his neck in the accident." Some would argue, by today's standards, an error in diagnosis was made. By the state's standards, care was appropriate.

Financial Constraints

Health-care costs nearly doubled in the 1980s. At the time this chapter was written, an overhaul of the health-care system was just beginning. Even though computer-based systems have been shown to decrease error

(Schwartz et al., 1985) and improve the quality of medical care, they are expensive—not only the initial outlay, but with maintenance, training, and ongoing personnel costs. The "enemy of good is better." Will this be the health-care motto for the 1990s?

Patient Confidentiality

By the nature of the patient record, a wide variety of individuals and services will have access to the patient's current chart and previous medical records. Not only will it be difficult to prevent very personal and potentially damaging information from reaching the wrong hands, but the identification and correction of erroneous information will also be very difficult (just consider the problems individuals have with national credit bureaus). Who will be liable if information is released that an individual has AIDS, or if an entry mysteriously (unintentional or not) appears on a record erroneously labeling the individual as "paranoid schizophrenic with homicidal ideation"? If we cannot prevent current computer pirates and hackers from entering sensitive financial and defense systems, then how can we guarantee protection for medical files?

Liability

The question of liability is an important one in today's litigious society. The use of computers in medicine and errors attributed to that use will raise a host of legal issues, the most visible being in the area of liability. Such liability can be applicable to both the medical care provider and the software designer.

Once a computer system is designed (whether it be an expert system or a computer-assisted diagnostic program), the question arises as to who will be responsible for testing and certifying the system. Will liability then be with the certifying individual, stay with the software programmer, or lie with the health-care provider who uses it? Comparing the advice of the expert system to that of a human consultant, Miller et al. (1985) concluded that the health-care providers could be considered negligent if it is determined that they knew or should have known that the advice given was substandard. In addition, Miller et al. pointed out that the potential for liability could exist if the physicians are found not to exercise reasonable care by their failure to consult an available expert system. Could a case of negligence also be made against health-care providers if their therapeutic or diagnostic plans deviated from those suggested by the expert system? What about a decision to use one of many available expert systems? Is there liability if the health-care provider chooses an expert system that is later deemed substandard (and recalled a few years later)?

HUMAN ERROR, COMPUTER ADVICE, AND HOW FAR
TO GO WITH COMPUTERS AND AUTOMATION

Turning to the subject of this book, human error in medicine, recall Alexander Pope's words, "To err is human, to forgive divine." Don't believe it, say the lawyers, with the insurance companies looking on hopefully. Can computers really help in reducing errors? It would seem so, because it appears that computers can perform many critical tasks far quicker and more reliably than human doctors and health-care professionals. Human-factors professionals have a pretty good working knowledge of what computers can do better than humans and what humans can do better than computers.

For starters, computers can solve large sets of equations and crunch numbers quickly and accurately, which humans simply cannot. Computers can store large amounts of information in precise form and recall it very quickly and reliably. People are poor at this, but humans can make all kinds of associations in memory, which computers, at least up to now, are not very good at. Computers and humans both can recognize patterns of light or sound, and both have to be programmed, but the human programming occurs "naturally" and on the job. Programming a computer has to be done laboriously by the same error-making humans that the computer can potentially surpass. Does this mean the computer is only as good as its programmer? That may be changing, for with neural network theory and massively parallel computation, computers are, in effect, programming themselves as they are trained in new situations.

The main point is that computers and humans are complementary in many ways. Potentially, this complementarity gives each a good position from which to check up on the other. If humans and computers were too much alike, they might collaborate in making the same error. At any rate, all errors are not the same, and if, as new and unforeseen circumstances arise, the human is allowed to have any margin of decision-making discretion (as contrasted to acting like a preprogrammed zombie), there will necessarily be some trial-and-error behavior. The computer's job, at least a big part of it, is to confine human error making to small trial excursions from the norm without getting into errors from which there is no recovery (i.e., bad *conse-quences*). Computers can do this, but there remains much work to be done. There are also several traps, as described in the following paragraphs.

Lack of Technical Knowledge and Fear of Computers.   Ordinary users of computers or computer-based automatic machinery are seldom very knowledgeable about the innards of the computer and the details of the software that is running on the computer at the moment. This is likely to be true of health-care professionals. Further, there may be some cultural problems getting physicians to type and use computers directly. To date, the record has not been so good, and many have felt that computer interaction is somehow an affront to their professional dignity, or at least not in keeping with their skills. This should change as new generations of physicians and health-care workers come into the ranks who have computers at home and are computer literate.

Multiple Modes and General Complexity.   Because computers can do many different kinds of things, they have to be switched into different modes, depending on the objective and the current conditions. It often happens that the current mode is not completely evident to even the experienced user, and the user assumes that the computer is operating in a different mode than it actually is. This is a big problem, for example, as commercial aircraft become more automated and computerized. In particular, the sophisticated new "flight management system" allows the pilot to perform a great variety of functions, from planning to navigation to autopiloting. In the hospital, patient monitoring equipment had been adjusted by a health-care worker to serve one need. When that monitor was used for another need, it was effectively inoperative to monitor what the worker thought it was monitoring. Because there was no alarm, either locally or at the central nurses' station, the patient died. One way out of this trap is to provide software and speech or text displays that continually remind the user what mode the machinery is in and make sure their expectations are compatible with what the machinery can do as currently set.

Vigilance and Overtrust.   Human-factors specialists have long worked with the problem of automatic machinery and the monitoring thereof. There is always the problem that the more reliable the machinery is, the more robust it is (able to work in a variety of conditions), and the longer is its period of automatic control, the less frequently the human need to check up on it, and the greater is the tendency to trust it to function on its own without human supervision. This, then, raises a question about how independent the machinery should be and whether some real demand for human intervention and collaboration is not a good idea.

Abdication of Responsibility.   Perhaps the greatest danger is that users of computers will come to feel that the computer knows best, that ultimately decisions are being made by the computer, and that they themselves have no responsibility in the matter.

A one-dimensional scale of "degrees of computerization" is given in the following list:

1. The computer offers no assistance, human must do it all.

2. The computer offers a complete set of action alternatives, and
3.   narrows the selection down to a few, or
4.   suggests one, and
5.   executes that suggestion if the human approves, or
6.   allows the human a restricted time to veto before automatic execution, or
7.   executes automatically, then necessarily informs the human, or
8.   informs him after execution only if he asks, or
9.   informs him after execution if it, the computer, decides to.
10. The computer decides everything and acts autonomously, ignoring the human (Sheridan, 1992).

Although a simplification (introduction of automation is really a multidimensional matter), this scale has been useful in considering "how far to go" in computerizing for different application contexts. It is clear that in many medical applications we are well beyond Step 1 in the list, but nowhere close to Step 10. Hopefully some degree of computerization provides reliable results and eliminates human errors, but more computerization may well complicate the human's understanding and produce the potential for human errors with serious consequences.

SOCIAL IMPLICATIONS

Government agencies, medical associations, and other well-meaning individuals or groups may recommend the adoption of various computer-based systems in the medical arena. However, this does not necessarily insure medical provider implementation or patient acceptance of these systems. Nor do the perceived medical advantages of such a system necessarily translate to improved total patient care. People still are social creatures, and as medical systems become more computerized and automated, the computer will mediate more of what interpersonal contact remains.

Eventually computers will become more powerful, think faster, have larger memories, interface better, learn more efficiently, and have more accurate and extensive algorithms. They may even become the standard of medical care. To some extent, computers may be augmented by robots and other automated devices. But to many writers there is a dark side to all of this: They envision the worst form of alienation, the worst tragedy, which occurs if the patient and the health-care worker are happy to accept roles in which they are made to feel secure by the computer, whereas in actuality, they are enslaved. Engelberger (personal communication, 1981) reminded us that "it will always be far easier to make a robot of a man rather than to make a robot like a man."

In regard to the impacts of the computer on a person's self-perception, we are fond of citing Mazlish (1967), who referred to the computer as the "fourth discontinuity" in this self-perception. At first, said Mazlish, humans saw themselves as the center of all things, but Copernicus jarred this perception by depicting humans as isolated dwellers of a tiny planet of a minor star (which turned out to be at the edge of but one galaxy). Darwin came along and rudely asserted that people were descended from the apes and they from lower creatures still. Later Freud suggested that we humans are not even consciously in charge of our own facilities, that our egos and ids drive us. Now the computer may be dealing us the ultimate affront—surpassing us intellectually—beating us in our best suit, and seemingly inviting us to give in.

Surely there are promises of greater product quality, better efficiency, assistance in error reduction, and improved patient care that motivate the new technology changes. Nevertheless, the potential negative impacts must be examined, and those impacts need to be reduced to achieve a satisfactory adoption of automation by a society.

CONCLUSION

Computers have contributed much to health care, and many new uses of computers show promise. The pressure to control costs will inhibit progress some but must not be allowed to curtail developments that ultimately will both enhance care and reduce costs . Doctors, nurses, and other health-care workers must be trained in the technology of computers as relevant to their jobs and be conditioned to be wary of mode errors (where they assume a system is operating in one programmed mode when actually it is operating in another). They must be cautious about overtrust of the computer's "intelligence" and appreciate the need for continuing human responsibility. The notion of "people versus computers" must be seen as an improper perspective. The more appropriate perspective is that of human–computer collaboration. The fundamental complementarity of the two bodes well for greater performance and reliability and reduction of errors in the health-care system.

SUGGESTED READING

Albright, R. G. (1988). *A basic guide to online information systems for health care professionals.* Arlington, VA: Information Resource Press.

Anderson, J. G. & Jay, S. J. (Eds.). (1987). *Use and impact of computers in clinical medicine.* New York: Springer-Verlag.

Arenson, R. L. (Ed). (1986). The use of computers in radiology. *The Radiology Clinics of North America, 24*(1), 1–133.

Ball, M. J. (Ed.). (1988). *Nursing informatics.* New York: Springer-Verlag.

Booth, P. (1989). *An introduction to human-computer interaction.* Hillsdale, NJ: Lawrence Erlbaum Associates.

Blum, B. I. (1986). *Clinical information systems.* New York: Springer-Verlag.

Fassett, W. E., & Christensen, D. B. (1986). Computer applications in pharmacy. Philadelphia: Lea and Febiger.

Gallant, S. (1993). *Neural network learning and expert systems.* Cambridge, MA: MIT Press.

Kurzweil, R. (1990). *The age of intelligent machines.* Cambridge, MA: MIT Press.

Orthner, H. F., & Blum, B. I. (Eds.). (1989). *Implementing health care information systems.* New York: Springer-Verlag.

McDonald, C. J. (Ed.). (1988). Computer-stored medical record systems [Special issue]. *M.D. Computing, 5*(5).

Miller, P. L. (1988). *Selected topics in medical artificial intelligence.* New York: Springer-Verlag.

Reiser, S. J., & Anbar, M. (Eds.). (1984) *The machine at the bedside: strategies for using technology in patient care.* Cambridge, England: Cambridge University Press.

Spackman, K. A., & Connelly, D. P. (1987). Knowledge-based systems in laboratory medicine and pathology. *Archives of Pathology and Laboratory Medicine, 111*, 116–119.

REFERENCES

Baxt, W. G. (1991). Use of an artificial neural network for the diagnosis of myocardial infarction. *Annals of Internal Medicine, 115*, 843–848.

Buharali, A., & Sheridan, T. B. (1982). Fuzzy set aids for telling a computer how to decide. *Proceedings of 1982 IEEE International Conference on Cybernetics and Society,* (pp. 643–647). 82-CH-1840-8, Seattle, WA.

Felsenthatl, E. (1993, May 3). Cookbook care: Maine limits liability for doctors who meet treatment guidelines. *The Wall Street Journal,* p. A1.

Hollnagel, E. (1989, July). *AI and HCI: Much ado about nothing?* Copenhagen: Computer Resources International.

Kingsland, L. C., Lindberg, D. A., & Sharp, G. C. (1986). Anatomy of a knowledge-based consultant system: AI/RHEUM. *M.D. Computing, 3*(5), 18–26.

Martin, J. F., Schneider, A. M., & Smith, N. T. (1987). Multiple model adaptive control of blood pressure using sodium nitroprusside. *IEEE Transactions on Biomedical Engineering, 34*, 603–611.

Mazlish, B. (1967). The fourth discontinuity. *Technology and Culture, 8*(1), 1–15.

McDonald, C. J., Hui, S. L., Smith, D. M., Tierney, W. M., Cohen, S. J., Weinberger, M., & McCabe, G. P. (1984). Reminders to physicians from an introspective computer medical record. *Annals of Internal Medicine, 100*, 130–138.

Meline, L. J., Westenskow, D. R., Pace, N. L., & Bodily, M. N. (1985). Computer controlled regulation of sodium nipride infusion. *Anesthesia and Analgesia, 64*, 38–42.

Miller, A. S., Blott, B. H., & Hames, T. K. (1992). Review of neural network applications in medical imaging and signal processing. *Medical and Biological Engineering and Computing, 30*(5), 449–464.

Miller, R., Schaffner, K., & Meisel, A. (1985). Ethical and legal issues related to the use of computer programs in clinical medicine. *Annals of Internal Medicine, 102*, 529–536.

Packer, J. S., Mason, D. G., Cade, J. F., & McKinley, S. M. (1987). An adaptive controller for closed loop management of blood pressure in seriously ill patients. *IEEE Transactions on Biomedical Engineering, 34*, 612–616.

Schwartz, W., Weiss, K. M., & Buchanan, A. V. (1985). Error control in medical data. *M.D. Computing, 2*(2), 19–25.

Sheridan, T. B. (1992). *Telerobotics, automation, and human supervisory control.* Cambridge, MA: MIT Press.

Shono, H., Oga, M., Shimomura, K., Yamasaki, M., Ito, Y., Muro, M., & Sugimori, H. (1992). Application of fuzzy logic to the Apgar scoring system. *International Journal of Biomedical Computing, 30*(2), 113–123.

Swor, R. A., & Hoelzer, M. (1990, March). A computer-assisted quality assurance audit in a multiprovider EMS system. *Annals of Internal Medicine, 19*, 286–290.

Weinstein, J. N., Rubinstein, L. V., Koutsoukos, A. D., Kohn, K. W., Grever, M. R., Monks, A., Scudiero, D. A., Welch, L., Chiausa, A. J., Fojo, A. T., Viswanadhan, V. N., & Paull, K. D. (1993). Neural Networks in the discovery of new treatments for cancer and AIDS. *Proceedings of the World Congress of Neural Networks, 1*, 111–116.

Westenskow, D. R. (1986). Automating patient care with closed-loop control. *M.D. Computing, 3*(2), 14–20.

Ying, H., McEachern, M., Eddleman, D. W., & Sheppard, L. C. (1992). Fuzzy control of mean arterial pressure in postsurgical patients with sodium nitroprusside infusion. *IEEE Transactions on Biomedical Engineering, 39*(10), 1060–1070.

9 Medical Devices, Medical Errors, and Medical Accidents

J. W. Senders
EMSCO, Inc.

Human error in medicine, and the adverse events that may follow, are problems of psychology and engineering, not of medicine. The following brief description of a serious problem stemming from a simple, well-established and very low-tech medical device shows that even in such technology there is much room for improvement. The case also reveals that there is even more room for improvement in the philosophy and general outlook of the people who design, manufacture, and market such devices—they need to know some psychology. They need to know about human behavior, ergonomics, and statistics.

The Anesthetic and Life Support Drugs Advisory Committee (ALSDAC) of the Food and Drug Administration (FDA) also needs to know these things. It is currently the case that medical devices may be approved by the FDA if they meet the criteria of being safe and effective when used as intended. It would be even better if the manufacturers added one more small criterion of their own before requesting approval for new medical devices: That devices be safe, even if not effective, when used in ways other than that which was intended.

The manufacturers' lack of interest in ensuring that devices be safe when used other than as intended opens the door to disaster. For example, it has led to a series of deaths and injuries over a 14-year period between the introduction and the withdrawal of preloaded syringes of concentrated Lidocaine. As described in the following case study, the intended use was that it be diluted by injection into bags or bottles in intravenous (IV) infusion systems.

159

A CASE STUDY: THE PRELOADED SYRINGE OF
CONCENTRATED LIDOCAINE

Lidocaine HCl has a wide variety of uses. One of the more common is "for acute management of life-threatening ventricular arrhythmias, such as occur during acute myocardial infarction," as stated in an information sheet published by Astra Pharmaceutical Products, Inc. The bulletin goes on to quote Danahy and Aronow (1976): "The incidence of ventricular fibrillation is particularly high in the first few hours after an acute infarction, before most patients reach the hospital. We have no ideal drug for this problem, but Lidocaine appears to be the best agent currently available." There were, until 1992, a great many sizes and formats of Lidocaine on the market. The events described all involve the erroneous use of either of two then-available 20% preparations in place of a 2% preparation. The concentrates were either a 5-mL syringe containing 1,000 mg of Lidocaine or a 10-mL syringe containing 2,000 mg of Lidocaine. The 2% preparation was a 5-mL syringe containing 100 mg of Lidocaine.

A common prescription is for a bolus dose of 100 mg of Lidocaine in a 5-mL preloaded syringe to be injected "directly" (not meaning through the skin into a vein, but almost invariably through a Y-port in an IV system). This is often followed by a level-sustaining drip administration of Lidocaine through the same IV system. The bag preparation for the drip is usually made up by injecting the contents of the 1,000- or 2,000-mg syringe into a half or a full liter bag of normal saline. The result is 2% Lidocaine. The 5-mL syringes are identical in diameter and length. The 10-mL syringe is the same length with a 40% larger diameter.

The preloaded syringes have been sold since 1975. In 1992 they were withdrawn from the market by the pharmaceutical firms that manufactured and sold them.

TABLE 9.1
Manufacturer ABC 1975–1982

Year	# Events Using: 1g	2g	# Deaths	# Permanent Inj.	# Recoveries
1975	0	1	1	0	0
1976	1	0	0	1	0
1977	1	0	0	1	0
1978	2	2	2	2	0
1979	5	3	3	1	4
1980	3	2 (+1 u/k)	4	1 (+1 u/k)	0
1981	4	0	3	0	1
1982	0	1	0	0	1
TOTALS	16	9 (+ 1u/k)	13	6 (+1 u/k)	6

TABLE 9.2
Manufacturer ABC 1982–1990

Year	Sales (in thousands)			# Events	Rate, Per Million	# Deaths & Injuries	Rate, Per Million
	100 mg	1 g	2 g				
1982	1,628	262	480	1	1.35	0	0
1983	1,654	203	416	4	9.61	2	4.81
1984	1,594	163	343	4	7.91	3	5.93
1985	1,776	143	347	2	4.08	0	0
1986	2,237	161	355	2	3.88	1	1.94
1987	1,778	115	248	1	2.75	0	0
1988	1,811	86	216	1	3.31	1	3.31
1989	1,603	75	158	2	8.58	2	8.58
1990	1,458	56	109	2	12.12	1	6.06
TOTALS	15,539	1,264	2,672	19	4.83	10	2.54

From the very year of introduction of the preloaded syringes, 1975, there has been repeated accidental misadministration of the concentrate into the Y-port of an IV set with serious adverse consequences. The number of incidents is difficult to know exactly, due both to the inadequacies of the reporting system and to the litigiphobia that afflicts hospitals and medical personnel. It appears that among the four manufacturers, there have been about 100 recorded. More than half of the patients who received the overdose died or suffered irreversible neurological damage. Table 9.1 presents data from one pharmaceutical manufacturer for the period from 1975 through 1982.

From 1982 through 1990, sales data were available, and a more comprehensive and revealing set of data is shown in Table 9.2.

The number of deaths in this period was six and the number of permanent injuries four. The outcome of two events is unknown. The remainder recovered.

Note that the sales of the concentrates fell off in time so that the event rate and the injury rate both climbed, even though the absolute numbers may have appeared (to some optimists) to drop. There was no peak due to inexperience with a subsequent settling down as has from time to time been claimed.

Data from another pharmaceutical manufacturer are presented in Table 9.3.

The overall rate of 6.65 per million is surely low, given that event data for 1978–1980 are missing, and it seems unlikely that there were none in that 3-year period. Data on death and injury rates were not available, but they are unlikely to be very much different from the data in Tables 9.1 and 9.2.

The number of events in the three tables is 69. The reporting rate has

TABLE 9.3
Manufacturer XYZ 1977–1990

| | Sales (in thousands) | | | # | Rate, Per | # Deaths |
Years	100 mg	1 g	2 g	Events	Million	& Injuries
1977	653	106	N/A	N/A		
1978	686	107	108	N/A		
1979	827	157	178	N/A		
1980	980	244	272	N/A		
1981	1,011	253	315	2	3.52	N/A
1982	986	194	252	4	8.98	N/A
1983	1,212	154	240	1	2.54	N/A
1984	1,520	156	266	0	0	N/A
1985	1,754	129	235	2	5.48	N/A
1986	1,651	101	214	6	19.10	N/A
1987	1,352	65	172	3	12.6	N/A
1988	1,968	81	186	4	15.0	N/A
1989	2,270	68	225	0	0	N/A
1990	2,910	N/A	N/A	1	N/A	
TOTALS	19,780	1,815	2,663	23	6.65 (1981–1989 inc. 22 events)	

been estimated to be somewhere between 10% and 50% of the actual rate. Those events that did not lead to reportable injuries were generally not reported, and there were many events discussed in the proceedings of the ALSDAC that clearly had not found their way into the data. The declining sales figures suggest that informal word was circulating.

The Psychology of the Manufacturers

From the very first incident, the manufacturer's practice has been to attribute responsibility to the person, usually a nurse, who administered the drug. There was a concentration of attention on the labeling, on the packaging, and on the words used for warnings, all to no avail. Not one of the remedies used by the manufacturers was ever demonstrated to alter the probability of the precipitating error that led to the accident. Nor, so far as could be determined, was there even any thought given to the idea of *assessing* the efficacy of the various package changes that were made. Routine letters were sent to the hospitals where the events occurred pointing out that the syringes were designed so as to make such misuse difficult (despite the evidence). The fact that the labels said what should and should not be done was brought forth as proof that if only people would read the labels the accidents would not happen.

Both internally and in reports to the FDA the events were identified as overdose rather than as package design problems. The result of this was that they were entered into the *medical* rather than the *engineering* information channels.

Within two of the manufacturers' files were reports showing that informal committees had been organized and had met briefly. The committees decided that it was impossible to redesign the syringe so as to interfere with, to interrupt, or to interdict the execution of the error in order to prevent injury to patients or make it less likely.

In November 1979, the FDA wrote to the manufacturers and suggested, after reviewing the sad events, two courses of action:

Toward this end, we suggest the following for your consideration:

a. Discontinuance of the marketing of the 1000 mg. and 2000 mg. dosages in 5 ml. and 10 ml. syringe systems, and making these dosages available in larger volumes only, for example 25 ml. and 50 ml.

b. Change the configuration of the 5 ml. and 10 ml. vials, 1000 mg. and 2000 mg., making them incompatible with the direct injection apparatus.

We invite your comments.

One manufacturer met with the FDA in response to the letter and discussed labeling primarily. There was also discussion (XYZ Pharmaceutical, 1980) of the possibility of making the product "foolproof," of the fact that "many of the nurses do not even speak English," and of the fact that "people in the hospital setting are not reading labeling but rather are reacting to color codes and packaging." So far as could be determined, none of the persons in the group had any experience in engineering, design, or psychology. None of them had any knowledge of why people make errors, or whether people could be prevented from making errors by adding warnings and the like. Nor was there evidence showing that any effort was made to find out why the accidents occurred except to identify the error that led to the accident and the person involved, usually a nurse.

One internal report, chilling in its to-the-point brevity said, "Physically redesigning the Lidocaine (syringe) *at this time* (emphasis in original) was determined not to be required. This is not to say it won't be required sometime in the future, but we will stick with labeling improvements in the near term" (ABC Pharmaceutical, 1980)—this while acknowledging that people were not reading labels! The report went on to suggest the addition of more labels. None of these was ever assessed for efficacy.

Many people died or were seriously injured subsequent to that decision. It seems clear that the writer of that report knew that it was possible to redesign for safety but elected to recommend that it not be done until it was required by the FDA.

The annual sales of the preparation dropped in the mid-1980s. Although there is no documentary evidence, the reduction in sales suggests that the word had begun informally to spread and the number of hospitals purchasing the product fell off. It was claimed by a representative of one of the manufacturers at a meeting of the ALSDAC that the number of incidents (those that found their way into the reporting system at any rate) decreased also. The claimed reduction in frequency was produced in support of an untested assertion that there was usually a rash of incidents when a new preparation or drug was introduced and that this was followed by a reduction as people gained experience in the use of the new product. However, in terms of number of incidents per million units sold there was no reduction, and possibly even an increase, in incident rate. (See the tables presented earlier.) Basically neither industry nor government made, or at any rate made public, any systematic analysis of the available data.

The attitude described here was maintained throughout the entire period that this preparation of the drug was marketed.

The Psychology of the Anesthetic and Life Support Drug Advisory Committee

The Lidocaine problem was well known by 1985. There was discussion in the ALSDAC meetings bearing on the topic in the meetings of 1985, 1986, and 1987.

In the meeting on December 13, 1984, the following appears in the minutes: "You will see, over time, that we hit a peak of case reports of these Lidocaine overdoses in 1979 and '80, and slowly have been tapering off since then." Yet for one manufacturer alone there were 4 cases in 1983 and 4 in 1984 with declining sales of the product.

The record continues: "Even though the frequency for reporting Lidocaine overdoses and complaining about Lidocaine labeling seems to be declining, that may be a reflection ... of a certain acceptance of the risk, and it no longer is a reportable event." Also appearing in the minutes is: "I think, if we do anything, we need to recommend a change in the packaging. ... The labeling will just not solve the problem"; and from another speaker: "Nobody is reading the labeling in the first place." There was then a unanimous vote to change the packaging rather than changing the labeling. This was never done. It is of interest to note that one of the members of the ALSDAC stated that the dangerous syringe did not have a needle on it. It does. He was corrected by another who said that the needle was hidden "way back inside" a shield. It is not. The needle is permanently attached and projects well beyond a shield that has only one function: It prevents direct injection into a vein through the skin. The needle fits admirably on the IV port. Neither of these ALSDAC members had their facts in order or the object of discussion in front of them.

Like the manufacturers' groups, the members of ALSDAC made assertions that the syringe could not be redesigned so as to prevent the injury or death that might follow from an initial error. So far as could be determined, none of the persons in ALSDAC had any experience in engineering, design, ergonomics, or psychology. None of them appeared to have any knowledge of why people make errors or whether people could be prevented from making errors by adding warnings and the like. Nor was there any evidence showing that any effort was made to find out why the accidents occurred except to identify the error that led to the accident. There was, as there was in the manufacturers' groups dealing with the problem, talk about foolproofing and idiotproofing the packaging. The only significant difference between ALSDAC and the manufacturers was that ALSDAC also considered some members' suggestions that the preparation be withdrawn from the market. However, no action was taken to accomplish this.

In the ALSDAC meeting of April 1985, a motion was passed unanimously to limit the contents of a preloaded syringe to 100 mg. This is cited in the record of the meeting in December 1986 when a question was asked as to what was being done. Later in the meeting a motion was made to require that the 20% concentrate, instead of being removed from the market, should be packaged so that it could not be injected into an IV port. After some parliamentary discussion, the motion was withdrawn and a new motion was made to refer the matter to a subcommittee to look into it and report at the next meeting. It was passed unanimously.

In the meeting of May 1987 it was decided that ALSDAC could not mandate that something be made idiotproof, and the idea was dropped in favor of improved labeling.

It is clear that the ALSDAC recognized the existence of a serious problem but was unable to address the issues. There was a lack of technical competence with respect to data analysis, theory of human errors, ergonomic design, and mechanical engineering. Ultimately, labeling was seen as the only solution, even though it was simultaneously recognized that it would not solve the problem because it had become eminently clear that people did not read labels in many critical medical settings. It was easier to accept two quite incompatible ideas about the accidents than to seek competent external advice.

The following year ALSDAC passed unanimously a motion to recommend that the preloaded syringes of concentrated Lidocaine be withdrawn from the market. Nothing was done about it.

BACKGROUND AND THEORY OF ERROR AND ACCIDENT

The nature of error has been discussed in other chapters of this book. Any proposal to do something about errors in medicine depends directly and strongly on some theory of what an error is, how an error is generated, what happens after an error comes into being, how it is that an error is detected, and what it is that causes injury or death to a patient. I must, therefore, define error (in medicine and elsewhere) in my way in order to be able to discuss what can be done about errors and their consequences in a medical setting.

To cure medical error, one must know something about the disease. What is an error? Why do errors occur? Why do medical errors occur? The causal mechanisms of medical errors, if there are any, must be the same as those of errors in nuclear power plants or aircraft, or in the kitchen. Although most people use the term medical errors, what they talk about are medical accidents. What is the difference between error and accident? I first try to give answers to these and other questions.

Definitions, Theories, Models, and Taxonomies

What is an error? From the viewpoint of an external observer, an error is a failure to perform an intended action that was correct given the circumstances. An error can occur only if there was or should have been an *appropriate intention to act* on the basis of a perceived or a remembered state of events. Then, if the action finally taken was not the one that was or should have been intended, there has been an error.

An error must *not* be defined by an adverse or serious outcome. An adverse outcome or accident may happen with no antecedent error. This may occur if the intention was the proper one, the action was properly executed, and the outcome was not certain. Examples are playing a game, deciding whether to carry an umbrella, administering a medication, or performing an operation known to be risky.

What Is an Accident?

An accident is an *unplanned, unexpected, and undesired event,* usually with an adverse consequence. An adverse outcome after an error, by this definition, must be construed to be an accident. No one plans, expects, or desires an accident.

An error is a psychological event with psychological causes, if errors are caused at all (there is always the possibility that causes of all or some errors can not be identified). An error may have any of a (possibly large) number of causes. A defined causal mechanism can give rise to a taxonomy of errors.

As for almost anything, there are an almost unlimited number of taxonomies of error.

Some Taxonomies of Error

Of the many possible taxonomies of error, here are a few that have been found useful.

Errors can be classified according to a hypothetical internal causative process.

Input Error or Misperception.   The input data are incorrectly perceived, then an incorrect intention is formed based on that misperception, and the wrong action is performed predicated on the incorrect intention. Thus an action was committed other than that which would have been intended had the input been correctly perceived. For example, a person may be confronted by the phrase "1000 mg" and see it as "100.0 mg." The person decides that it should be administered as a bolus into a Y-port and successfully does so. A fatal overdose results.

Intention Error or Mistake.   The input data are correctly perceived, an incorrect intention is formed, and the wrong action is performed predicated on the incorrect intention. Thus an action was committed other than that which should have been intended given that the input was correctly perceived. For example, a person may be confronted by the phrase "1000 mg" and see it as "1000 mg." That person incorrectly decides that it should be administered as a bolus into a Y-port and successfully does so. A fatal overdose results.

Execution Error or Slip.   The input data are correctly perceived, the correct intention is formed, and the wrong action is performed. Thus an action was committed other than that which was intended. For example, a person may be confronted by the phrase "1000 mg" and see it as "1000 mg." The person correctly decides that it should be administered as a drip after dilution in a drip bag. That person becomes distracted while approaching the patient and, from habit, injects the contents as a bolus into a Y-port. A fatal overdose results.

One can classify errors according to the assumed locus of the causal process.

Endogenous Error.   This is an error that arises from processes inside the actor. The elimination or reduction of such errors must involve psychology, physiology, or neurology. The error resulting from distraction cited previously is endogenous. It probably results from the capture of the lower probability process of injection into a bag by the higher probability process of injection into a Y-port. It should be noted that about four times as many of the bolus doses than of the concentrates are sold and the two task sequences have common elements of action. Such a situation is an opportunity for a capture error.

Exogenous Error.   This is an error that arises from processes outside the actor. The elimination or reduction of such errors involves design of objects and work environments and correction of policies, protocols, and procedures. For example, the inconsistent use of an extraneous ".0" in the quantity "100" induces the false interpretation of "1000" as "100.0" and could lead to an overdose accident. The unnecessary custom of keeping both bolus and dilution syringes in the same area permits the substitution error.

Errors can be classified according to the observable nature of the error. If an error actually results in an action, then there is a phenomenon that can be observed. The particular appearance of the error may be called its *mode*. An example of a phenomenological taxonomy of error (by modes) follows.

Error of Omission.   This is an error characterized by the leaving out of an appropriate step in a process.

Error of Insertion.   This is an error characterized by the adding of an inappropriate step to a process.

Error of Repetition.   This is an error characterized by the inappropriate adding of a normally appropriate step to a process.

Error of Substitution.   This is an error characterized by an inappropriate object, action, place, or time instead of the appropriate object, action, place, or time.

Why Are There So Many Taxonomies?

The various taxonomies serve different purposes. The internal cause taxonomy may provide a theoretical basis for a program of behavioral or neurological research. The locus of causal process taxonomy divides the universe of error into those that can be analyzed and cured by engineering, design, societal, and procedural changes, and those that can be analyzed and cured through psychological intervention and modification. The observable nature of the error taxonomy provides a basis for the analysis of the consequences that will follow on the expression of the error in a particular working environment (an operating room or a nursing station, for example).

It is obvious that the various taxonomies are sometimes in conflict with regard to the ways in which errors may be differentiated. The taxonomy organized in terms of hypothetical internal causal mechanisms conflicts with the taxonomy organized in terms of the phenomena. For example, the three differently classified examples of input error, intention error, and execution error constructed for the internal causal process taxonomy all lead to a single class of substitution error in terms of the observable nature of the error.

The conflicts serve to emphasize that for the most part each taxonomy serves a special purpose. The conflicts arise because of the differences of the goals.

How to Talk About Errors

Much of the discourse about errors, whether in medicine or in other domains, is confusing and confused because of the differences in terminology used by the discussants. One person's error may be another's accident. One person's slip may be another's mistake. A generally accepted, standard terminology has yet to be established.

The *mode* of an error will result in some kind of *expression*, that is, something wrong will be done in the particular environment in which the action occurs. The expression must depend on what is available to be done in the environment. In a medical setting, an *error of substitution* (its *mode*) may result in a nurse's picking up a 2-g prefilled Lidocaine syringe (its *expression*) instead of a 100-mg syringe. In a nuclear power plant, the same error might be expressed by turning off the wrong pump. The error is the same in both cases: A wrong act has been substituted for the right act.

Finally the *expression* may lead to a negative *consequence*, that is, something adverse will happen as a result of what was erroneously done; that is an accident caused by an error. For example, the syringe substitution error can result in a massive overdose of Lidocaine. As has been indicated previously, attention is usually directed to the negative consequence because that is (usually) reported. Attention is also directed to the identifiable error that led to the consequence because it usually identifies the person who can be assigned responsibility.

What Is a Medical Error?

It is common to discuss errors in medical settings in terms of their expressions, that is, what was done wrong. It is common to report errors in medical settings only in terms of their adverse consequences, that is, what happened to the patient. This habit has serious shortcomings. Only those consequences that result in injury or death are noted. The number of incidents involving Lidocaine overdose, for example, is an interesting statistic, but it does not indicate much about the number of errors or why they occurred.

In reality, what are reported are not medical errors but medical accidents consequent on errors for which someone might be held responsible. What is not seen are those errors that occurred and were caught before they were completed. There is no good estimate of the probability of substitution errors on the night shift, or by physicians, or by pharmacists.

It would be of great importance to know the modes of the errors that were not harmful. The information could, for example, help in estimating the risk of the introduction of a new drug, a new package, or a new device into a health-care setting. Some kind of better data collection process is needed.

A medical error is an error that happens in a medical setting and is made by someone who is engaged in a medical activity. Some small fraction of medical errors lead to accidents that usually happen to a patient—the very person the medical establishment is dedicated to protecting and helping. It is usually the case that if there is no adverse outcome—no accident—the error is not reported and does not become part of the experience base of the practice of medicine.

It is clear that to understand what accidents are likely to happen to patients, information must be accumulated about all errors: those that injure or kill, and the near misses or those that have not yet done so. Then appropriate protective measures can be in place to wait for the error and interdict the accident rather than to be put into place after the accident.

Why Is a Medical Error Different from Any Other Error?

The high level of sophistication of modern medicines and medical devices makes the effect of misadministration more likely to be harmful than would be the case, for example, if the medicines were pharmaceutically inert (homeopathic remedies, perhaps).

Because of the very personal nature of medicine, the myth and mystique that surrounds it, and the image of the doctor in today's society, error in medicine is always seen as a special case of medicine rather than as a special case of error. Because of the frequency of litigation against hospitals and physicians in consequence of real or imagined medical error, the other class of persons studying error are risk managers, insurance specialists, and legal counsel. The consequence is that medical errors are usually analyzed exclusively by persons qualified in medicine, risk analysis, or law rather than by persons qualified to study error.

The unfortunate result has been the isolating of medical errors from much, though not all, of the body of theory, analysis, and application that has been developed to deal with error in other fields such as aviation and nuclear power. Because of the intensely personal nature of medicine and because of the ostensibly curative, helping, and ameliorative nature of the medical process, the consequences of medical error are viewed with more alarm than those in many other enterprises. That all of us, sooner or later, will receive the benefits, and be exposed to the risks, of modern medicine exacerbates the sense of alarm. The adverse outcomes are directly opposed to the ostensive nature of the whole enterprise.

It is difficult to convince politicians and administrators, patients and physicians, judges and lawyers, and juries, that one need not necessarily know anything particularly deep about medicine to be able to identify, analyze, and rectify the mechanisms that led to a medical error. The problems that arise and remain unsolved because of this difficulty have led to repeated injuries and deaths to patients and to subsequent expensive litigation. The solution lies in the application of nonmedical knowledge to the cure.

Much as human behavior in a medical setting is still behavior and not medicine, human error in a medical setting is still error and not medicine. Medical error must be considered to be the result of the expression of error in a situation in which there are medically significant things to be done and done wrong. If an error that was expressed in a medical setting instead happened to occur in a nuclear power plant, the only differences would be in the words used.

The Biases in the Data

A major problem for the collector of medical error data is that so little is reported. Reporting biases stem from hospitals' fear of litigation and from the fact that the interest of administrators and regulators is directed to the adverse outcomes rather than to the mechanisms that led to the outcomes. Underdoses rarely kill or injure and therefore do not get reported. Yet an underdose is the same error as an overdose but with the opposite sign.

Pharmaceuticals are frequently inherently dangerous. Anesthetics, in particular, can be considered as designed to bring patients to a state between life and death. As a consequence, the potential for catastrophic outcome of minor overdose errors is very much increased. In general, that which is beneficial in small amounts may be of no avail in lesser amounts and malign in greater amounts.

The result is that overdoses are more frequently reported than underdoses, so that the statistics arc distorted. There is no generally accepted error-reporting system designed to reveal all those errors that occur but have not (on that occasion at any rate) led to adverse consequences. Such a universal reporting system is necessary because there are so many identifiable errors that are benign except under special circumstances. If they were identified and appropriate actions taken they might still occur but would effectively be disarmed through design changes.

Fear of litigation is one of the major reasons for the underreporting of inconsequential errors. The argument appears to be that error reports might be used to prove at some later date that a hospital knew that its staff made errors of a particular kind. Then, the logic seems to go, if at some later time the same kind of error resulted in an accident to a patient, negligence could be alleged. The negligence could have been in not disciplining or discharging personnel, in not increasing training or motivation, or in not changing something (no matter what).

The Mental Act of God (MAOG)

Should people, the actors, be blamed for their errors? Should they be held responsible? Blame implies a theory that incipient errors can be perceived by the actor before they are executed and voluntarily controlled to prevent their execution. Responsibility implies a theory that adverse consequences arise because of flaws in behavior.

An act of God is defined thus: "In law; a direct, sudden, and irresistible action of natural forces, such as could not humanly have been foreseen or prevented" (The American College Dictionary), and is also defined thus: "Operation of uncontrollable natural forces" (Concise Oxford Dictionary).

Errors, to the extent that we have data, are random; the moment when an error will occur cannot be predicted. There is no aura that signals to an actor that an error is about to occur even though people can be trained to be aware of a heightened probability of error. From the point of view of the actors, the errors that they commit are MAOGs. The actors are the victims of the error. The patient is the victim of the expression of the error in a badly designed medical setting that permits the error to be completed and to produce an injury. Blaming people for making errors is like blaming them for breathing. They will do both willy-nilly. Leape et al. (1991) said: "Preventing medical injury will require attention to the systemic causes and consequences of errors, an effort that goes well beyond identifying culpable persons" (p. 383).

Why Do Accidents Happen?

An accident, as outlined earlier, can be the result of an error that is made, that is not detected and interrupted by anyone, and that can be completed in an action that injures or kills a patient. It is my belief that with the best of personnel selection and training, the highest of dedication, and the greatest of motivation, errors will still be made. An error-free environment is an illusory goal or, worse, a deliberate abdication of responsibility on the part of those who should do something about the consequences of error but prefer the easier route of blaming the people for MAOGs.

WHAT IS THE SOLUTION?

An error can occur, and can be self-detected and (sometimes) corrected, at many points in the sequence of *mental* and *physical* events encompassing perception, decision, and the initiation of action. An error can occur, and can be self-detected and (sometimes) corrected at many points between the beginning and the end of an action.

Such detection can be of the mode of an error, of its expression, or of its consequence. For example, a nurse may start to reach for a 2-g Lidocaine syringe and change the motion toward its correct goal, the 100-mg syringe. This correction might be a conscious act or not; little has been done on the analysis of such barely expressed errors. For example, a nurse may actually pick up the wrong syringe and immediately return it to be replaced with the correct one. There has been an error, but the detection and correction occurred in the same moment as the action. There are many opportunities for self-detection. Experience indicates that the probability and frequency of such self-corrected errors will be high. Some recent data gathered by Senders and Cohen (1992) suggest that for every error completed by nurses (e.g., the actual use of a wrong syringe) there are about 10 that were caught before they actually were completed, so that there was no accident and therefore no report. Errors are much more likely than might be thought on the basis of data from present reporting systems. Even the depressing report of Leape et al. (1991) on hospital errors probably greatly underestimates their probability.

What Are the Remedies?

To reduce the probability of error, one can work on people, on procedures, or on the work environment. A reduction of error probability may be achieved by application of personnel factors aimed at improving human performance such as more powerful motivation, improved and more frequently renewed training, and the use of scientifically designed work-rest schedules.

Error may also be reduced and the probability of self-detection increased through the use of standard vocabularies and symbol sets, by standardized labeling and packaging, and by standardized design of the controls, displays, and interfaces of devices. Medical materials and devices should always be stored in a standard way and in standard locations in any medical environment.

It is an unfortunate fact that the process of the administration of pharmaceuticals, from the writing of a prescription to the preparation of the dose to the administration of the medication, is full of problems. Some of these stem from appalling handwriting and ambiguous abbreviations, poorly designed packaging, and nonstandard labeling. The gruesome catalog of medication errors presented by Davis and Cohen (1981) demonstrates, among other things, the effect of the absence of a controlled vocabulary with an absolute and enforced rejection of deviations from it for everyone involved.

To improve self-detection of errors, there must be two kinds of remedies. Some must be psychological. Sellen (1990), for example, suggested the possibility of training in the use of tactics leading to improved probability of self-detection of error. Other remedies must depend on the redesign both of the elements of the system and of the system as a whole. In an ideal world, a prescription would say in clear and unambiguous words what medication was to be given to which patient, when, how, how much, and so on. A medication container would tell the person holding it what its name is, what the appropriate dose is, how it should be used, and what the consequences will be if it is used in any of a variety of improper ways. It would say all this in multiple ways, clearly and unambiguously, as if on the assumption that the person holding it was blind, stupid, and ignorant. Medical things should announce their identities to the user through many independent and redundant perceptual routes.

The probability of error detection can be increased through the use of variation of shape, texture, color, mass, and size, all of which actively engage human perception and memory. This would provide parallel channels of information feedback about the identity of what has been grasped, whether a container of pharmaceutical or a control on a medical device.

Detection can also be improved by providing more active feedback to the user. The use of a bar-code reader and voice synthesizer on the cabinet in which medicines are stored would allow immediate confirmation by a watchful chip of what was picked up. Such equipment is in widespread industrial use, and costs are low. A cabinet so equipped that announced to nurses what they had just picked up would be an inexpensive and powerful collaborator against error.

Finally, to interdict the completion of the error in action, such as the bolus injection into an IV Y-port of a preloaded syringe of concentrated Lidocaine, the syringe should be designed so as not to fit where the contents should not be injected. What is needed is a design aimed either at physical interdiction of consummation of an error, in the manner of a lock-and-key pair, or at slowing or interfering with the consummation of the error. For example, the requirement of additional preparatory acts to be performed in order to use the syringe containing the concentrate would have offered renewed opportunity to detect the error in syringe selection. Ideally the device used not in accord with its intended use should simply not work at all—and should complain loudly about it.

Preventive Analysis

As Senders and Norman (1983) pointed out, it is always possible to find a reason for an accident and equally impossible to find a cause. The argument rests on the fact that (virtually without exception) accidents can be traced back to reasons masquerading as (apparent) causes, but prediction of the accident on the basis of knowledge of those same causes never occurs. There is an enormous number of chains of events antecedent to an accident. It can be said of each event that without that event, the accident would not have happened. These must be considered as reasons for the accident. They are the result of rationalization after the fact. Knowledge of the putatively causal events would not have led a rational observer to predict the accident. Nor, in my view, would such knowledge have led even a sophisticated observer/analyst to alter the estimated probability of the accident.

It must therefore be accepted that errors will inevitably occur and that the times of accidents cannot be predicted. The best approach is not to try exclusively only to *prevent* errors but also to reduce the probability of making errors and to increase the probability of self-detection and interruption of errors. In addition, failure mode analysis and fail-safe design must be used to interrupt or interdict the consummation of those errors that are executed, thereby minimizing the consequences even if the error occurs and is not detected.

Failure Mode Analysis

It must be accepted that errors will inevitably occur and that the times of accidents cannot be predicted. However, it is possible to predict the forms that errors will take. This makes preventive design possible. The remedy is to prevent the translation of the error into an accident. The manufacturers and the FDA must stop expecting nurses and physicians to use things correctly every time. Each medication package and each medical device must be subjected to failure mode analysis. In brief, the designer of anything to be used in the medical arena must ask:

What incorrect actions can people do?
What adverse result can arise from these incorrect actions?
How can those actions be prevented from being completed?

The application of failure mode analysis means that the possible ways in which each package or device can be misused should be exhaustively tabulated. Then the outcomes of each misuse are identified and evaluated, and those that are unacceptable must be designed out. That is, if a possibility is undesirable, then the possibility must be eliminated unless there is overwhelming reason to waive the requirement. This is like locking the barn door before the horses gallop out; the more usual approach has been to use a running horse detector to shut the door.

For example, the IV system with tubing, bags, Y-ports, and the like can be designed so that a prefilled syringe that is supposed to be injected into a drip bag and used as a drip in dilute form must be incapable of being injected into the arm of a Y-port. This requires, of course, meticulous analysis of even such simple systems as a prefilled syringe or the IV set. It also requires the enforcing of standards industry wide to provide assurance of incompatibility where it is necessary. Because the greater concentration of technical knowledge rests in the manufacturing establishment, it should be the responsibility of the manufacturers to perform the analysis and present the results for evaluation by the FDA.

CONCLUSIONS

There are few or no medical errors; there are many errors that occur in medical settings. Those that are not prevented from running their courses can lead to accidents. These are the ones that come to our attention; the others are lost in the course of time.

The lost errors are those that have not or have not yet injured or killed a patient. The lost information could have been used to predict what errors are likely to injure or kill a patient in the future.

There must be a system for voluntary and anonymous reporting of *all* errors that occur in the medical system. The reports must be collected into a single database that will permit search and collation of data by multiple attributes. Such a system is needed to predict the kinds of incorrect things people are likely to do in the future.

Many errors stem from the absence of a controlled vocabulary for use in the medical setting. Such errors could be eliminated by appropriate design as well as promulgation and enforcement of the use of a controlled vocabulary. All communication of medical orders and all names of medical preparations and devices should conform to the standards of the controlled vocabulary. This is especially true of prescriptions that are an errorgenous link between the medical system and the patient. Other errors stem from poor handwriting and the use of improper abbreviations (Davis, 1988), especially in the writing of prescriptions.

Improvements in training and work schedules will be effective in reducing, but will not eliminate, errors of medical personnel.

Because not all errors can be eliminated, it is necessary to increase the probability of self-detection of errors before they are expressed in action through training, through design of devices and containers, and through consideration of the error-inducing aspects of the social and physical environment in which errors occur.

Finally, because not all errors will be self-detected, it is necessary to eliminate or diminish the consequences of error in the medical setting through design. Human failure mode analysis is the necessary tool.

REFERENCES

Danahy, D. T., & Aronow, W. S. (1976). *Drug Therapy, 6,* 80–84.

Davis, N. (1988). *Medical abbreviations: 7000 conveniences at the expense of communications and safety.* Huntingdon Valley, PA: N. M. Davis.

Davis, N., & Cohen, M. (1981). *Medication errors: Causes and prevention.* Huntingdon Valley, PA: N. M. Davis.

Leape, L., Brennan, T. A., Laird, N., Lawthers, A. G., Localio, A. R., Barnes, B. A., Hebert, L., Newhouse, J. P., Weiler, P. C., & Hiatt, H. (1991). The nature of adverse events in hospitalized patients. *New England Journal of Medicine, 324*(6), 377–384.

Sellen, A. (1990). *Mechanisms of human error and human error detection.* Unpublished doctoral dissertation, University of California at San Diego.

Senders, J., & Cohen, M. (1992). *Near misses and real accidents.* Unpublished survey of nurses' errors.

Senders, J. W., & Norman, D. A. (1983). *On the dual nature of error causation.* Paper presented at the Second Clambake Conference on the Nature and Source of Human Error, Bellagio, Italy.

10 Radiopharmaceutical Misadministrations: What's Wrong?[1]

Dennis I. Serig
United States Nuclear Regulatory Commission

This chapter describes nuclear medicine and radiopharmaceutical misadministrations. It then discusses the conventional wisdom as to the cause of radiopharmaceutical misadministrations (i.e., what's wrong) and notes that a human factors perspective might lead to a more useful understanding of what's wrong. Development of a database designed to provide that more useful understanding is described. Data summaries are reviewed in terms of their support for the conventional wisdom versus the human factors perspective as to what's wrong.

BACKGROUND

Nuclear medicine involves the purposeful injection, ingestion, or inhalation of material containing a small amount of radioactivity (i.e., a radiopharmaceutical). The patient's exposure to radiation must be justified by a medical need. Most often the medical need is diagnostic, and the selection of radiopharmaceutical, dosage, and route of administration depends on the bodily structure or function about which information is required. Following administration of a diagnostic radiopharmaceutical, the patient is scanned or counted with electronic devices to obtain the required information. Examples of diagnostic nuclear medicine procedures are a

---

[1]Views expressed are the author's and not necessarily those of the U.S. Nuclear Regulatory Commission.

179

thyroid uptake using Iodine 131 sodium iodide and bone imaging using Technetium 99m labeled medronate or oxidronate (i.e., a bone scan).

The patient's medical need may also be therapeutic. In such cases, the administered radiopharmaceutical is intended to correct the patient's condition through destruction of specific body cells. Examples of therapeutic nuclear medicine include the use of Iodine 131 for treatment of Grave's disease or thyroid carcinoma.

When all goes well, a nuclear medicine patient's need is met without difficulty. However, on some occasions there are problems in meeting patient needs. Radiopharmaceutical misadministrations, exemplified by the following four incidents, are such occasions.

At a suburban clinic, a technologist injected a patient with what he believed to be a radiopharmaceutical used for bone imaging. Several hours later the patient was scanned. There was no evidence of bone uptake. Instead, the patient appeared to have been injected with a radiopharmaceutical used for brain and kidney imaging.

At a nearby large city hospital, the need for a lung ventilation study was written into a patient's chart. In preparing the patient for the study, the technologist selected and administered a vial of Xenon 133 with twice the prescribed radioactivity.

Across the state, a patient was referred to a university hospital for a renal ultrasound scan. Instead, the patient was injected with Technetium 99m Gluceptate and given a renal nuclear medicine scan.

At a government-owned medical center near the university hospital, a technologist scanned the nuclear medicine request form for a patient and noted that it involved Technetium 99m DTPA. She drew a standard dose of that radiopharmaceutical and injected it before noting that the requested study required inhalation of the radiopharmaceutical in aerosol form.

Wrong radiopharmaceutical, wrong dose, wrong patient, wrong route of administration—what's wrong? The conventional wisdom appears to be that random human error is what's wrong. Random? Random as in occurring without definite reason or pattern? Although such a view is common, it is not very useful to those who work to reduce the number of radiopharmaceutical misadministrations. What, after all, can be done to prevent events that occur without definite reason or pattern?

The perception of random human error as the cause of radiopharmaceutical misadministrations stems largely from within the nuclear medicine community. From that perspective, the perception may be very reasonable. Based on the commonly accepted estimate of one radiopharmaceutical misadministration per 10,000 administrations, the average nuclear medicine facility will experience less than one such event per year. Chances are that the errors leading to any two misadministrations at such a facility and the circumstances surrounding those errors will be different. From that perspective, it is easy for the nuclear medicine community to conclude that the human errors leading to radiopharmaceutical misadministrations occur without definite reason or pattern.

Could viewing radiopharmaceutical misadministrations from another perspective lead to a different perception? Several years ago, the Nuclear Regulatory Commission (NRC) initiated an effort to establish a human factors perspective on radiopharmaceutical misadministrations. The goal was to see whether that perspective might lead to a more satisfying answer to the question "what's wrong?" and, if so, to help rethink the challenge of reducing radiopharmaceutical misadministrations.

The concepts of human–machine system, task, and human error are crucial to a human factors perspective of nuclear medicine. Meister (1987) defines a human-machine system as "an organization of men and women and the machines they operate and maintain in order to perform assigned jobs that implement the purpose for which the system was developed" (p. 18). McCormick (1976) defines tasks as "units of human performance that are system related" (p. 24). Finally, Kinkade (1988) defines human errors as "simply deviations in performance from a specified standard" (p. 10).

Nuclear medicine can be viewed as a human–machine system. It involves an organization of people and equipment and requires successful performance of a number of tasks in order to meet a patient's medical need. Moreover, failure to perform a nuclear medicine task successfully (i.e., an error in Kinkade's terms) can lead to a radiopharmaceutical misadministration.

Not all errors lead to misadministrations. One reason is that nuclear medicine, like many other systems, is tolerant of some errors. For example, the specified standard of performance for dosage of some radiopharmaceuticals is ± 50% of the prescribed dose. An error leading to a smaller deviation from the prescribed dose would not be considered a misadministration. Another reason that not all errors lead to misadministrations is that some errors are detected and corrected. For example, timely double-checking of patient identification in the nuclear medicine clinic can detect that a ward has sent the wrong patient prior to administration of a radiopharmaceutical.

Whether or not an error leads to a misadministration and what, if any, type of misadministration occurs depends on the task being performed and the specific error. Human factors practitioners frequently develop a task inventory (i.e., a list of the tasks humans perform in a system) in order to help determine what type of errors can occur and what their system conse-

Task No.	Task Description
1.	Identify the need for diagnostic information or therapy.
2.	Make a record of the need.
3.	Transmit the need to the nuclear medicine department and to the patient.
4.	Schedule a clinical procedure.
5.	Prepare the radiopharmaceutical.
6.	Prepare the dose.
7.	Transport patient to clinic or prepared dose to patient.
8.	Conduct the clinical procedure.
9.	Process the results of the clinical procedure to obtain diagnostic information or therapy results.
10.	Interpret the information.
11.	Transmit the information to the requester.

FIG. 10.1.   A general inventory of nuclear medicine tasks.

quences might be. Figure 10.1 provides a general inventory of nuclear medicine tasks.

Typically the inventory of tasks would be further broken into subtasks and task elements. The goal of such a detailed breakdown is to describe required human performance at a level where all potential errors important to system performance can be identified. Figure 10.2 provides a typical sequence of subtasks for the task of preparing the radiopharmaceutical.

Figure 10.2 also illustrates the breakdown of subtask 5.1, elute Technetium 99m from a generator to obtain sodium pertechnetate, into task elements. The list of subtasks associated with a nuclear medicine task varies depending on, among other things, the intended clinical procedure. For example, preparation of a radiopharmaceutical for injection would differ from preparation of a radiopharmaceutical for inhalation.

Another way that human factors practitioners view systems is in terms of the degree of human versus machine control. Human–machine systems with the greatest degree of human control are termed manual systems. McCormick (1976) stated that manual systems "consist of hand tools and other aids which are coupled together by the human operator" (p. 13) and suggested that such systems were exemplified by a cook with his or her utensils.

Consistent with McCormick's (1976) picture of manual systems, much of nuclear medicine involves human performance with paper forms, charge cards, various vials, vial shields, syringes, syringe shields, and other small devices and simple pieces of equipment. Humans perform the individual tasks, subtasks, and task elements associated with the various equipment items, and, like a good cook, they also orchestrate the overall performance to achieve system goals.

At least implicit in every cooking and nuclear medicine task are subtasks and task elements that reflect the need not just to use the equipment but to control and monitor that use in a way that ensures success of the entire performance. Errors on control subtasks and task elements (e.g., on selection of a kit vial while preparing a radiopharmaceutical—see Fig. 10.2, subtask 5.3) can lead to radiopharmaceutical misadministrations as easily as errors on subtasks and task elements involving equipment use. Errors on monitoring subtasks and task elements (e.g., on verification of patient identification) can result in failure to prevent a radiopharmaceutical misadministration.

To derive value from a human factors perspective of a system, the reasons for performance deviations from a specified standard (i.e., human errors) must be identified. In order to do this, human factors practitioners evaluate

Task, Subtask/Task Element No.*	Task, Subtask, or Task Element Description
5.	Prepare the radiopharmaceutical
5.1.	Elute (i.e., remove by dissolving) Technetium 99m from a generator (i.e., a device where Technetium 99m accumulates as a decay product of Molybdenum 99) to acquire a vial of sodium pertechnetate (i.e., Technetium 99m in a saline solution).
5.1.1.	Select fresh elution vial (i.e., an evacuated vial for collecting sodium pertechnetate from a generator).
5.1.2.	Place elution vial in generator.
5.1.3.	. . .
5.2.	Place the vial of sodium pertechnetate in a radiation shield (i.e., a device to shield personnel from radiation).
5.3.	Select the appropriate radiopharmaceutical kit (i.e., a vial for mixing sodium pertechnetate with another material to produce a particular radiopharmaceutical) for the scheduled procedure.
5.4.	Place the kit vial into a radiation shield.
5.5.	Draw an appropriate amount of sodium pertechnetate from the elution vial.
5.6.	Inject the sodium pertechnetate into the kit vial.

*Tasks are identified by one-digit numbers, subtasks by two-digit numbers and task elements by three-digit numbers.

FIG. 10.2.   Breakdown of nuclear medicine task number 5 into subtasks and task elements.

tasks for mismatches between what is required of humans within the system and what those persons can reasonably be expected to do. They then search for factors that contribute to such mismatches.

An example of a performance requirement might be that a human must always correctly identify a particular material by its written name. A factor contributing to a mismatch might be that the name looks very similar to that of another material. The result is that the two materials are occasionally confused. Other factors, such as the name being handwritten or being printed on poor-quality paper by a nine-pin dot matrix printer with a well-used ribbon, can increase the mismatch and the error rate.

Once factors contributing to mismatches between what is required of a person and what that person can reasonably be expected to do are identified, they can be addressed. Common approaches to eliminating mismatches involve modifying the system to reduce human performance requirements or to enhance human performance capabilities. Modification of equipment design, procedures, and organizational structure and function are some ways to reduce human performance requirements. Modification of personnel selection and training are ways to improve the performance capabilities of people within the system.

Another approach to eliminating mismatches is to modify task allocation. Human tasks may be eliminated or reallocated to nonhuman elements of the system (e.g., machines or computers), which might perform them more reliably. Such an approach may be necessary if neither the performance required of people in the system nor the performance capabilities of those people can be modified sufficiently to eliminate the mismatch (Callan et al., in press).

A number of radiological health consequences are possible for patients receiving radiopharmaceutical misadministrations. The near-term consequences can range from negligible to a loss of thyroid function, which requires lifelong medication. The long-term radiological health consequences of some radiopharmaceutical misadministrations can be an increased risk of cancer. There are also nonradiological consequences of radiopharmaceutical misadministrations that should not be ignored. Among such consequences are delayed diagnosis, additional costs for repeated medical procedures, increased staff workload, and reduced confidence in the medical system.

METHOD

For a number of years the NRC required its medical licensees to submit reports concerning radiopharmaceutical misadministrations. Each report was like a small, not too clear, window on an isolated event. Often the view provided by the report was "ho-hum" or even frustrating. For example: "Our most senior and reliable technologist, Joe/Josephine, was in a hurry this morning and injected a patient with radiopharmaceutical A instead of radiopharmaceutical B. We are sorry. It won't happen again."

Sometimes the view provided by a report showed a flash of something interesting. For example: "Label color and wording on the intended kit vial and on the kit vial actually used were very similar." Even reports that provided interesting views raised more questions than they answered. How common is the reported error? Can it occur at other facilities? What factors can contribute to it?

In an effort to get a better view than was provided by the individual reports, the NRC built a computerized database of information about radiopharmaceutical misadministrations. It was hoped that the database would provide a larger, clearer window on radiopharmaceutical misadministrations. It was further hoped that the view through that window would reveal error-likely tasks, subtasks, and task elements and the factors that contributed to those error-likely situations.

The first step was to structure a dBASE III PLUS™ database that would capture the information from Diagnostic Misadministration Reports (see Fig. 10.3) submitted by licensees. Information from about 850 reports of misadministrations occurring in 1989 and 1990 was then entered into the database.

Early sorts and listings of information in the database indicated that it could serve as a larger, clearer window on radiopharmaceutical misadministrations. The window was not, however, as clear as desired. For example, looking at the data sorted according to the exact terminology used by licensees would have led an observer to believe that more than 50 chemical forms of radiopharmaceuticals containing Technetium 99m were used in nuclear medicine. Cleaning the window by adding a field to record standardized terms for the chemical forms of radiopharmaceuticals resulted in a data sort that reduced that number to less than 20—a number much more consistent with reality.

Other areas of the window were also cleaned. For example, a field containing standardized terms for the radioactive isotopes used in preparing radiopharmaceuticals was added. Original fields were retained to allow quality review of the cleaning process.

Early review of the data also indicated that the window could be widened. An example was the addition of fields to record the day of the week on which a misadministration occurred and the relationship of that day to weekends and holidays.

At this writing, each record in the database can contain information in 116 different fields. Information for 59 of those fields came directly from

NRC FORM 473
(5-88)
10 CFR § 35.33

U.S. NUCLEAR REGULATORY COMMISSION
Approved by OMB
3150-0140
Expires 5/31/91

DIAGNOSTIC MISADMINISTRATION REPORT

(N1) LICENSEE NAME

(N2) LICENSE NUMBER

(N3) CITY

(N4) STATE

(N5) EVENT DATE
MONTH DAY YEAR

(N6) REPORT DATE
MONTH DAY YEAR

(N7) TYPE OF MISADMINISTRATION

(01) WRONG RADIOPHARMACEUTICAL
(02) DOSAGE DIFFERING FROM PRESCRIBED BY 50%
(03) WRONG PATIENT
(04) WRONG ROUTE?

(N8) DID THE MISADMINISTRATION INVOLVE AN ISOTOPE OF IODINE
(999) YES   (111) NO

(N9) NUMBER OF PATIENTS WHO RECEIVED A MISADMINISTRATION UNDER THIS REPORT

(N10) INTENDED
(05) NO CLINICAL PROCEDURE
(06) NUCLEAR MEDICINE STUDY (Complete (N10A) INTENDED and (N11) GIVEN)
(07) X-RAY STUDY
(08) ULTRASOUND STUDY
(09) CT STUDY
(10) NMR STUDY
(11) OTHER:

(N10A) INTENDED
MILLICURIES | ISOTOPE | CHEMICAL FORM | STUDY

(N11) GIVEN
MILLICURIES | ISOTOPE | CHEMICAL FORM | STUDY

(N12) PRECIPITATOR

(71) REFERRING PHYSICIAN
(72) WARD NURSE
(73) WARD CLERK
(74) NUCLEAR PHARMACY
NAME OF NUCLEAR PHARMACY | CITY | STATE

(75) AUTHORIZED USER

(76) HOT LAB TECHNOLOGIST
(77) IMAGING TECHNOLOGIST
(78) CLINIC RECEPTIONIST
(79) SCHEDULING TECHNOLOGIST
(80) PATIENT
(81) OTHER:

(N13) ERROR

HOT LAB
(11) MISLABELED A SYRINGE
(12) MISLABELED A VIAL OR VIAL SHIELD
(13) RECONSTITUTED WRONG REAGENT KIT
(14) PLACED RECONSTITUTED VIAL IN WRONG SHIELD
(15) SELECTED WRONG VIAL WHEN DRAWING DOSAGE
(16) SET DOSE CALIBRATOR IMPROPERLY
(17) MISREAD DOSE CALIBRATOR
(18) MISUNDERSTOOD RADIOPHARMACEUTICAL OR DOSAGE ORDER

REFERRAL
(20) MISUNDERSTOOD REFERRING PHYSICIAN'S REQUEST
(21) REQUESTED WRONG STUDY
(22) REQUESTED STUDY FOR WRONG PATIENT

ADMINISTRATION
(30) SELECTED WRONG PATIENT
(31) ANSWERED WAITING ROOM PAGE INTENDED FOR OTHER PATIENT
(32) BROUGHT WRONG PATIENT TO CLINIC
(33) SELECTED WRONG SYRINGE FROM DOSAGE CART

OTHER
(40) Specify

(N14) CONTRIBUTING FACTORS
(80) STUDENT TECHNOLOGIST
(81) NEW EMPLOYEE
(82) FOREIGN LANGUAGE
(83) PATIENT INCOHERENT OR UNCONSCIOUS
(84) ID BRACELET NOT CHECKED
(85) REQUISITION NOT CHECKED
(86) PATIENT CHART NOT CHECKED
(87) NEW PROCEDURE
(88) HEAVY WORKLOAD
(89) OTHER:

(N15) ACTION TAKEN TO PREVENT RECURRENCE
IMPLEMENT NEW PROCEDURES FOR
(C1) VERIFICATION OF REQUEST
(C2) RADIOPHARMACEUTICAL LABELING AND HANDLING
(C3) VERIFICATION OF PATIENT IDENTIFICATION
(C4) REINSTRUCT PERSONNEL
(C5) REPRIMAND PERSONNEL
(C6) IMPROVE SUPERVISION OF PERSONNEL
(C7) NO ACTION
(C8) OTHER:

(N16) EFFECT ON PATIENTS | NONE APPARENT | SEE ABSTRACT

(N17) ABSTRACT (If more space is required, attach additional sheets.)

RADIATION OFFICER (Printed Name) | SIGNATURE | TELEPHONE | DATE

NUCLEAR REGULATORY COMMISSION USE
(N18) (999) YES (111) NO | (N19) AS | (N20) REGIONAL LOG NUMBER | (N21) ACCESSION NUMBER | (N22) INITIALS

NRC FORM 473 (5-88)

FIG. 10.3.   An NRC diagnostic misadministration report form.

NRC Form 473. Information in the other fields came from efforts to clean and widen the window provided by the database.

Following completion of the database, the capabilities of dBASE III PLUS™ were used to do simple sorts, counts, sums, and listings of data. The purpose was to find answers to questions such as how many of the various types of misadministrations occurred in 1989 and 1990, with what errors various types of misadministration were associated, and what contributed to those errors.

RESULTS

Table 10.1 shows the percentage of wrong radiopharmaceutical, wrong dose, wrong patient, and wrong route of administration misadministrations reported for 1989 and 1990.

The majority (68.9%) of the misadministrations involved the wrong radiopharmaceutical. That is, they involved patients whose physicians ordered nuclear medicine procedures but who received radiopharmaceuticals different from those required for the ordered procedure. A sizable minority (24%) of the misadministrations involved the wrong patient, that is, patients for whom nuclear medicine procedures were not ordered but who received radiopharmaceuticals anyway.

A small percentage (6.5%) of misadministrations involved the wrong dose, that is, patients for whom nuclear medicine procedures were ordered but who received radiopharmaceutical doses differing from their prescription by more than 50%. Finally, a very small percentage (0.6%) of the misadministrations involved the wrong route of administration; that is, patients who received radiopharmaceutical doses by injection, ingestion, or inhalation when one of the other two routes was appropriate to the intended nuclear medicine procedure.

Still, what was wrong? Why the wrong radiopharmaceutical, wrong dose, wrong patient, or wrong route of administration? Table 10.2 shows that, during 1989 and 1990, licensees attributed each type of misadministration to a number of different errors.

Table 10.2 also shows that a few types of error accounted for a large percentage of all the errors for each type of misadministration. For example, "reconstituted wrong reagent kit," "selected wrong vial when drawing dosage," and "selected wrong syringe from dosage cart" together accounted for almost 60% of the wrong radiopharmaceutical misadministrations. "Re-

TABLE 10.1
Number of Misadministrations by Type for 1989 and 1990

Type	Number	%
Wrong radiopharmaceutical	585	68.9
Wrong dose	55	6.5
Wrong patient	204	24.0
Wrong route	5	0.6
Totals	849	100.0

TABLE 10.2

First Error by Type of Misadministration for 1989 and 1990

Errors	Type 1	Type 2	Type 3	Type 4	Totals
None reported	65	8	17	—	90
Mislabeled syringe	19	1	—	—	20
Mislabeled a vial or vial shield	13	—	—	—	13
Reconstituted wrong reagent kit	52	—	—	—	52
Placed reconstituted vial in wrong shield	11	—	—	—	11
Selected wrong vial when drawing dosage	167	—	—	—	167
Set dose calibrator incorrectly	—	2	—	—	2
Misread dose calibrator	—	3	—	—	3
Misunderstood radiopharmaceutical or dosage order	15	14	—	—	29
Misunderstood referring physician's request	26	2	12	—	40
Requested wrong study	8	2	22	1	33
Requested study for wrong patient	—	—	47	—	47
Selected wrong patient	10	—	57	—	67
Answered waiting room page intended for other patient	6	—	12	—	18
Brought wrong patient to clinic	2	—	21	—	23
Selected wrong syringe from dosage cart	131	—	—	—	131
Other (specify)	60	23	16	4	103
Totals	585	55	204	5	849

Codes for misadministration types

Type 1 = Wrong radiopharmaceutical
Type 2 = Wrong dose
Type 3 = Wrong patient
Type 4 = Wrong route

quested wrong study," "requested study for wrong patient," "selected wrong patient," and "brought wrong patient to clinic" together accounted for more than 70% of the wrong patient misadministrations.

As a group, the seven errors that most often led to wrong radiopharmaceutical or wrong patient misadministrations accounted for more than 58% of all misadministrations reported during 1989 and 1990. It is clear that, although humans could make a number of errors leading to a misadministration, misadministrations were attributed to some errors much more often than to others.

What about the causes of the more frequently reported (excluding "none reported" and "other") errors associated with various types of misadministration? The diagnostic misadministration report form allowed licensees to mark boxes indicating which of several factors contributed to an error resulting in a misadministration. Some licensees marked none, some marked one, and others marked more than one such factor.

Table 10.3 shows the factors that licensees marked as contributing to the three most frequent errors leading to wrong radiopharmaceutical misadministrations.

The three most frequent errors leading to wrong radiopharmaceutical misadministrations were reconstitution of the wrong radiopharmaceutical kit, selection of the wrong vial when drawing a dose, and selection of the wrong syringe from the dosage cart. For each of those errors, licensees most often marked heavy workload as a contributing factor followed by the group of factors including student technologist, new employee, and new procedure. For selection of the wrong syringe from the dosage cart, licensees also

TABLE 10.3
Factors Contributing to the Most Frequent Errors Associated with Wrong
Radiopharmaceutical Misadministrations in 1989 and 1990

Contributing Factors	Error 13	Error 15	Error 33	Total
Student technologist	5	14	5	24
New employee	6	20	12	38
New procedure	1	—	—	1
Foreign language	—	—	—	—
Patient incoherent or unconscious	—	—	1	1
ID bracelet not checked	—	—	—	—
Requisition not checked	—	—	8	8
Patient chart not checked	—	—	4	4
Heavy workload	16	58	46	120

Codes for error types

13 = Reconstituted wrong reagent kit
15 = Selected wrong vial when drawing dosage
33 = Selected wrong syringe from dosage cart

TABLE 10.4

Factors Contributing to the Most Frequent Errors Associated with Wrong
Dose Misadministrations in 1989 and 1990

Contributing Factors	Error 18
Student technologist	1
New employee	2
New procedure	1
Foreign language	—
Patient incoherent or unconscious	—
ID bracelet not checked	—
Requisition not checked	—
Patient chart not checked	—
Heavy workload	1

Codes for error types

18 = Misunderstood radiopharmaceutical or dosage order

marked patient incoherent or unconscious and the group of factors including requisition not checked and patient chart not checked.

Table 10.4 shows the factors that licensees marked as contributing to the most frequent error leading to wrong dose misadministrations.

The most frequent error leading to wrong dose misadministrations was misunderstood radiopharmaceutical or dosage order. For that error, licens-

TABLE 10.5

Factors Contributing to the Most Frequent Errors Associated with Wrong
Patient Misadministrations in 1989 and 1990

Contributing Factors	Error 21	Error 22	Error 30	Error 32	Totals
Student technologist	—	—	—	—	—
New employee	1	2	1	—	4
New procedure	1	—	—	—	1
Foreign language	—	—	—	1	1
Patient incoherent or unconscious	—	1	3	5	9
ID bracelet not checked	—	1	39	14	54
Requisition not checked	2	—	3	—	5
Patient chart not checked	9	28	11	1	49
Heavy workload	2	4	9	3	18

Codes for error types

21 = Requested wrong study
22 = Requested study for wrong patient
30 = Selected wrong patient
32 = Brought wrong patient to clinic

ees marked heavy workload and the group of factors including student technologist, new employee, and new procedure as contributing factors.

Table 10.5 shows the four factors that licensees marked as contributing to the most frequent errors leading to wrong patient misadministrations.

The four most frequent errors leading to wrong patient misadministrations were requested wrong study, requested study for wrong patient, selected wrong patient, and brought wrong patient to clinic. For each of those errors, licensees most often marked the group of errors including identification bracelet not checked, requisition not checked, and patient chart not checked. Licensees also marked heavy workload as a factor contributing to all four errors. Licensees marked the group of factors including student technologist, new employee, and new procedure and the group of factors including foreign language and patient incoherent or unconscious a few times for some, but not all, of the four errors.

Table 10.6 shows the factors that licensees marked as contributing to an error leading to a single wrong route of administration misadministration.

The error leading to the wrong route misadministration shown in Table 10.6 was requested wrong study. The licensee marked new procedure as a factor contributing to the requested wrong study error.

The data in Tables 10.3–10.6 can be summarized in terms of groups of contributing factors. The group of contributing factors that included student technologist, new employee, and new procedure suggested that associated errors were due to inadequate familiarity or experience with some task, subtask, or task element. Members of this group of factors contributed to the most frequent errors leading to all four types of misadministration.

TABLE 10.6

Factors Contributing to the Most Frequent Errors Associated with Wrong Route Misadministrations in 1989 and 1990

Contributing Factors	Error 21
Student technologist	—
New employee	—
New procedure	1
Foreign language	—
Patient incoherent or unconscious	—
ID bracelet not checked	—
Requisition not checked	—
Patient chart not checked	—
Heavy workload	—

Codes for error types

21 = Requested wrong study

The group of contributing factors that included foreign language and patient incoherent or unconscious suggested that associated errors were due to difficulties in obtaining necessary information from the patient. Members of this group of factors contributed to the most frequent errors leading to wrong radiopharmaceutical and wrong patient misadministrations.

The group of contributing factors that included identification bracelet not checked, requisition not checked, and patient chart not checked suggested that the error was failure to perform a control or monitoring subtask or task element. Members of this group contributed to the most frequent errors leading to wrong radiopharmaceutical and wrong patient misadministrations.

The sole member of the fourth group of contributing factors was heavy workload. It suggested that performance of some tasks, subtasks, or task elements was influenced by some aspect of the quantity of work to be done. Heavy workload contributed to the most frequent errors leading to wrong radiopharmaceutical, wrong dose, and wrong patient misadministrations.

DISCUSSION

Random human error? Together, wrong radiopharmaceutical and wrong patient misadministrations accounted for more than 92% of the reports for 1989 and 1990. Moreover, despite the fact that a wide range of errors can lead to misadministrations, only a small number actually accounted for the majority of the data. Which misadministrations occurred and which errors led to those misadministrations looked a lot less random when information from a large number of reports was pooled.

So what is wrong? Like cooking, nuclear medicine is a manual system. If you know one of your grandmother's wonderful recipes, you also know that what "a pinch of this" and "a rounded spoonful of that" mean is knowledge gained only through long experience. The measures used in radiopharmaceutical preparation are not so personal, but successful routines to meet goals and to avoid problems are developed and assimilated into the larger process only through experience. Sometimes such routines are recognized in formal procedures to be followed by everyone. Often they remain the idiosyncratic practices of the person who developed them.

Inadequately developed or assimilated routines may explain the contribution that the group of factors including student technologist, new employee, and new procedure makes to the most frequent errors leading to all four types of misadministration. Student technologists and new employees may simply not have had the experience necessary to develop routines for avoiding errors on all the various nuclear medicine tasks. Even new employees with previous work experience may need to learn routines appropriate to their new situation. At the same time, experienced new-hires may be affected by negative transfer of training (i.e., the need to unlearn aspects of previous routines that are no longer appropriate).

New procedures can require changes in everyone's routines. In such situations, long-term employees are placed in much the same situation as new employees with previous work experience. Learning the new procedure and assimilating it into a long-standing routine may take time. At the same time, aspects of a long-standing routine may have to be unlearned.

Experience gaps and negative transfer of training, whether suffered by student technologists, new employees, or long-term employees, can affect any nuclear medicine task. The specific task, subtask, or task element that is not performed properly will vary with the person and facility. The wide range of possibilities results in the appearance of random human error. The actual root cause is, however, inadequate familiarity or experience with specific tasks, subtasks, or task elements.

What else is wrong? The process of nuclear medicine is guided by a flow of information. That information is used to control the process (e.g., to select the correct radiopharmaceutical or patient). It is also used to monitor performance (e.g., to verify correct radiopharmaceutical or patient selection). The group of factors including foreign language and patient incoherent or unconscious and the group of factors including identification bracelet not checked, requisition not checked, and patient chart not checked are both related to the flow and verification of information. Both groups also contribute to the most frequent errors leading to wrong radiopharmaceutical and wrong patient misadministrations.

Patient identification is vital to a number of nuclear medicine tasks. Some nuclear medicine subtasks/task elements require that patients themselves provide this information (e.g., when the patient does not have an identification bracelet or when verification of identity is required). Responses from speakers of a foreign language or incoherent patients may seem to satisfy the information requirement. Recognition that a response did not actually provide the required information may not come until after a misadministration has occurred.

The nuclear medicine process may continue even when the speaker of a foreign language or patient who is incoherent or unconscious fails to provide identification information. Nuclear medicine personnel may, for example, use name labels posted at a patient's bedside or door in lieu of information directly from the patient. If, as sometimes happens, the information from other sources matches the expectations of the task performer but does not match the patient actually present, a misadministration can occur.

Checking requisitions, patient charts, and identification bracelets uses information flowing through the system to verify the results of certain subtasks and task elements. For example, checking the patient's chart in the nuclear medicine clinic is a means of confirming that the request for a nuclear medicine procedure received from the patient's ward is correct. Failure to perform such checks does not halt the nuclear medicine process, but it does increase the likelihood that previous errors will not be detected.

Licensee reports and the current study considered failures to check requisitions, charts, and identifications to be factors contributing to errors and misadministrations. However, from the human factors perspective, failure to perform a check could itself be considered an error. The current study did not allow for identification of factors that might have contributed to such an error. However, heavy workload, the final factor identified as contributing to errors leading to misadministrations, may be part of the answer.

Heavy workload contributed to the errors most frequently leading to wrong radiopharmaceutical, wrong dose, and wrong patient misadministrations. But how? One way might be by affecting the performance of subtasks and task elements intended to monitor performance (e.g., verification of the outcome of previous subtasks and task elements).

Heavy workload may cause conscious or inadvertent omission of subtasks and task elements intended to monitor performance. Other factors, such as distractions, may have a similar effect. As noted previously, failure to perform such subtasks and task elements does not halt the nuclear medicine process. Such failures do, however, increase the likelihood that errors on previous subtasks and task elements will go undetected and will result in a misadministration.

Misadministrations caused by random human error do not invite many solutions. In contrast, misadministrations caused by mismatches between what a system requires a person to do and what that person can reasonably be expected to do can be addressed in a number of ways.

Perhaps the most obvious approach to reduce mismatches due to lack of familiarity or experience with tasks is improved training. At the same time, simplification and standardization of procedures might increase training effectiveness, reduce the number of opportunities for error, and reduce training costs. An effort to simplify and standardize procedures might also lead to improvement of human interfaces with equipment and to reallocation of some tasks from humans to equipment items that can perform them more reliably.

One approach to reducing the omission of monitoring subtask and task elements might be to reduce workload during performance of particular tasks. One approach to reducing workload involves changes to staffing and scheduling. Such an approach is not always possible. Changes to procedures and human–equipment interfaces that make monitoring task performance easier may be a viable alternative.

Approaches for reducing misadministrations by resolving mismatches between what the nuclear medicine system requires a person to do and what that person can reasonably be expected to do must, of course, be tailored, coordinated, and tested in the real world. Doing so is, however, much less of a challenge than dealing with random human error.

REFERENCES

Callan, J. R., Kelly, R. T., Quinn, M. L., Gwynne, J. W., Moore, R. A., Muckler, F. A., Kasumovic, J., Saunders, W. M., LePage, R. P., Chin, E., Schoenfeld, I., & Serig, D. I. (in press). *Human factors evaluation of brachytherapy using remote afterloaders* (NUREG/CR-6125, Vol. I). Bethesda, MD: United States Nuclear Regulatory Commission.

Kinkade, R. G. (1988). *Human factors primer for nuclear utility managers* (NP-5714) Palo Alto, CA: Electric Power Research Institute.

McCormick, E. J. (1976). *Human factors in engineering and design.* New York: McGraw-Hill.

Meister, D. (1987). System design, development, and testing. In G. Salvendy (Ed.), *Handbook of human factors* (pp. 17–42). New York: Wiley.

11 Human Error in Dynamic Medical Domains[1]

David M. Gaba
Stanford University School of Medicine

This book addresses Human Error in Medicine. Each of the two aspects of the title are vast areas of concern. *Medicine* (with a capital *M*) as used here encompasses an enormous array of activities associated with health care, ranging from the self-administration of over-the-counter medications to skilled nursing care, chronic care of the elderly, high-technology diagnostics such as imaging and nuclear medicine, and the equally technological interventions of surgery and intensive care.

In this chapter we consider human error in anesthesiology and its close relative, intensive care. This domain is representative of the subset of Medicine that deals with complex, dynamic, high-risk, acute patient care. As this chapter describes, the combination of complexity, time pressure, and dynamism makes the domain distinctly different from much of the rest of the health-care system. Although the chapter focuses on the behaviors of operational personnel, it is appropriate to address briefly the issue of organizational structure and managerial decision making.

HUMAN ERROR: BY THE MANAGERS OR THE OPERATORS?

Human error, for its part, is a rather ambiguous term encompassing a variety of human behaviors, wherein the definition of *error* is controversial. Suboptimal performance is not necessarily the same as making an error. Furthermore, many investigators now believe that the traditional concept of error in

[1]Portions of this chapter have previously appeared in whole or in part in D. M. Gaba, K. J. Fish, & S. K. Howard (1994). *Crisis Management in Anesthesia.* New York: Churchill Livingstone, and are used by permission.

real-world work situations must be replaced with a more sophisticated understanding of the performance of the entire organization relative to its explicit and implicit goals. Complex, high-risk industries, like anesthesia, require an organizational structure that provides the proper foundation for successful operations in combination with appropriate cognitive and technical performance of the operational personnel.

In some cases, when evaluating the performance of the organization as a whole, it is apparent that the operational personnel themselves are really the "victims" of problematic decisions further back in the system. The results of these decisions are termed by Reason (1990a, 1990b) as "latent errors." These are

> errors whose adverse consequences may lie dormant within the system for a long time, only becoming evident when they combine with other factors to breach the system's defenses. [They are] most likely to be spawned by those whose activities are removed in both time and space from the direct control interface: designers, high-level decision makers, construction workers, managers and maintenance personnel. (p. 173)

Richard Cook and David Woods discuss these issues in detail in their chapter in this book.

This chapter focuses on the details of the performance of anesthesia personnel. Skilled performance is vitally important to patient safety and is usually what makes the difference between a near miss and an accident. Thus, understanding the elements of ideal performance will reveal the possible ramifications of suboptimal performance. In exploring how skilled anesthetists conduct their work and the complex cognition required of them, we are able to pinpoint the many different ways in which less than perfect performance can result in the generation of accidents or in the failure to recover from them before harm comes to the patient. Many requirements of the anesthetist's performance are shared by all segments of the health-care system, but this chapter concentrates on those aspects of the domain that are relatively different from the rest of Medicine.

ANESTHESIA IS AN UNIQUE SUBSET OF MEDICINE

What makes anesthesia different from the rest of Medicine? Why do we in anesthesia, and not those in other specialities (e.g., internal medicine or pediatrics), seek models for our work from arcane fields far removed from the care of the sick, such as aviation, space flight, process control (nuclear power and chemical manufacturing), shipping, military command, and fire fighting? The reason is that characteristics of the work are analogous. The dominant features of the anesthetist's environment include a combination of extreme dynamism, intense time pressure, high complexity, frequent uncertainty, and palpable risk. This combination is considerably different from that encountered in most medical fields. Thus, anesthetists have had to turn away from much of the research on decision making in medicine and look instead at other human activities that share these features.

Anesthetists share an aphorism with pilots that describes the work environment: "hours of boredom, moments of terror." It is largely the occasional moments of terror, and not the routine hours of boredom, that define the anesthetist's role in the operating room and the mental attitudes required to perform the job successfully. This is one aspect of anesthesia that sets it (and a few related fields such as intensive care and emergency room medicine) apart from most other branches of medical care, certainly apart from primary care and chronic care. Another aspect is the direct physical involvement of the anesthetist in the tasks of patient care. This includes the performance of invasive procedures, the administration of rapidly acting, potentially lethal medications, and the operation of increasingly complex devices. In all likelihood, the emphasis on direct action and the aura of "danger" lurking just below the surface are key factors that attracted many anesthetists to this line of medical work rather than others.

INTRODUCTION TO ANESTHESIOLOGY

A brief overview of the world of anesthesia may put some of these issues into focus. Anesthesiology is the branch of medicine that deals with the processes of rendering patients insensible to pain during surgery (surgical anesthesia) or when faced with acute or chronic pain states (pain management). Critical care medicine is another specialty of anesthesiology. The anesthetized patient can be thought of as one who has been purposefully made critically ill. Such patients typically are dependent on mechanical ventilation and under the influence of powerful drugs. This chapter primarily deals with surgical anesthesia.

An *anesthetist* is anyone who administers anesthesia. Anesthesiology is a medical specialty, and an *anesthesiologist* is a physician who has specialized in anesthesiology. In the United States (and in a few other places), there are *nurse anesthetists* who are registered nurses with several years of training in anesthesia. There is a mandatory certification process for nurse anesthetists, such that they are often known as "Certified Nurse Anesthetists" or CRNAs. They can administer general anesthesia and monitor patients under the supervision of a physician, who may or may not be an anesthesiologist. Fifty percent of anesthetics in the United States are conducted with nurse anesthetists. In this chapter, the term *anesthetist* refers generically to whoever is the anesthesia provider.

Surgery and Anesthesia

Every patient who has surgery requires some kind of anesthesia. There are three types of anesthesia: local, regional, and general. For local anesthesia, drugs that block nerve conduction (local anesthetics) are injected immediately around the site of surgery before and during the surgical procedure. Local anesthesia is often provided by surgeons without anesthetists present, but when patients have diseases that need to be monitored, when they may require carefully managed sedation, or when a more extensive operation might ensue necessitating general anesthesia, an anesthetist is asked to conduct *monitored anesthesia care* and is present throughout the case.

For regional anesthesia, the anesthetist injects local anesthetics in the region of specific nerves (nerve blocks), or near the spinal cord (epidural or spinal anesthesia). The local anesthetics then numb whole regions of the body, and the patient is monitored by the anesthetist.

General anesthesia consists of unconsciousness, amnesia, inhibition of bodily responses to pain, and often temporary muscle paralysis. The patient's breathing is often taken over artificially by the anesthesiologist and mechanical ventilators. Even more careful monitoring may be required for the patient under general anesthesia because the patient is unconscious and many intrinsic bodily functions are altered.

Preoperative Evaluation

Each patient must be evaluated by an anesthetist prior to beginning anesthesia. For patients who are already in the hospital, this is typically done the afternoon or evening before surgery. The anesthetist interviews the patient and checks the history and notes in the patient's chart. Appropriate physical examinations are made and any relevant laboratory tests are reviewed or ordered. For patients with significant medical problems, additional tests or consultations may be requested, and in some cases the surgery is postponed or canceled because of such problems. A balance must be struck between the urgency and benefit of the surgery and the risks of the underlying problems and the likelihood of improving them prior to surgery. Hard data on risks and benefits that can be applied to specific cases are relatively scarce. The subjective perception of risks and benefits will differ between medical specialties, between schools of thought within a specialty, and between individual practitioners. Anecdotally, it seems that surgeons overestimate benefits and underestimate risks relative to anesthetists, although there are few data documenting this phenomenon. These differing perceptions are sometimes a source of conflict between anesthetists and surgeons. A preliminary report of a survey of anesthesiologists' experiences with *production pressures* (the pressure to place efficiency and throughput as the top priorities rather than safety) suggests that a significant fraction of them feel strong pressures that sometimes cause them to perform in an unsafe manner (Howard, Gaba, & Jump, 1993).

It is now common for patients to have surgery on an outpatient basis. This complicates performing a preoperative anesthetic evaluation. Patients with significant medical problems usually are not handled as outpatients, but there are growing pressures to expand the outpatient system to include a larger fraction of the patient population.

General Anesthesia

Anesthesia is initially induced with an intravenous induction agent that acts very rapidly and places the patient in a relatively deep level of anesthesia. However, these drugs typically have a very short duration of action (5–10 min or less) so that anesthesia must be maintained using one or more other drugs. A variety of drugs and techniques can be used to maintain anesthesia. When anesthesia is induced, the patient will need respiratory support. Even if the patient can breathe spontaneously, special tools and techniques may be needed to keep the airway open. For the majority of cases, it is advantageous to provide a direct route of gases into the lungs, so an *endotracheal tube* is placed through the mouth (or occasionally the nose) into the windpipe (trachea) and connected to the anesthesia system. A cuff on the tube provides an airtight seal. To place the tube usually requires that the patient's muscles be paralyzed (at least for a short time) with a drug like Curare. Placing the endotracheal tube is usually easy, but in some patients due to their anatomy, it is very difficult or even impossible using standard techniques. Most cases involve artificial ventilation by using a breathing bag or ventilator.

The drugs that induce general anesthesia usually have some effects on the cardiovascular system, so the anesthetist carefully monitors these effects and may have to give other drugs to modify blood pressure, heart rate, and so forth. There is not a fixed dose of most drugs used in anesthesia; rather, they are titrated by their effects on the patient. While the patient is unconscious, their body will still respond to stimuli. Thus, anesthesia is a balancing act between the level of painful stimuli (both the surgery and the presence of the tube in the trachea) and the depth of anesthesia. The balance is continually adjusted by the anesthetist. For many types of surgery, muscle paralysis is required, and the anesthesiologist will administer Curare-like drugs to accomplish this. The patient's ventilation must be mechanically controlled—failure to do so will result in asphyxia, cardiac arrest, and death.

Throughout the operation, the anesthetist is engaged in a large number of tasks. These include (but are not limited to):

- Monitoring the patient and the life-support equipment by direct and electronic observation.

- Adjusting the anesthetic level and administering other medications as needed.
- Recording vital signs and other parameters at least every 5 min.
- Responding to changes in the patient's condition and taking emergency actions if needed.
- Evaluating blood loss and urine output.
- Administering IV fluids and, when needed, blood.
- Adjusting the position of the OR table as needed by the surgeons.

End of the Operation

Toward the end of the operation the anesthesiologist prepares to "wake up" the patient (although some patients are taken to the intensive care unit still deeply anesthetized and allowed to wake up slowly). Waking up the patient consists of reducing the level of anesthesia to the minimum necessary, reversing muscle-paralysis drugs, and allowing the patient to breathe, spontaneously. When the surgery is completed, the anesthesiologist administers 100% oxygen and when the patient is breathing satisfactorily and shows signs of regaining cough reflexes, the endotracheal tube is removed. The patient is given oxygen by mask and the ventilation carefully watched. The patient is then taken to the postanesthesia care unit (PACU, or recovery room), where specially trained nurses oversee the patient's further emergence from anesthesia. Postoperative problems with cardiovascular, pulmonary, and other systems are managed by the PACU nurses under the immediate supervision of anesthesiologists.

Regional Anesthesia

The course of events for regional anesthesia is similar except that usually the regional block is established at the beginning of the procedure, and the appropriate numbing effect is evaluated.

THE COGNITIVE PROCESSES OF THE ANESTHETIST

To assess suboptimal performance in anesthesia, it is necessary to understand the complex cognitive processes that the anesthetist uses to perform the job successfully. This will allow the identification of a variety of pathways in which skilled performance is necessary, and a number of key points at which suboptimal performance (or overt errors) may lead to adverse patient outcomes.

The anesthetist has two concurrent goals for anesthetic care: (a) to execute a planned sequence of actions to induce and maintain anesthesia, followed by a planned sequence to terminate the anesthetic, and (b) to detect and correct perturbations from the desired course.

Anesthetic Plans and Their Adaptation

An anesthetic plan is developed for each case. In constructing the plan, the anesthetist conducts a preoperative evaluation of the patient to: (a) identify the technical requirements of the proposed surgery; (b) assess the existence

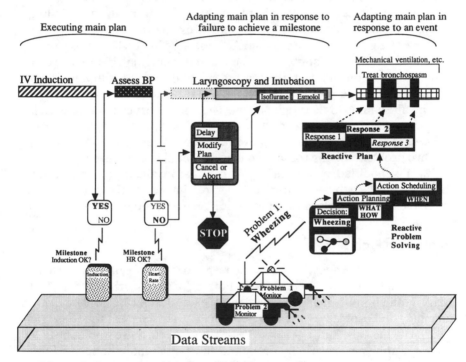

FIG. 11.1.   Real-time adaptation of anesthetic plans. As the anesthetic plan is executed it must be adapted to real-time contingencies. The horizontal bars across the top of the diagram show a small portion of the original plan dealing with the induction of anesthesia. At various points the anesthetist checks to see that certain *milestones* are met before proceeding to the next stage of the plan. In this example the milestones are satisfactory induction of anesthesia, and satisfactory control of blood pressure. If a milestone is not achieved the next step of the plan can be delayed, it can be modified, or the anesthetic can be suspended or aborted. In addition to checking for milestones the anesthetist observes the data streams, watching for problems (the "police cars"). If a problem is detected a process of reactive planning is invoked in order to adapt the original plan to the changed situation. In this example wheezing (bronchospasm) is detected, which necessitates the superimposition of several treatments for bronchospasm on top of the planned institution of mechanical ventilation.

and severity of underlying medical problems; and (c) match these requirements to the mental, physical, and technological resources available. Most
surgical procedures are standardized, so the technical requirements are well
known. Many patients have few, if any, medical problems that could significantly alter the anesthetic plan. Therefore, for many operations a routine,
off-the-shelf anesthetic plan will suffice. When a procedure is new or challenging, the patient has significant underlying diseases, or standard resources are not available, creative planning may be needed. Most commonly
this involves adapting one or more standard plans (singly or in combination)
to produce a compromise plan that best fits the mix of goals and constraints.

Plans are themselves dynamic resources. Once formed, they facilitate
anticipation and preparation for upcoming events or contingencies. The
plan acts as a guide for future action. As illustrated in Fig. 11.1, as the plan is
executed, the anesthetist observes the data streams available to assess the
patient's condition and the progress of the plan. Either the failure to achieve
a critical milestone on time or the occurrence of a serious problem will
require alterations to the original plan. The anesthetist can decide to delay
the next action in the sequence, to modify the plan, or to suspend or abort
the case. The existing plan may have to be modified through a process of
reactive problem solving (described in more detail later in this chapter). As
the plan is modified, actions may be taken that invalidate other aspects of
the original plan, thus requiring even more changes of the plan. In some
cases, the original goals of the anesthetic plan may need to be reconsidered.

Reaction to Problems

How often does the anesthetist have to respond to a problem? The available
research data (Cooper et al., 1987; Forrest, Cahalan, & Rehder, 1990)
suggest that, in spite of all the planning, at least 18% of cases will involve an
unanticipated problem that could harm the patient and that requires intervention by the anesthetist. At least 3%–5% of cases in these studies involved
a *serious* unplanned event that required substantial intervention, "with or
without full recovery of the patient" (Forrest et al., 1990, p. 263). Because
logistical aspects of these studies required them to exclude critically ill and
emergency patients, these figures are probably lower bounds on the actual
frequency of serious events.

Estimates of the frequency of adverse *outcomes* related to anesthesia care
are very difficult to obtain. Except in situations where a healthy patient
suffers a major injury that cannot be related to the surgery, it is hard to
separate the contribution of underlying surgical disease, underlying medical
disease, surgical care, and anesthesia care as discrete factors in the patient's
outcome. Unlike aviation, in which planes are never supposed to crash,

human beings do become ill and die. Anesthesia and surgery both involve major alterations to the patient's normal condition, which makes some incidence of complications inevitable. The available data, although few in quantity, suggest that on the order of 1 in 2,000 cases results in the patient's death due at least in part to anesthesia factors (Lunn & Devlin, 1987). Only 1 in 100,000 to 200,000 cases result in death due *solely* to anesthesia. The frequency of brain damage and other serious permanent injury is not well established at present.

These data for anesthesia yield a picture that is similar in some respects to that seen in other industries in that the incidence of fatal outcomes is very low in routine settings. For example, fatal accidents during routine commercial airplane flights occur at a rate of 0.36 per 100,000 departures. However, the analogy between pilots and anesthetists is not perfect. Anesthetists face problem situations during their work much more frequently than do pilots. Only a few flights require an unforeseen intervention, yet nearly 20% of anesthetics require an unforeseen intervention, and 5% require *major* unplanned interventions. Thus, it is important for the anesthetist to understand how problems occur, how to prevent them from occurring, and how to detect and correct them before they can harm the patient. This process of intervening in the chain of accident evolution is one of the key features of skilled performance by the anesthetist.

Sources of Problems

The chain of accident evolution is schematically presented in Fig. 11.2. Problems can arise from a variety of sources. One prominent source is the patient. Underlying diseases can become manifest during an operation independent of the stresses of surgery and anesthesia. These stresses will interact with underlying conditions, making acute exacerbations even more likely. The surgery itself often causes additional specific problems, including the surgeons compressing vital organs or lacerating significant arteries or veins. Other difficulties may be triggered when equipment fails (although these failures are rarely the sole cause of the problem), or when the anesthetist or surgeon makes an overt error. Several strategies are used to *prevent* problems from occurring. Precase evaluation of the patient and checking the anesthesia equipment disclose immediate hazards. Appropriate planning (as discussed earlier) will make the likelihood of error on the part of the anesthetist or surgeon less likely, but human fallibility cannot be entirely eliminated. Therefore, a substantial part of the anesthetist's task, and a critical arena for optimal versus suboptimal performance, is the reactive detection and correction of the problems that will inevitably occur.

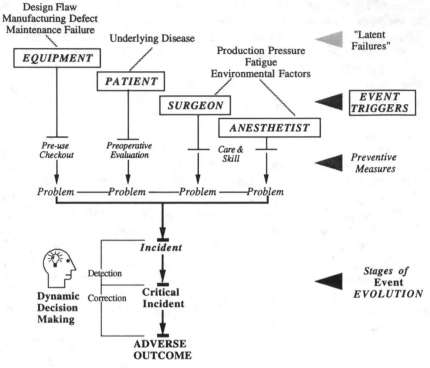

FIG. 11.2.   The chain of evolution of adverse events during anesthesia. The proximate causes of problems, termed *event triggers,* are shown in boxes. The likelihood of a problem being triggered is influenced by the underlying latent errors relevant to each event trigger. Problems can be prevented by specific actions such as pre-use checkout of equipment and preoperative evaluation of the patient. Once a problem occurs it can evolve further, becoming an *incident,* then a *critical incident* (which can directly harm the patient), or finally an *adverse outcome* for the patient. The anesthetist's task is to interrupt this chain of evolution at the earliest possible stage using a process of dynamic decision making. (From Gaba, Fish, & Howard, 1994. Reprinted by permission.)

Evolution of Problems

Once a problem occurs, there are various possibilities for its future evolution. These are shown schematically in Fig. 11.3, and include the following:

- A single problem worsens, and by itself evolves into an adverse outcome.
- The problem begins to evolve but remains self-limited without any intervention.
- Multiple small problems combine to trigger a problem that can evolve into an adverse outcome. The original problems by themselves would not have evolved further.

Symbol Key:

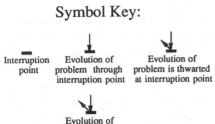

▬
Interruption point

Evolution of problem through interruption point

Evolution of problem is thwarted at interruption point

Evolution of problem should be thwarted but is not

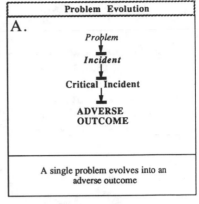

Problem Evolution

A.

Problem

Incident

Critical Incident

ADVERSE OUTCOME

A single problem evolves into an adverse outcome

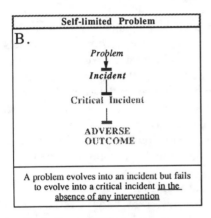

Self-limited Problem

B.

Problem

Incident

Critical Incident

ADVERSE OUTCOME

A problem evolves into an incident but fails to evolve into a critical incident <u>in the absence of any intervention</u>

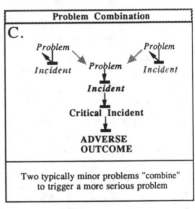

Problem Combination

C.

Problem   *Problem*

Incident   *Problem*   *Incident*

Incident

Critical Incident

ADVERSE OUTCOME

Two typically minor problems "combine" to trigger a more serious problem

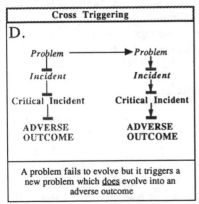

Cross Triggering

D.

Problem ———→ *Problem*

Incident   *Incident*

Critical Incident   **Critical Incident**

ADVERSE OUTCOME   **ADVERSE OUTCOME**

A problem fails to evolve but it triggers a new problem which <u>does</u> evolve into an adverse outcome

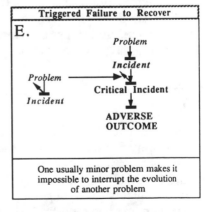

Triggered Failure to Recover

E.

Problem

Incident

Problem ———→ **Critical Incident**

Incident

ADVERSE OUTCOME

One usually minor problem makes it impossible to interrupt the evolution of another problem

FIG. 11.3. Example pathways of the evolution and interaction of adverse events. Many pathways involve complex interactions between multiple problems. Other pathways of accident evolution are possible but not shown. (From Gaba, Fish, & Howard, 1994. Reprinted by permission)

- A single problem triggers another problem that evolves into an adverse outcome.
- An evolving problem should be able to be stopped, but the recovery pathway is faulty.

The complexity and rapid time course of the possible pathways of accident evolution present a major challenge to anesthetists and account for the unique cognitive profile of anesthesia and related medical fields.

A PROCESS MODEL OF THE ANESTHETIST'S PROBLEM SOLVING

A process model has been developed to describe formally how the anesthetists react to intraoperative events (Fig. 11.4). It is consistent with models described by investigators studying other complex dynamic domains (Klein, 1989; Klein, Orasanu, Calderwood, & Zsambok, 1993; Rasmussen, 1986; Reason, 1990b). At its core, the model contains a repeated loop of *observation, decision, action,* and *reevaluation* (shown by the heavy black arrows in Fig. 11.4). This loop is under the control of two higher levels of cognition, termed *supervisory control* and *resource management*. Because they involve "thinking about the anesthetist's own thinking," these higher levels are examples of *metacognition*.

In the following section, the individual elements in the model are described in detail to demonstrate how the different behaviors can have different consequences for the patient and how each element may require different strategies to prevent or handle possible error pathways. For each element, the possible types of human errors associated with it are discussed. Refer to Fig. 11.4 for the relationship between these elements.

Observation (Fig. 11.4, Procedural Level and Sensory/ Motor Level)

Management of dynamic situations depends on responding to changing information, of which the anesthetist has many sources. These include clinical observation of the patient; a multitude of displays from electronic monitors; visual inspection of the surgical field, the activities of nurses, and suction canisters and sponges; normal and abnormal sounds of the patient and equipment; and reports of laboratory test results or X-ray film interpretations. The human mind can only pay *close* attention to one or two items at a time. The anesthetist's metacognitive process of supervisory control must decide what information to attend to and how frequently to observe it.

Human errors occurring at this point include:

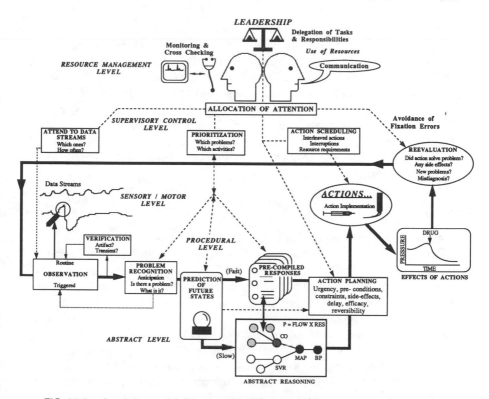

FIG. 11.4.   A process model of the anesthetist's real-time decision making and actions. Five levels of cognition are shown operating in parallel. The process involves a core process loop (heavy black line) comprised of observation, decision, action, and reevaluation. The core process is under the direction of higher levels of supervisory control and resource management. Each element of the model involves different cognitive skills, and each element is the locus of a different set of suboptimal behaviors or errors. A challenging aspect of anesthesia is the dynamic coordination of all these functions by the individual anesthetist, by the anesthesia crew, and by the OR team as a whole. (From Gaba, Fish, & Howard, 1994. Reprinted by permission.)

- Not observing a data stream at all.
- Not observing a data stream frequently enough.
- Not observing the optimum data stream given the existing situation.

Why might these errors occur? One possibility is a loss of overall *vigilance*, the capacity to *sustain* attention. It plays a crucial role in the observation and detection of problems and thus is a necessary prerequisite for meaningful care of the patient. Vigilance can be degraded in a number of ways, because human beings are susceptible to impairment due to several performance-shaping factors, including fatigue, illness, drugs, lack of motivation, and attitudes of invulnerability.

Although vigilance is *essential*, it alone cannot interrupt the chain of accident evolution. It thus is a necessary, but not sufficient, component of decision making and action. Failures of observation can still affect the vigilant practitioner. One is the failure to attend to all relevant information, resulting from excessive attention being paid to only a subset of the data stream. Even perfect allocation of attention can be overwhelmed by the sheer amount of information and the rapidity with which it is changing. In this circumstance, even the perfect individual cannot switch attention between multiple data sources fast enough to keep up with the dynamism of the environment.

Verification (Fig. 11.4, Procedural Level)

Not only is there a great deal of information in the OR environment, but that information is not always reliable. Most patient monitoring uses relatively weak signals detectable at the surface of the body (to avoid invading the body more than is necessary). This makes it especially prone to *artifacts* (false data). Even direct clinical observations such as vision or listening with a stethoscope can be ambiguous. Brief *transients* (true data of short duration) can also occur. These will typically correct themselves quickly. If either artifacts or transients are incorrectly interpreted as indicating a major problem requiring a strong response, unnecessary and risky actions might be taken. In most cases, critical observations can be verified before they are acted on. Verification uses a variety of methods, including:

- Repeating the observation or observing the short-term trend.
- Observing an existing redundant channel (e.g., blood pressure measured directly in an artery—an "invasive" technique—versus that measured with a blood pressure cuff).
- Correlating multiple related variables (e.g., heart rate, heart rhythm, and blood pressure).
- Activating a new monitoring modality (e.g., placing a special catheter inside the heart).
- Recalibrating an instrument or testing its function.
- Replacing an entire instrument with a backup device.
- Asking for a second opinion.

Knowing when and how to verify data is an important metacognitive skill. For example, the anesthetist must decide under what conditions it is useful to invest time, attention, and energy in establishing a new data source, such as placing a special heart catheter in the middle of a case, as opposed to relying on more indirect data sources that are already in place. Conversely, failure to verify data properly can result in devoting excessive attention to a nonproblem or to taking actions that are more dangerous than the underlying event.

Problem Recognition (Fig. 11.4, Procedural Level)

Having made and verified observations, the next step is to decide if they indicate that the patient's course is "on track" or if a problem is occurring. If a problem is found, a decision must be made as to its nature and importance. This process of problem recognition is a central feature of the theories of cognition that apply to complex dynamic worlds (Klein, 1989; Klein, Calderwood, & Macgregor, 1989; Klein et al., 1993; Orasanu, 1993; Reason, 1990b). Problem recognition involves matching sets of cues (data) in the OR to patterns that are known to represent specific types of problems. A common type of human error is to observe the signs of a problem but fail to *recognize* them as being problematic. This error is made more likely by the fact that the existing cues are dynamically changing and are ambiguous.

Even when a problem *is* detected, the cues may not specify an unique cause. Therefore, one part of supervisory control involves what to do when a clear-cut match or "diagnosis" cannot be made. Anesthetists, like other dynamic decision makers, use approximation strategies to handle these ambiguous situations; psychologists describe these strategies as *heuristics* (Tversky & Kahneman, 1974). One heuristic is to categorize an event as an instance of one of a variety of *generic problems*, each of which encompasses many different underlying conditions. For example, generic problems during anesthesia might include "difficulty with ventilation," "low blood oxygen concentration," or "low blood pressure." Although the exact diagnosis underlying the generic problem may not be immediately apparent, there are generic therapies that are appropriate for all diagnoses in the class (see later discussion on *precompiled responses*).

Another heuristic that is used when there are multiple possible diagnoses is to gamble on a single specific diagnosis, betting initially on the occurrence of the candidate that is most frequently encountered (Reason, 1987, terms this *frequency gambling*). In preparing for a case, the anesthetist may adjust a mental "index of suspicion" for recognizing certain specific problems that can be anticipated for that patient or procedure. The anesthetist must also decide when a single underlying diagnosis explains all of the data or whether the data indicate multiple simultaneous causes (Woods, Roth, & Pople, 1987). This decision is important because excessive attempts to refine the diagnosis can be very costly in terms of allocation of attention and may prevent the real problem from being adequately addressed. Experienced clinicians often use heuristics to yield considerable time saving in dealing with problems. These shortcuts are a two-edged sword. Both frequency

gambling and inappropriate allocation of attention solely to expected problems can seriously derail problem solving when these gambles do not pay
off.

Prediction of Future States (Fig. 11.4, Procedural Level and Abstract Level)

Problems must be assessed in terms of their significance for the *future*
condition of the patient. Predicting future states based on the occurrence of
seemingly trivial problems is a major part of the anticipatory behaviors that
characterize dynamic decision makers (Sarter & Woods, 1991). These predictions should influence the priority given to problems because those that
are immediately threatening or are highly likely to evolve into critical
incidents deserve the highest priority. Prediction of future states should also
influence action planning by defining the time-frame available for required
actions. Therefore, possible errors include:

- Failure to predict the evolution of catastrophic consequences from the
  early stages of a problem.
- Failure to use predictions in assigning priorities to tasks during action
  planning.

Precompiled Responses and Abstract Reasoning (Fig. 11.4, Procedural Level and Abstract Level)

Having recognized a problem, how does the expert anesthetist respond?
The classical paradigm of decision making postulates a careful comparison
of the evidence with various causal hypotheses that could explain them
(Orasanu & Connolly, 1993). This is supposed to be followed by a careful
analysis of all possible actions and solutions to the problem. This approach,
although powerful, is relatively slow and does not work well with ambiguous
or scanty evidence. It is rarely seen in real-world problem-solving situations
(Orasanu & Connolly, 1993). Many perioperative problems faced by anesthetists require quick action to prevent a rapid cascade to a catastrophic
adverse outcome, and for these problems, deriving a solution through formal deductive reasoning from "first principles" is just too slow.

The initial responses of experienced practitioners to perioperative events
arise from *precompiled responses* to that type of event (DeAnda & Gaba,
1991; Rasmussen, 1986, 1987; Reason, 1987, 1990b; Schwid & O'Donnell,
1992). This has been referred to as "recognition-primed decision making"
(Klein, 1989, p. 47). Experienced practitioners can rapidly retrieve and
execute precompiled responses to common problems. During precase planning, they may mentally reorder and rehearse these responses based on the
patient's condition, the surgical procedure, and the problems to be expected

(Xiao, Milgram, & Doyle, 1992). Failure to do so can be considered an error because it renders suboptimal the anesthetist's ability to respond quickly and accurately to that patient's needs.

Precompiled responses usually are acquired primarily through personal experience; only a few that involve major catastrophes (e.g., for cardiac arrest) have been explicitly codified and taught systematically. Most anesthetists appear to have compiled their responses haphazardly, making them less than optimally prepared to respond to complex or highly critical events. Research using anesthesia simulators has demonstrated that anesthetists, both novices and experts, varied widely in their ability to respond to critical incidents (DeAnda & Gaba, 1990, 1991; Gaba & DeAnda, 1989; Schwid & O'Donnell, 1992).

Although well thought out precompiled responses can provide quick, if partial, solutions to problems, they are destined to fail when the problem is not due to the suspected cause or when it does not respond to the usual measures. Conducting anesthesia purely by precompiled "cookbook" procedures is not desirable. Therefore, even when quick action is required, careful reasoning about the problem utilizing fundamental medical knowledge also takes place in the background while precompiled responses are being executed. This kind of abstract reasoning may involve a search for high-level analogies (Reason, 1990b) or true deductive reasoning from the deep knowledge base and careful analysis of all possible solutions (the classical decision paradigm).

When anesthetists were observed while managing simulated crises they appeared to respond primarily with precompiled plans while linking their course of action to abstract medical concepts (DeAnda & Gaba, 1990, 1991; Gaba & DeAnda, 1989). Whether this merely represented explanation or justification or whether it was true abstract reasoning is unclear, in part because the particular simulated crises they faced did not *require* novel abstract solutions. Thus, the degree to which abstract reasoning is mixed in with procedural reasoning during the management of critical incidents during anesthesia is not known at this time. Clearly, performance will be suboptimal if the anesthetist cannot successfully merge precompiled responses with abstract reasoning when the standard approaches are not succeeding.

Coordination of Activities via Supervisory Control (Fig. 11.4, Supervisory Control Level)

The importance of allocating attention between different data streams has already been emphasized in the section on observation. The anesthetist's attention must also be distributed among many different tasks, and possibly among many simultaneous problems. In fact, empirical analyses have been conducted of the tasks performed by novice and experienced anesthetists

(Boquet, Bushman, & Davenport, 1980; Dallen, Nguyen, Zornow, Weinger, & Gaba, 1990; Drui, Behm, & Martin, 1973; Dzwonczyk, Allard, McDonald, Block, & Yablock, 1992; Gaba & DeAnda, 1991; Herndon, Weinger, Paulus, Zornow, & Gaba, 1991; Herndon, Weinger, Zornow, & Gaba, 1993; McDonald, Dzwoncyzk, Gupta, & Dahl, 1990; McDonald & Dzwonczyk, 1988), and the mental workload these tasks imposed (Dallen et al., 1990; Gaba, Herndon, Zornow, Weinger, & Dallen, 1991; Gaba & Lee, 1990; Loeb, 1993), during actual surgical cases. These studies disclosed that not only are there a plethora of tasks but, periodically, they generate enough mental workload to degrade the anesthetist's ability to respond to secondary tasks. Anesthetists must therefore modulate their level of attention as the task demands change.

Workload Management Strategies.   One aspect of modulating attention is the *active* management of workload. Several authors (Gaba, 1992; Gopher, 1991; Schneider & Detweiler, 1988; Xiao, Milgram, & Doyle, 1992) describe a variety of strategies for workload management. Failure to carry out these strategies will almost certainly lead to difficulties in managing complex and dynamic situations. The strategies are:

• *Distributing workload over time.* The anesthetist prepares for future tasks when the current load is low (*preloading*), and will delay or shed low-priority tasks when the workload is high (*offloading*). The interleaving of tasks together (*multiplexing*) is another example of distributing workload over time.

• *Distributing workload over resources.* When workload cannot be distributed over time, and when additional noncompeting resources are available, task loads can be distributed to them. Some resources are internal to the individual anesthetist. For example, a single anesthetist *can* simultaneously ventilate the patient by hand, assess the cardiac rhythm, and discuss patient care with the surgeon. The single anesthetist *cannot* simultaneously insert an intravenous line and ventilate the patient by hand. If they are to be performed at the same time, these two tasks must be distributed to different individuals. The anesthetist coordinates this distribution of tasks and supervises their proper execution at the resource management level.

• *Changing the nature of the task.* The nature of task is not fixed. Surgery and anesthesia can be postponed or aborted. Tasks can be executed to different standards of performance. As the standards are loosened, the workload required to perform them is reduced. For example, during periods of massive blood loss, the anesthetist will focus primarily on administering blood and fluids and on monitoring the blood pressure. The acceptable limits of blood pressure will be widened. In such cases, less critical tasks such as writing on the anesthesia record will be foregone (offloading) to lessen the workload.

The supervisory control level determines how frequently different data sources are observed, what priorities are given to routine tasks versus problem solving, and how to schedule actions so that the necessary attention and motor resources will be available to execute them. The intensive demands on the anesthetist's attention can easily swamp the available mental resources. Therefore, the ideal anesthetist strikes a balance between acting quickly on every small perturbation (which requires substantial attention) and adopting a more conservative "wait and see" attitude. This balance must be constantly shifted between these extremes as the situation changes. However, it has been observed that during true crisis situations, some practitioners show a great reluctance to switch from "business as usual" to "emergency mode" even when serious problems are detected (DeAnda & Gaba, 1991; Howard, Gaba, Fish, Yang, & Sarnquist, 1992; Schwid & O'Donnell, 1992). Erring too far in the direction of "wait and see" is an error that can be particularly catastrophic.

Action Planning and Action Scheduling (Fig. 11.4, Procedural Level and Supervisory Control Level)

Supervisory control also involves the optimum planning of *actions* and the scheduling of their efficient execution. At any given time there are multiple tasks, each of which is intrinsically appropriate, yet they cannot all be done at once. Each action must be interleaved with the myriad other concurrent activities. The expert anesthetist considers many factors in planning and adapting optimum action sequences, including:

- *Preconditions* necessary for carrying out the actions (e.g., it is impossible to measure the blood flow generated by the heart if the appropriate catheter needed to make the measurement is not already in place inside the heart).
- *Constraints* on the proposed actions (e.g., it is impossible to check the size of the patient's pupils when the head is fully draped in the surgical field).
- *Side effects* of the proposed actions.
- *Rapidity and ease* of implementing the actions.
- *Certainty of success* of the actions.
- *Reversibility* of the action and the "cost of being wrong."
- *Cost* of the action in terms of attention and of resources.

Experts in other complex dynamic domains (specifically tank commanders and fire chiefs) have been observed to conduct a mental simulation of the actions they are contemplating to determine whether there are hidden flaws in their plans (Klein, 1989; Klein et al., 1989). Although this practice would make sense for anesthetists, they have not yet been observed doing so. Perhaps this is because anesthetists are able to execute their actions incrementally, dynamically changing their plans as they go along. Suboptimal action planning is common, especially with novices. Primarily, the failures involve making poor choices for the set of actions to be performed or attempting to perform actions that are incompatible with each other.

Action Implementation (Fig. 11.4, Sensory / Motor Level)

A particular hallmark of anesthesia is that the decision maker does not just "write orders" but is involved directly and physically in implementing the actions decided upon. Executing these actions requires substantial attention and may in fact impair the anesthetist's physical ability to perform other activities (e.g., when an action requires a sterile procedure). When performing actions, a variety of errors of execution (*slips*), may occur (Norman, 1981). Some of the risks due to slips in anesthesia have been addressed through the use of *engineered safety devices* that physically prevent incorrect actions. For example, newer anesthesia machines have interlocks that physically prevent the simultaneous administration of more than one volatile anesthetic agent. Other interlocks physically prevent the selection of a gas mixture containing less than 21% oxygen.

Reevaluation (Fig. 11.4, Supervisory Control Level)

Successful dynamic problem solving in anesthesia requires the supervisory control level to initiate frequent reevaluation of the situation. The initial diagnosis and situation assessment can be incorrect, especially when the available cues do not precisely identify a problem. Even actions that are appropriate for the problem are not always successful, and they sometimes cause serious side effects. Furthermore, there is often more than one problem to deal with at a time. Only by frequently reassessing the situation can the anesthetist adapt to dynamically changing circumstances. The reevaluation process returns the anesthetist to the "observation" step of the process model, but with specific assessments in mind:

- Did the actions have any effect (e.g., did the drug reach the patient, or could there have been a problem with the intravenous infusion)?
- Is the problem getting better, or is it getting worse?

- Are there any side effects of the actions?
- Are there any additional problems that were missed before?
- Was the initial situation assessment or diagnosis correct?

The process of continually updating the situation assessment and of monitoring the efficacy of chosen actions is termed *situation awareness*, a concept that has been used extensively in aviation (Endsley, 1988; Gaba, Howard, & Small, in press; Sarter & Woods, 1991; Tenney, Adams, Pew, Huggins, & Rogers, 1992).

Fixation Errors.   Faulty reevaluation, inadequate plan adaptation, or loss of situation awareness will result in a type of human error termed a "fixation error" (DeKeyser & Woods, 1990; DeKeyser, Woods, Masson, & Van Daele, 1988). Incorrect initial diagnoses are to be expected given the complexity and uncertainty of the data available to the anesthetist. What makes a fixation *error* is the *persistent* failure to revise a diagnosis or plan in the face of readily available evidence that suggests a revision is necessary. There are three main types of fixation error (DeKeyser & Woods, 1990; DeKeyser et al., 1988):

- *This And Only This.* The persistent failure to revise a diagnosis or plan despite plentiful evidence to the contrary.
- *Everything But This.* The persistent failure to commit to the definitive treatment of a major problem.
- *Everything's OK.* The persistent belief that no problem is occurring in spite of plentiful evidence that it is.

This type of error is extremely common in dynamic situations. Fixation errors were reported among the errors observed in laboratory studies of anesthesiologists' responses to simulated critical incidents (Botney, Gaba, Howard, & Jump, 1993; DeAnda & Gaba, 1990, 1991; Gaba & DeAnda, 1989; Howard et al., 1992; Schwid & O'Donnell, 1992). Such errors occurred for both novice residents and experienced practitioners. In some cases, fixation errors prevented anesthesiologists from correctly managing poten-tially lethal situations.

Resource Management

The elements of the process model described previously deal only with the activities of the anesthetist as an individual. The final element, the resource management level (a concept borrowed from aviation human-factors re-searchers), describes the anesthetist's requirement to command and control *all* the resources at hand. This is, in essence, the ability to translate the *knowledge* of what needs to be done into effective *activity* in the real world. The resources available to the anesthetist include self, other operating room personnel (including other anesthetists), medical equipment, cognitive aids (written checklists, manuals), and external resources (laboratory service, radiology, medical consultants, phone hotlines).

Managing the *self* resource is the main thrust of the supervisory control level. But managing the other real-world resources requires other cognitive skills, of which interaction with the rest of the operating room team is particularly important. There has been little investigation of team behavior in the OR, but the research on commercial air crews (Foushee & Helmreich, 1988; Helmreich & Foushee, 1993; Helmreich, Foushee, Benson, & Russini, 1986; Orasanu, 1993; Orlady & Foushee, 1987) appears to apply well to anesthesia (Gaba, Howard, & Fish, 1994; Howard et al., 1992). The hallmarks of resource management derived from these studies are:

- Prioritization of tasks.
- Distribution of workload.
- Monitoring and cross-checking; utilization of all available data.
- Leadership.
- Communication.
- Mobilization and use of all available resources.
- Team building, including explicit briefings and sharing of mental models.

Studies of simulated anesthetic crises in which the entire OR team was present have shown examples of suboptimal resource management (Botney, Gaba, & Howard, 1993; Botney, Gaba, Howard, & Jump, 1993; Howard et al., 1992). These included:

- *Poor communication.* Commands were ambiguous and spoken "into thin air." The individuals addressed did not realize they were being spoken too, or could not understand what was desired. In some cases, anesthetists inadvertently misled colleagues who had been called in to help by not explaining clearly what had already happened and what actions they had already taken.
- *Poor workload management.* Task preloading and offloading to match the workload demand did not always occur. Nurses and surgeons were not utilized as resources to help in patient management, and the anesthetists sometimes failed completely to inform the surgeons of major problems even when continuing the surgery complicated the situation.
- *Fixation errors.* Surprisingly, in spite of the hypervigilance typically

induced by the simulator environment, all three types of fixation error have been observed.

• *Poor leadership.* Complex emergencies deteriorated because no one assumed overall command authority. There was decision by committee, and no attempts were made to unify the mental models of the participants. In some cases, less experienced practitioners took over the case (appropriately) from a more experienced practitioner.

SOCIAL PSYCHOLOGY OF THE OR ENVIRONMENT

Unlike teams in aviation, the military, and police or fire services, the OR team does not have an official command structure. The physicians (surgeon and anesthesiologist) are simultaneously coequally responsible for the patient during the perioperative period. Each has an area of primary responsibility, but there is considerable overlap between them. In this regard, the OR team is quite unusual.

Strictly speaking, a *team* is: "a distinguishable set of two or more people who interact, dynamically, interdependently, and adaptively toward a common and valued goal/objective/mission, who have each been assigned specific roles or functions to perform, and who have a limited life-span of membership" (Salas, Dickinson, Converse, & Tannenbaum, 1992, p. 4). A team is distinct from a *group* in that a group is typically an ad hoc collection of individuals without a specific mission and without specific roles. The degree to which the various members of the OR team agree on common objectives is debatable. Clearly, all agree that a good outcome for the patient is the ultimate goal. However, there can be considerable disagreement on how to achieve this goal and which elements of patient care have the highest priority.

These differences are probably traceable to the fact that the OR team is itself made up of several *crews* (nursing, surgery, anesthesia). Each crew is comprised of one or more individuals representing a profession with its own culture, traditions, and norms. There is no guarantee of coordinated action between the crews, although if the same crews work together frequently, they are likely to evolve more coordinated patterns of behavior at the local level.

Status and hierarchy effects are important in team performance. Especially in crisis situations, the lower status crew member will tend to defer to the higher status individual, even if that individual is performing poorly. In aviation, a number of airplane crashes have occurred in which overbearing captains were combined with unassertive subordinates (copilot and flight engineers). The team was not able to respond effectively even when the subordinates knew that something was wrong (Foushee & Helmreich, 1988).

As yet, there is primarily anecdotal information about such phenomena in the operating room, both within the anesthesia crew and between the anesthesia and surgical crew. However, conflicts between surgeons and anesthetists can result in production pressures on anesthetists to proceed with anesthesia even when they believe it is unsafe to do so. The production pressure survey found that anesthetists do feel this kind of pressure from surgeons, and the anesthetists also reported internal pressures as well (Howard et al., 1993). The authors hypothesized that repeated exposure to external pressures led anesthetists to internalize the pressures so that they then feel impelled to go ahead with cases against their better judgment, even in the absence of overt external coercion. Investigating or changing these aspects of teamwork will be difficult because these relationships are driven by economic considerations and are part of complex organizational and interpersonal networks linking the different medical cultures.

Difficulties in crew coordination *within* the anesthesia crew are most frequent in training situations. In the cockpit, on each leg of a flight, captains and first officers traditionally alternate the explicitly defined roles of "pilot flying" and "pilot nonflying," each of which involves a multitude of interrelated tasks. The roles of the anesthesia trainee and the faculty member during patient care are rarely made explicit. The trainee is often expected to do all tasks with only occasional assistance from the supervisor. Who is responsible for what tasks in a crisis is not predefined. Interestingly, two factors frequently found associated with anesthesia critical incidents have been "teaching in progress" and "inadequate supervision" (Cooper, Newbower, & Kitz, 1984).

SUMMARY

The consideration of performance and error in this example of a dynamic medical specialty goes well beyond classical notions of error. Yes, there are overt slips and mistakes due to lack of knowledge and skill, but much of the suboptimality that can be identified resides not in the decisions and actions themselves, but in the complex way in which they are linked together. The dynamic environment demands an enormous number of trade-offs that must be continuously balanced and refined. The failure modes associated with these dynamic tradeoffs are different from those seen in other areas of Medicine.

This review of the complex tasks of the anesthetist might leave the impression that anesthesia is markedly unsafe, but this is not true. There is an increasingly good safety record of anesthesia both for young, healthy patients and for the critically ill. That outcomes have improved over time is really a testament to the dedication of the anesthesia professionals who must work in this challenging arena. The goal of the profession is to make anesthesia even safer, especially in view of an increasingly aged patient population that presents ever more difficult challenges to the anesthetists.

Existing research has exposed some of the weak spots in the training of anesthetists and in their operational environment, yet these weak spots are in the process of being addressed. Increased attention has been focused on the ergonomics of the anesthetist's work environment to reduce the possibility of slips (Loeb, Weinger, & Englund, 1993; Weinger & Englund, 1990). Other investigators are beginning to address the complex sociology of the OR and its impact on teamwork, crew coordination, and production pressure (see the chapter by Helmreich and Schaefer in this book). New training modalities that target problem solving, supervisory control, and resource management have been developed. These modalities typically make extensive use of recently developed anesthesia simulation environments. Some training courses have been explicitly modeled on the types of training in resource management that were pioneered in aviation (Holzman et al., 1993; Howard et al., 1992). The profession of anesthesiology has been a leader within Medicine in addressing issues related to the performance of practitioners. Anesthestists are confident that success in their domain will be continued and that anesthesiology will be able to help lead other fields of Medicine as they address comparable problems in their own domains.

REFERENCES

Boquet, G., Bushman, J.,& Davenport, H. (1980). The anaesthetic machine, a study of function and design. *British Journal of Anaesthesia, 52,* 61–67.

Botney, R., Gaba, D., & Howard, S. (1993). Anesthesiologist performance during a simulated loss of pipeline oxygen (abstract). *Anesthesiology, 79,* A1118.

Botney, R., Gaba, D., Howard, S., & Jump, B. (1993). The role of fixation error in preventing the detection and correction of a simulated volatile anesthetic overdose (abstract). *Anesthesiology, 79,* A1115.

Cooper, J. B., Cullen, D. J., Nemeskal, R., Hoaglin, D. C., Gevirtz, C. C., Csete, M.,& Venable, C. (1987). Effects of information feedback and pulse oximetry on the incidence of anesthesia complications. *Anesthesiology, 67,* 686–694.

Cooper, J. B., Newbower, R., & Kitz, R. (1984). An analysis of major errors and equipment failures in anesthesia management: Considerations for prevention and detection. *Anesthesiology, 60,* 34–42.

Dallen, L., Nguyen, L., Zornow, M., Weinger, M., & Gaba, D. (1990). Task analysis/ workload of anesthetists performing general anesthesia (abstract). *Anesthesiology, 73,* A498.

DeAnda, A., & Gaba, D. (1990). Unplanned incidents during comprehensive anesthesia simulation. *Anesthesia and Analgesia, 71,* 77–82.

DeAnda, A., & Gaba, D. (1991). The role of experience in the response to simulated critical incidents. *Anesthesia and Analgesia, 72,* 308–315.

DeKeyser, V., & Woods, D. (1990). Fixation errors: Failures to revise situation assessment in dynamic and risky systems. In A. Colombo & A. Bustamante (Eds.),

Systems reliability assessment (pp. 231–251). Dordrecht, The Netherlands: Kluwer Academic.

DeKeyser, V., Woods, D., Masson, M.,& Van Daele, A. (1988). *Fixation errors in dynamic and complex systems: Descriptive forms, psychological mechanisms, potential countermeasures.* Technical Report for NATO Division of Scientific Affairs, 2–10.

Drui, A., Behm, R.,& Martin, W. (1973). Predesign investigation of the anesthesia operational environment. *Anesthesia and Analgesia, 52,* 584–591.

Dzwonczyk, R., Allard, J., McDonald, J., Block, F., & Yablock, D. (1992). The effect of automatic record keeping on vigilance and record keeping time (abstract). *Anesthesia and Analgesia, 74,* S79.

Endsley, M. (1988). Design and evaluation for situation awareness enhancement. *Proceedings of the Human Factors Society 32nd Annual Meeting* (pp. 97–101). Santa Monica, CA: Human Factors Society.

Forrest, J., Cahalan, M., & Rehder, K. (1990). Multicenter study of general anesthesia. II. Results. *Anesthesiology, 72,* 262–268.

Foushee, H., & Helmreich, R. (1988). Group interaction and flight crew performance. In E. Wiener & D. Nagel (Eds.), *Human factors in aviation* (pp. 189–228). San Diego: Academic Press.

Gaba, D. (1992). Dynamic decision-making in anesthesiology: Cognitive models and training approaches. In D. Evans & V. Patel (Eds.), *Advanced models of cognition for medical training and practice* (pp. 122–147). Berlin: Springer-Verlag.

Gaba, D. M., & DeAnda, A. (1989). The response of anesthesia trainees to simulated critical incidents. *Anesthesia and Analgesia, 68,* 444–451.

Gaba, D., Herndon, O., Zornow, M., Weinger, M., & Dallen, L. (1991). Task analysis, vigilance, and workload in novice residents (abstract). *Anesthesiology, 75,* A1060.

Gaba, D., Howard, S., & Fish, K. (1994). *Crisis management in anesthesiology.* New York: Churchill-Livingstone.

Gaba, D., Howard, S., & Small, S. (in press). Situation awareness in anesthesiology. *Human Factors.*

Gaba, D., & Lee, T. (1990). Measuring the workload of the anesthesiologist. *Anesthesia and Analgesia, 71,* 354–361.

Gopher, D. (1991). *Mental workload—A tutorial.* Conference on Human Error in Anesthesia, Pacific Grove, CA, February, 1991. Rockville, MD: United States Food and Drug Administration.

Helmreich, R., & Foushee, H. (1993). Why crew resource management? In E. Weiner, B. Kanki, & R. Helmreich (Eds.), *Cockpit resource management* (pp. 3–45) San Diego: Academic Press.

Helmreich, R. L., Foushee, H. C., Benson, R., & Russini, W. (1986). Cockpit resource management: Exploring the attitude-performance linkage. *Aviation, Space, and Environmental Medicine, 57,* 1198–1200.

Herndon, O., Weinger, M., Paulus, M., Zornow, M., & Gaba, D. (1991). Analysis of the task of administering anesthesia: Additional objective measures (abstract). *Anesthesiology, 75,* A47.

Herndon, O., Weinger, M., Zornow, M., & Gaba, D. (1993). The use of automated record keeping saves time in complicated anesthetic procedures (abstract). *Anesthesia and Analgesia, 76,* S140.

Holzman, R., Cooper, J., Gaba, D., Philip, J., Small, S., & Feinstein, D. (1993).

Participant responses to realistic simulation training in anesthesia crisis resource management (ACRM) (abstract). *Anesthesiology, 79,* A1112.

Howard, S., Gaba, D., Fish, K., Yang, G., & Sarnquist, F. (1992). Anesthesia crisis resource management training: Teaching anesthesiologists to handle critical incidents. *Aviation, Space, and Environmental Medicine, 63,* 763–770.

Howard, S., Gaba, D., & Jump, B. (1993). A survey of anesthesiologists attitudes towards production pressures (abstract). *Journal of Clinical Monitoring, 9,* 231.

Klein, G. (1989). Recognition-primed decisions. *Advances in Man-Machine Systems Research, 5,* 47–92.

Klein, G., Calderwood, R., & Macgregor, D. (1989). Critical decision method for eliciting knowledge. *IEEE Transactions on Systems, Man, and Cybernetics, 19,* 462–472.

Klein, G., Orasanu, J., Calderwood, R., & Zsambok, C. (Eds.). (1993). *Decision making in action: Models and methods.* Norwood, NJ: Ablex.

Loeb, R. (1993). A measure of intraoperative attention to monitor displays. *Anesthesia and Analgesia, 76,* 337–341.

Loeb, R., Weinger, M., & Englund, C. (1993). Ergonomics of the anesthesia workspace. In J. Ehrenwerth, & J. Eisenkraft (Eds.), *Anesthesia equipment: Principles and applications* (pp. 385–404). St. Louis: Mosby.

Lunn, J., & Devlin, H. (1987, December 12). Lessons from the confidential inquiry into perioperative deaths in three NHS regions. *Lancet, 2,* 1384–1386.

McDonald, J., & Dzwonczyk, R. (1988). A time and motion study of the anaesthetist's intraoperative time. *British Journal of Anaesthesiology, 61,* 738–742.

McDonald, J., Dzwoncyzk, R., Gupta, B., & Dahl, M. (1990). A second time-motion study of the anaesthetist's intraoperative period. *British Journal of Anesthesia, 64,* 582–585.

Norman, D. (1981). Categorization of action slips. *Psychology Review, 88,* 1–15.

Orasanu, J. (1993). Decision making in the cockpit. In E. Wiener, B. Kanki, & R. Helmreich (Eds.), *Cockpit resource management* (pp. 137–172). San Diego: Academic Press.

Orasanu, J., & Connolly, T. (1993). The reinvention of decision making. In G. Klein, J. Orasanu, R. Calderwood, & C. Zsambok (Eds.), *Decision making in action: Models and methods* (pp. 3–20). Norwood, NJ: Ablex.

Orlady, H., & Foushee, H. (1987). *Cockpit resource management training. NASA Conference Publication 2455.* Washington, DC: National Aeronautics and Space Administration.

Rasmussen, J. (1986). *Information processing and human–machine interaction: An approach to cognitive engineering.* New York: Elsevier.

Rasmussen, J. (1987). Cognitive control and human error mechanisms. In J. Rasmussen, K. Duncan, & J. Leplat (Eds.), *New technology and human error* (pp. 53–61). Chichester, England: Wiley.

Reason, J. (1987). Generic error-modeling system (GEMS): A cognitive framework for locating common human error forms. In J. Rasmussen, K. Duncan, & J. Leplat (Eds.), *New technology and human error* (pp. 63–83). Chichester, England: Wiley.

Reason, J. (1990a). The contribution of latent human failures to the breakdown of complex systems. *Philosophical Transactions of the Royal Society of London, B327,* 475–484.

Reason, J. (1990b). *Human error.* Cambridge, England: Cambridge University Press.

Salas, E., Dickinson, T., Converse, S., & Tannenbaum, S. (1992). Toward an understanding of team performance and training. In R. Swezey, & E. Salas (Eds.), *Teams: Their training and performance* (pp. 3–29). Norwood, NJ: Ablex.

Sarter, N., & Woods, D. (1991). Situation awareness: A critical but ill-defined phenomenon. *International Journal of Aviation Psychology, 1,* 45–57.

Schneider, W., & Detweiler, M. (1988). The role of practice in dual-task performance: Toward workload modeling in a connectionist/control architecture. *Human Factors, 30,* 539–566.

Schwid, H., & O'Donnell, D. (1992). Anesthesiologists' management of simulated critical incidents. *Anesthesiology, 76,* 495–501.

Tenney, Y., Adams, M., Pew, R., Huggins, A., & Rogers, W. (1992). *A principled approach to the measurement of situation awareness in commercial aviation (NASA Contractor Report 4451).* Washington, DC: National Aeronautics and Space Administration.

Tversky, A., & Kahneman, D. (1974). Judgment under uncertainty: Heuristics and biases. *Science, 185,* 1124–1131.

Weinger, M., & Englund, C. (1990). Ergonomic and human factors affecting anesthetic vigilance and monitoring performance in the operating room environment. *Anesthesiology, 73,* 995–1021.

Woods, D., Roth, E., & Pople, H., Jr. (1987). *Cognitive environment simulation: An artificial intelligence system for human performance assessment.* Washington, DC: U.S. Nuclear Regulatory Commission.

Xiao, Y., Milgram, P., & Doyle, D. (1992). Off-loading, prevention, and preparation: Planning behaviors in complex system management. *25th Annual Conference of the Human Factors Association of Canada* (pp. 193–200). Mississauga, Ontario: Human Factors Association of Canada.

12 Team Performance in the Operating Room

Robert L. Helmreich
The University of Texas at Austin

Hans-Gerhard Schaefer
University of Basel/Kantonsspital

Surgical procedures require the coordinated efforts of doctors and nurses working under time pressure. In this setting, human error can result in death or permanent damage to a patient. This chapter discusses the multiple factors that influence the performance of the operating room teams of a European teaching hospital. Results of a survey of operating room personnel in this hospital are presented along with observations of team behavior during operations.

Although the most obvious goal of research into operating room (OR) behavior is to reduce the incidence of human error, this is not a sufficient outcome. Teams that function well in the operating room serve their patients better, but optimal performance also includes the timely and efficient accomplishment of procedures and the maintenance of staff morale and job satisfaction. By gaining a better understanding of how medical teams perform, it may become possible to develop strategies to enhance team performance and hence to reduce the likelihood of error.

The history of concern with interpersonal human factors in the OR is a short one. The seminal work of David Gaba and his colleagues on anesthetic mishaps, decision making, and situation awareness and simulator training has been a major contribution (Gaba, 1989, 1992; Gaba, Maxwell, & DeAnda, 1987). The concerns of anesthesiologists have focused attention on human factors and group dynamics in this environment (Caplan, 1989; Chopra, Bovill, Spierdijk, & Koornneef, 1992; Ewell & Adams, 1993; Howard, Gaba, Fish, Yang, & Sarnquist, 1992; Kumar, Barcellos, Mehta, & Carter, 1988; Schaefer, Scheidegger, & Helmreich, 1993). Weinger and Englund (1990) provide an excellent overview of the multiple factors influencing the behavior of anesthesiologists and the intersection of individuals and groups with equipment under varying time pressures and degrees of fatigue. Investigations of causal factors in anesthetic mishaps suggest that between 75% (Chopra et al., 1992) and 80% (Kumar et al., 1988) of them involve human error. The critical role of human failure is also demonstrated by a survey in which 24% of a sample of anesthesiologists reported committing an error with lethal results (McDonald & Peterson, 1985).

The dynamics of the OR are complex because it is a point of intersection for multiple groups with their own agendas and requirements. The environment contains four identifiable teams during preoperative, operative, and postoperative periods—surgeons, anesthesiologists, surgical nurses, and anesthesia nurses. Anesthesiologists and surgeons are classified as either consultants or residents. Consultants are specialists responsible for resident training as well as the bulk of patient treatment. Residents are in training to become qualified as specialists. Whereas there are strong hierarchical separations between physicians and nursing staff, there is not a clear division of authority between the senior surgeon and the senior anesthesiologist.

Team composition is continually changing with consultants, for example, coming and going to discharge other duties. Depending on the complexity of the operation and the condition of the patient, there may be between 4 and 15 people working in the OR at any time. The team functions within the framework of a support system containing laboratories, orderlies, and other staff. The primary tasks, of course, are managing the patient's condition so as to stabilize it and to alleviate pain (if the patient is conscious) and accomplishing the necessary surgical procedure. Actions of one team can influence behavioral requirements for the other. For example, some incisions by surgeons may cause large shifts in blood pressure or heart rate, requiring intervention by the anesthesia team. Similarly, patients too lightly anesthetized may cough and cause problems for the surgical team. Figure 12.1 defines the groups that intersect and interact in the service of the patient.

As Fig. 12.1 indicates, the operating room represents a task-oriented situation where a number of people perform a variety of activities toward a common goal, the well-being of the patient. Human failure in this setting involves errors of commission (for example, severing an artery or administering an inappropriate drug) or errors of omission (for example, failing to note falling blood pressure). However, little is gained by placing blame for error on a single individual or team. Errors (as well as superior performance) have their roots in the backgrounds of the participants, the dynamics of the group, and the environment in which the activity occurs. To understand the causes of error, it is necessary to consider the organizational, physical, and social context as well as the specific behavior. Such analysis is often called a *systems approach*. System-level analyses have been employed

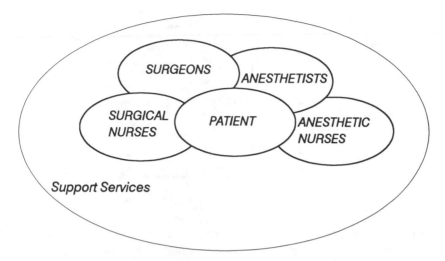

FIG. 12.1.   Team composition of the operating room.

to understand catastrophic failures in a number of demanding environments including aviation, marine shipping, and nuclear and petrochemical plant operations (Helmreich, 1992, in press; Moshansky, 1992; Perrow, 1984; Rasmussen, Duncan, & Leplat, 1987; Reason, 1990).

There is a critical difference in reactions to human error in medicine and in environments such as nuclear power plants and aviation. When a patient dies in the operating theater, it is a regrettable occurrence but one involving a single, ill individual. When a nuclear reactor fails because of operator error or an airworthy plane crashes because of crew failure, the outcome is highly visible and results in public demands for investigation and remediation. As a result, resources are made available for research, and empirical data accumulate. Perhaps in part because of anxieties about personal vulnerability on the part of legislators who fly and appropriate funds, research into human error in aviation has been relatively well supported by the government. In addition, the fact that cockpit voice recorders and flight data recorders (the so-called "black boxes") provide an objective record of actions and communications has facilitated and contributed to development of a substantial literature on pilot error. Recently, several anesthesiologists concerned with human error and performance in the OR have noted strong parallels between behavioral issues on the flightdeck and observed behavior in the OR and have embraced the systems approach to analysis of this environment (Ewell & Adams, 1993; Howard et al., 1992; Schaefer & Scheidegger, 1993). Data from research into flight crew behavior have guided the present investigation, including the derivation of a conceptual model of operating room behavior (Helmreich & Foushee, 1993; Helmreich, Wiener, & Kanki, 1993).

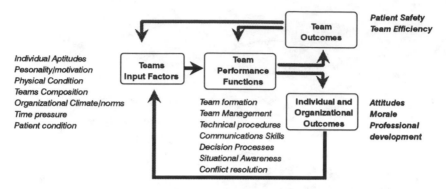

FIG. 12.2.   A model of operating room performance.

The model places the OR in the context of an organizational system. It specifies external influences on performance and defines the group dynamics that can influence outcomes. The general form of the model is based on an *input, process, outcome* conceptualization (Helmreich & Foushee, 1993; McGrath, 1964). Organizational, environmental, and individual factors set the stage for group processes that in turn lead to multiple outcomes. Both processes and outcomes form a recursive system that influences subsequent interactions. Figure 12.2 shows the model that can serve as a framework in which to place research findings and a guide for further investigation. The Department of Anesthesia in the European hospital, with responsibility for the preparation and maintenance of surgical patients, serves as the reference group.

INPUT FACTORS

Organizational Factors

Culture and Structure.   The departments of anesthesia and surgery in this European hospital are the administrative homes of the physicians involved in the OR. Each is directed by a *professor* in the specialty and *consultant specialists*, who provide day-to-day supervision and instruction, and *residents*, who are completing a 4–7 year program to become certified as specialists in either surgery or anesthesia. The two nursing specialties, directed by Anesthesia and Surgery, report to the Chief Administrator for Surgical Nursing, although they are expected to work closely with their medical counterparts. Each nursing component is responsible for the supervision and training of nurses into the two specialties and is staffed by diplomats (fully qualified nurses) and nurses in training.

Two types of patients receive services from the hospital. The first, National Health Service, publicly supported patients, have fees paid by the state at a fixed rate. The second, private patients, have some form of personal insurance and choice of physician. They pay fees to the departments of surgery and anesthesia. Some of the fees collected are placed in a departmental discretionary fund to support research and augment salaries.

The Department of Anesthesia is led by a professor who is motivated to keep the department in the forefront internationally and is willing to commit resources to new programs. However, the structure of the department and organizational policies that are described in the following sections limit the chair's ability to change direction rapidly.

Departmental Efficiency.   Residents and consultants are scheduled for fixed duty periods in the OR and are on call for scheduled periods, including nights and weekends. Both groups are salaried; hence, variations in time worked are a function of situational constraints or personal motivation. Even though there is considerable wasted time in waiting for patients, and so on, there would be no *financial* gain for OR personnel if they were to complete their OR duties more rapidly and with less lost time. Thus, attempts to increase efficiency need to motivate participants by providing gains in job satisfaction and individuals' sense of personal efficacy.

Training.   Staff training requirements create problems and conflict and influence patient treatment. For example, some procedures that are more complex or more stressful for patients may be employed to provide training for residents (for example, fiber optic intubations with the patient awake, insertion of Swan-Ganz catheters, epidural catheters, and transesophageal probes). In addition, residents may be excluded from some activities (for example, intubation of patients) so nurses can acquire experience.

Efforts to master new equipment or to participate in new types of training, such as team coordination and communications, are voluntary. Introducing new training programs within the organizational structure requires either increasing the efficiency of present activities, getting personnel to volunteer nonpaid time, or obtaining additional operational funds for more staff. Only the first option, increasing efficiency, seems viable under current economic conditions. However, as noted earlier, this approach would require providing tangible incentives for staff to work more effectively and to dedicate time gained to additional training.

Evaluation and Quality Control.   The organizational norm regarding performance evaluation is to critique substandard behavior but not to single out above-average performance for recognition. There is not a formal system of performance feedback either for consultants or residents. There is no regulatory requirement in this country for a quality control program to measure the quality of patient treatment. However, the current chair instituted such a program on assuming his position 4 years ago. Detailed records of each case are maintained, and a quality control staff collects follow-up data on the condition of each patient and any adverse reactions. Aggregated, these data provide reliable information on departmental performance and allow detection of trends in performance across time. At present, the data are not used to provide feedback to individual residents or consultants. There is considerable anxiety on the part of staff members about the possible use of quality control data in a professionally threatening manner.

Work Norms. Each anesthesia consultant and resident is assigned to regularly scheduled duties in the OR and, in addition, is on call several times a month. During the on-call period there is responsibility for all emergency cases requiring anesthesia. Residents are scheduled for a 55-hour week (recently reduced from 65 hours). Consultants are less formally scheduled but work between 65 and 75 hours a week. In addition to their clinical and training duties, consultants, like their faculty counterparts in other university departments, are expected to maintain a program of research and publication.

Anesthesiologists' job pace and requirements are extremely uneven. There is a high level of demand and coordination required during the patient induction and emergence phases. This also includes coordination with the support staff who deliver the patient and prepare the OR. During the maintenance phase, while surgical procedures are accomplished, there are often relatively low demands except for vigilance and monitoring. However, in the event of medical emergency, the workload during this period can be extremely high. According to Chopra et al. (1992), 45% of incidents involving error occurred during this phase, which suggests that problems in monitoring may be critical. Multitasking requirements are also present at both the individual and group level. The anesthesiologist must both monitor patient condition and conduct procedures such as drug administration or catheterization (Weinger & Englund, 1990). Consultants may be responsible for supervision of anesthesia in up to three operating theaters. This means as many as six patients may be active at any time (three in operating theaters and three undergoing induction or waking up).

When residents or nursing staff are inexperienced or if difficulties arise with one or more patients, the workload can become extremely high and can demand effective workload distribution and prioritization throughout the patient's stay on the surgical floor. Patients are normally sent to a recovery room as soon as they emerge from anesthesia. However, this room is closed at 5:00 p.m. daily and is not open on weekends or holidays. If there are unexpected difficulties with a patient or if it is an emergency procedure, there may be difficulties in getting space for the patient in intermediate or intensive care units—a responsibility of the anesthesiologist.

Role Strain and Conflict. Both anesthesiologists and surgeons play multiple roles during the discharge of their duties in the OR and elsewhere. Both groups are responsible for patient care and staff training, and, as noted, these requirements can produce conflict between the need to train and the desire to use the most efficient and least demanding procedures. There are also multiple demands on staff for clinical work, teaching, research, and publication.

A more fundamental conflict in roles can be caused by differing perceptions of the surgeon and anesthesiologist role. In the model posed here, the four primary groups were seen as part of a single, superordinate operating room team. In this view, the patient is the implicit customer for the service activity, and the team's efforts are directed toward safe and effective management of the patient's condition and needs.

Another model of OR activity has been proposed that defines suppliers and customers differently (Schisler, 1992). It focuses on the *surgeon* as the customer of the services provided by the anesthesia teams. The pattern of interactions in the OR is undoubtedly impacted by the views held by participants. The dynamics of the situation where the surgeon is the customer ("who is always right") would necessarily differ from environments where all perceive themselves to be part of the same team with shared responsibilities. Probably the most dysfunctional situation is one where participants from the various subgroups hold differing views of the role structure.

Whatever the perceptions of roles, there are clearly shifts in role centrality that influence the relations of surgeons and anesthesiologists and can affect interactions. The primary purpose for the patient (and staff) to be in the OR is to accomplish some surgical procedure. Awareness of this function can lead the surgeon to a self-perception of being the de facto team leader. Nonetheless, responsibility for patient management is equally central to the endeavor, and awareness of this role can lead anesthesiologists to a similar perception of their role.

When the patient is taken to the recovery room, intensive care, or intermediate care, primary responsibility for the patient shifts to the anesthesiologist, who is charged with pain control and postoperative patient maintenance. In this setting, the role definition for anesthesiologists is clear and their status unequivocally defined.

Interdepartmental Conflict. The operation of the OR, with patient induction by anesthesiologists required before surgeons can perform their tasks, requires advance planning and coordinated scheduling of activities that are highly variable in duration. Anesthesiologists would like induction as shortly as possible before surgery to reduce the time spent monitoring unconscious patients. They would like coordination with activities in the theater so that a new patient's arrival is based on completion of previous surgery, allows enough time for induction, and minimizes waiting time for theater or surgeons.

Surgeons, on the other hand, would like the anesthetized patient ready in the theater when *they* desire to begin surgical procedures. These goals are not necessarily compatible and create the possibility of conflict between surgeons, anesthesiologists, anesthetic nurses, and surgical nurses. These conflicts play out in two settings—daily planning meetings in the departmental office and in the operating theaters. Several underlying factors create problems for the planning activity. One is the fact that unanticipated emergency work can take precedence and interfere with any schedule defined. In a study of surgical nurses in a U.S. hospital, Denison and Sutton (1990) found similar problems with scheduling. Another problem is a lack of advance planning and coordination among groups to maximize common outcomes for surgery, anesthesia, and nursing (for example, efficient use of each group's time). There is also an expressed concern by the surgical staff that the scheduling of surgical procedures must be flexible to be competitive with private hospitals that are more oriented toward serving paying patients.

At the daily planning meeting that is held mid-afternoon in the Department of Anesthesia, surgeons arrive and present a proposed schedule of operations for the following day (scheduled surgery is conducted Monday through Friday with the weekends reserved for emergency procedures). This is discussed in a forum that includes the consultant anesthesiologist on duty the following day and the head nurses (or delegates) from both anesthesia and surgery. The proposed schedule of surgery includes theater assignments and estimates of time required for the procedure. One of the persisting complaints by anesthesia personnel is that the surgical staff have not thoroughly verified patients' conditions and that estimates regarding time involved are inaccurate. At the conclusion of this meeting, the surgical, nursing, and anesthesia groups agree on a master plan, reconcile differences, and print and distribute the schedule for the following day. At the time of schedule creation, the anesthetic residents and consultants generally have not yet examined the patient to determine if there are possible complications that may require special procedures for induction (for example, coronary artery disease, restricted airways, or obesity) or if possible complications would indicate special assignment to particular consultants, residents, or nurses.

In the operating theaters, the anesthesiologist normally calls for the patient from the ward after consulting with the surgical nurse, based on the best estimate of when the theater will be free and the surgeon present. Occasionally, a senior surgeon instead of the anesthesiologist will call the patient from the ward. Although there is not a formal written procedure delegating this responsibility, deviation from the normal practice can and does create conflict and hard feelings. Once anesthetized, the patient must be continually monitored by an anesthesiologist or anesthesia nurse. Hence, if there is a long delay before the room or surgeon is ready, this creates ill-will among the anesthesia staff. Conversely, if the surgical staff is ready to proceed but the patient has not been called for or induction completed, the schedule is also disrupted and tempers can flare. Other sources of scheduling problems can come from laboratory delays in processing samples or missing information such as patient notes.

Communications regarding patient scheduling appear to be one of the major sources of conflict between the groups. Although there are many unpredictable factors that can render time estimates invalid (e.g., patient emergencies, etc.), there is a perception on both sides that many of the problems result from imperfect communication and coordination.

Environmental Factors

Operating Room Design.   Unlike most theaters in the United States, the operating facility in the European hospital is designed for processing multiple patients. Each theater is attached to two induction rooms. Patients are brought first to one of the induction rooms where they are anesthetized and wait for availability of the theater. While the first patient is receiving surgery, the next patient can be anesthetized. When the surgery is complete, the patient can be moved to the other induction room, the operating theater cleaned and prepared, and the second patient positioned for surgery. This design is undeniably efficient. However, it has the potential of placing great demands and a very high workload on the consultant anesthesiologist, residents, and nurses. The physical layout of the theater shows little recognition of design factors such as placement of equipment to improve usability.

Equipment.   Technology has advanced rapidly in anesthesia delivery and patient monitoring. Equipment is manufactured by several vendors and varies in sophistication. Many monitoring functions are automated and have visual and auditory warnings when patient parameters exceed normal ranges. The 12 theaters in this hospital have four different types of anesthesia monitoring equipment with different displays and levels of sophistication.

It was noted earlier that research data indicate that the highest percentages of anesthetic incidents occur during the maintenance phase of anesthesia when the primary task is monitoring. Monitoring is made more difficult by variations in equipment and the fact that many spurious alarms are presented in an already noisy environment.

One problem observed in this hospital was intermittent failure of newly installed monitoring equipment, requiring that old style equipment be brought in and hooked up. This resulted in a lack of monitoring capability while the equipment shift was accomplished. The causes of equipment failure have roots in the hardware and software involved and in operators unfamiliar with new operating procedures. As noted, there are no provisions for formal recurrent training of staff, including operational familiarization with new hardware.

Patients.   It is a truism that patient outcomes would be better if patients were not so ill. Much of the outcome variability is influenced by patient age, physical condition, and physique. A high proportion of patients are elderly with a variety of physical and mental disabilities associated with aging. Variability in patient reactions increases the need for vigilance in monitoring throughout surgical procedures.

Team Factors

Composition.   As mentioned earlier, four distinct subgroups function in the OR—surgeons, anesthesiologists, surgical nurses, and anesthetic nurses. Although it is becoming popular to refer to the "operating room team," it is not clear that most participants view the OR in this way. Observations suggest a tendency to define one's primary group in terms of function and status. Hence, there are two recognizable medical teams and two nursing teams with closer affiliation between the two anesthetic groups and the two surgical groups.

Climate.   The morale of workers and resultant group climate of an organization are recognized as critical determinants of performance (Steers & Porter, 1991). Groups operating with a positive group climate are more satisfying to their members and, by extrapolation, should produce high motivation for effective performance.

Norms.   All work groups develop norms governing appropriate behaviors. In the multigroup setting of the OR, team members are probably most strongly influenced by the norms of their primary group (e.g., anesthetic nurse or surgeon). One of the tasks facing researchers is to understand the differing sets of norms that govern OR behavior.

Intergroup Norms.   Groups also develop conventions for dealing with other groups. These include patterns of work communication and social interaction. With four distinct groups, the norms governing relations among the groups should be important determinants of process and outcomes.

Individual Factors

Attitudes.   Research has shown that attitudes regarding crew coordination, communication, leadership, and personal capabilities are highly predictive of crew performance in aviation (Helmreich, Foushee, Benson, & Russini, 1986). Attitudes aggregated by group can also provide information on group culture and norms. It is almost axiomatic that if subgroups differ in views of or there is no agreement among individuals on appropriate management and behavioral strategies, they are unlikely to function optimally as teams.

Aptitude/Intelligence.   Each of the groups working in the OR is composed of individuals with specialty training who have been chosen from an applicant pool. Although there is undoubtedly within- and between-group variance in aptitude and intelligence, in these selective groups it is unlikely that intelligence is a major determinant of process and outcomes. There are, however, highly visible differences among staff in the psychomotor skills involved in surgery, intubation, and catheterization.

Personality/Motivation.   In vocations that require interpersonal interaction and team efforts for success, personality factors are critical determinants of group processes. Research data also suggest that individuals with different personality types are differentially attracted to some professions and that this is true of medicine and nursing (Barberat, Goldschmid, Neirynck-Carton-de-Wiart, Nughes, & Ricci, 1990; Helmreich, Spence, Beane, Lucker, & Matthews, 1980; Spence & Helmreich, 1983). Anecdotal evidence in medicine suggests that surgeons and anesthesiologists differ normatively in personality. If there are personality differences between OR groups, these could influence the patterns of interaction and group climate.

There is considerable evidence showing that personality and achievement motivation are strong determinants of individual and team performance (e.g., Chidester, Helmreich, Gregorich, & Geis, 1991; Helmreich et al., 1980; Rose, Helmreich, Fogg, & McFadden, 1993; Spence & Helmreich, 1983). In medicine (although not broken down by specialty area), Barberat et al. (1990) found that successful doctors in Switzerland differed significantly from those with lesser professional attainment on several scales of the California Personality Inventory (Gough, 1987). This profile indicates that successful physicians are higher in intellectual concerns, management potential, and self-actualization, but lower in femininity or interpersonal orientation, although physicians as a group are elevated on the latter scale. Further research into personality factors within all OR teams is much needed.

Knowledge/Training.   As has been noted, one of the functions of this hospital is the training and qualification of nurses and specialists. As a result, most surgeries have a training as well as a patient-service component. The differential qualifications of individuals on the teams together with the need to train while serving the patient's needs influence the dynamics of the OR.

Physical Condition/Fatigue.   Because residents and consultants spend extended periods on call and are required to respond to emergencies as needed, they may find themselves working under conditions of considerable fatigue. Emergencies can require performance when the circadian cycle is disrupted. Gravenstein, Cooper, and Orkin (1990) surveyed anesthesiologists and anesthesia residents and found that the majority of respondents reported having made errors in the administration of anesthesia as a result of fatigue. Fatigue can reduce the capacity for vigilance and monitoring and can change patterns of interpersonal interaction (Graeber, 1988). The team can and should serve as a mechanism to reduce fatigue-induced error.

Perhaps the most critical issue regarding fatigue and the condition of individuals surrounds work transitions (National Research Council, 1993). Vigilance is assumed at all times, but it is difficult to maintain during long, slack periods. At any moment, whether stimulated by eternal alert or through monitoring, individuals on the team must be prepared to make a transition from a passive, vigilant state to active intervention, frequently with a life in the balance.

Emotional State.   Instances of serious interpersonal conflict have been observed in the OR, typically having to do with interteam conflict over patient preparation or management. Team dynamics and performance can either be facilitated by positive emotional states or compromised by negative states. There is a lack of research into the implications for safety and efficiency of conflict in the OR, but the issue warrants systematic investigation.

Preliminary Research on Attitudes as Inputs to the OR

The first approach to determining behavioral issues related to OR performance was to survey the attitudes of operating room personnel. The research was aided by investigations of flight crew attitudes using two research instruments that have been widely used in U.S. and foreign organizations, the *Cockpit Management Attitudes Questionnaire* (Helmreich, 1984) and the

Flight Management Attitudes Questionnaire (FMAQ; Helmreich, Merritt, Sherman, Gregorich, & Wiener, 1993). These surveys have been validated as predictive of crew performance and have been able to isolate group norms and cultures (Merritt & Helmreich, in press). The new survey, the *Operating Room Management Attitudes Questionnaire (ORMAQ;* Schaefer & Helmreich, 1993) contains questions that deal with team communication and coordination, leadership, work motivation, organizational climate, and recognition of the effects of stressors on personal capabilities.[1]

Like the *FMAQ,* the *ORMAQ* focuses on concepts derived from Hofstede's (1980) seminal research on cultural differences in work values. One of Hofstede's dimensions, *Power Distance* (PD), reflects the extent to which less powerful members of organizations and cultures *expect* and *accept* that power is distributed unevenly. PD is reflected in the decision-making style of superiors, ranging from participative to dictatorial, and the willingness of subordinates to disagree with superiors. Asian and Latin cultures were found by Hofstede (1980) to be high in PD, whereas the United States and many European cultures tend to be moderate on this dimension.

Survey Results.   The *ORMAQ* contains 64 multiple-choice questions answered on 5-point Likert scales. In addition, two open-ended items ask how team effectiveness and job satisfaction could be increased. Data were collected on a German translation of the *ORMAQ* from 156 staff members including 53 surgeons, 45 anesthesiologists, 32 surgical nurses, and 22 anesthesia nurses in the European hospital. This represented a response rate of more than 60% from each subgroup, an excellent rate of return for survey research. Overall, the results show that there are significant differences in opinion between the subgroups and large variability in attitudes within subgroups. Respondents generally agreed that communications and coordination are as important as technical proficiency for safety and efficiency and that the team concept is applicable in the operating room. However, there was much less consensus on items reflecting how optimal team coordination should be achieved. For example, anesthesiologists and anesthesia nurses were much more accepting than surgeons and surgical nurses of the idea that a preoperative briefing is important for team effectiveness.

Items reflecting leadership and the power distance concept showed large and highly significant differences ($ps < .01$ by analysis of variance). Surgeons and surgical nurses are more accepting than the other groups of the notion that junior team members should not question the decisions of senior staff.

---

[1]Copies of the instrument, available in German and English, may be obtained from the authors.

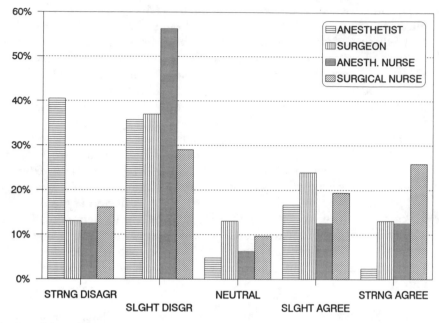

GROUP DIFFERENCE P<.007

FIG. 12.3. Responses to the question "Operating room team should not question decisions of senior staff."

However, as shown in Fig. 12.3, there is a broad range of opinions in each group.

Attitudes regarding the style of leadership desired and actually experienced cast further light on the dynamics of the OR. Four leadership styles were described in the following way on the questionnaire:

1. *Autocratic*—superior makes decisions and communicates them firmly, expects loyal obedience and no questions.
2. *Mild Autocratic*—superior makes decisions promptly, but explains them fully, provides reasons to subordinates and answers questions.
3. *Consultative*—normally consults with subordinates when important decision is to be made, listens to advice, considers it, and then makes decision.
4. *Democratic*—usually meets with subordinates when decision is to be made, puts problem before group and invites discussion, and then accepts majority viewpoint as decision.

Respondents were then asked to indicate which style they preferred and which they encountered most frequently in the OR. Figure 12.4 shows the preferred and actual leadership styles for each group. Anesthesiologists, surgeons, and anesthesia nurses most often preferred the consultative leadership style, whereas surgical nurses selected the democratic style most, followed by the consultative. Many surgeons, on the other hand, endorsed the mild autocratic style. All groups report a great discrepancy between the style they prefer and the one they encounter in the OR. The majority of each group reports encountering autocratic or mild autocratic leadership. These findings suggest that the team concept may be impaired in this setting because many participants do not experience the type of leadership they consider optimal.

Recognition that one is susceptible to the effects of stress and fatigue is important, because an individual who is not functioning optimally may be more prone to make errors and in more need of support from co-workers. A notable finding from aviation research on attitudes was that pilots normatively held unrealistic attitudes regarding the effects of stress and fatigue on their personal capabilities (Helmreich, 1984; Helmreich & Wilhelm, 1991). Aviators reported that they performed as well when fatigued as rested, that their decision making was as good in emergencies as in normal operations, and that as "true professionals," they could leave behind personal problems when flying. To our surprise the attitudes of medical professionals were equally unrealistic, although there was a great deal of variability in re-

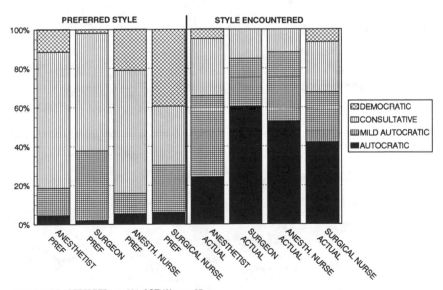

CHI SQUARE PREFERRED p <.001, ACTUAL p <.05

FIG. 12.4.   Distribution of preferred leadership style and leadership style encountered in the operating room.

sponses. Figure 12.5 shows the distribution of responses regarding performance when fatigued. The modal response is to agree slightly with this proposition. Similarly, the majority of medical personnel endorse the proposition that their decision making is as good in emergencies, as shown in Fig. 12.6. The notion of being impervious to the effects of stress can undermine reliance on other team members to maintain effectiveness under stressful conditions.

The survey findings suggest that operating room personnel do not have a consensus view of how team activities should be led and coordinated. This probably reflects the fact that the training of medical personnel focuses on technical aspects of the job and not on team activities.

Two thirds of those who completed the survey took the additional time required to write open-ended responses to items asking for recommendations for improving operating room efficiency and staff morale. This response rate indicates a high level of concern on the part of respondents. Regarding increased efficiency, the majority of the comments stressed the need for better communication and team coordination. Other common themes included a need for more feedback on performance and better leadership in the OR. Only a few respondents mentioned a need for more technical training. Suggestions for improving morale also centered on improving communications and attitudes regarding OR team functions. Many

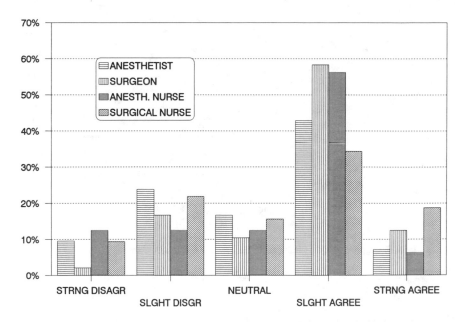

FIG. 12.5.   Responses to the question "Even when fatigued, I perform effectively during critical phases of operations."

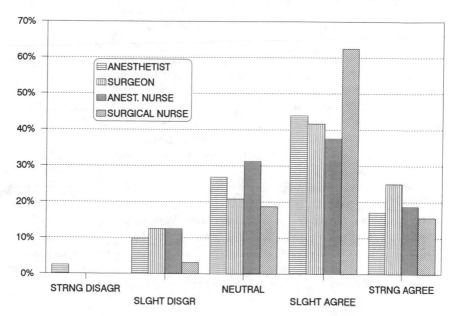

FIG. 12.6. Responses to the question "My decision making is as good in emergencies as under normal conditions."

stressed the need to better understand the work of other teams. A number pointed out the frustrations resulting from poor scheduling of operations, but only a few mentioned traditional concerns with pay and work time.

GROUP PROCESS FACTORS

Group processes in the OR can be classified on several dimensions. Anesthesiologists may interact with their peers (for example, the consultant and the resident) or with their extended team that includes the anesthetic nurse or nurses. Interactions can also take place at the interteam level with surgeons, surgical nurses, and support staff (i.e., orderlies). There may be less willingness to ask for assistance or to communicate problems encountered at the interteam level.

The second classification is in terms of function. Two broad types of tasks can be defined in the OR team setting—those dealing with cognitive and interpersonal issues and those dealing with technical aspects of patient management and surgery. Both are critical and combine to define group processes. These tasks can be further broken down into specific behaviors that can be evaluated in terms of their presence or absence and quality.

One set of cognitive and interpersonal tasks involves forming and maintaining the team (considered at the macro level as representing all four subgroups). Activities include establishing a favorable group climate and effective leadership. A second set of tasks is related to communications processes and decision making. It includes seeking and transmitting information, sharing mental models regarding needed actions, and determining the optimal course of action. A third consists of activities required to maintain situation awareness and to manage workload. It includes monitoring patient condition and the course of actions taken by the other team (for example, anesthesiologists maintaining awareness of surgeon actions that may impact patient vital signs). Strategies to maintain vigilance and to avoid distractions fall under this cluster.

The technical domain consists of two groups of tasks. The first includes the activities that are seen as primary professional responsibilities. For surgeons, this includes the procedures planned and required. For anesthesiologists, it involves management of anesthetic level along with physiologic support. The second consists of procedures employed to maintain accurate, required records of activities. This includes all of the paperwork for patient records and quality control required by the organization.

Research into Group Processes

Organizational researchers have recognized the importance of team function in the operating room. For example, Denison and Sutton (1990) observed operating room nurses in a U.S. teaching hospital. Their focus, however, was on the nursing subgroup rather than the larger team.

Understanding operating room behavior and performance presupposes the ability to recognize and evaluate observable behaviors reliably. This can best be achieved by defining specific behaviors that reflect effective and ineffective performance. Behaviors that are essential for safe and efficient function can be defined as shown in the following list. The goal of this categorization is to develop a rating methodology that can be employed reliably by trained observers.

1. Briefings (conduct and quality)
2. Inquiry/Assertion/Advocacy
3. Team self-critique (decisions and actions)
4. Communications/Decisions
5. Leadership-Followership/Concern for tasks
6. Interpersonal relationships/Group climate
7. Preparation/Planning/Vigilance
8. Workload distribution/Distraction avoidance
9. Teaching function (if applicable)
10. Conflict resolution (if present)

The authors verified the behavioral issues defined by observing team interactions in the anesthetic induction rooms and operating theaters of the European hospital. Numerous delays and frustrations caused by failures to coordinate activities among groups were noted. In the induction room, scheduling problems caused unnecessary stress in patients brought too early to theater to await anesthetic induction. Unclear or missing instructions on the written OR program regarding how to position patients on the table caused delays and often required many phone calls for clarification. A frequent scene in the OR was teams waiting with an anesthetized patient for the surgeon to appear. Such delays induced an atmosphere of anger and hostile resignation. In one instance, surgical residents and the anesthesia team waiting for the chief surgeon expressed their frustration to us, but on the surgeon's arrival no comments were made by either group—there was no apology or explanation from the surgeon as to why he was 30 minutes late, and there were no complaints or comments from any team member waiting. This created a tense and unhappy atmosphere during the operation, which could impact performance.

A near total absence of preoperative briefings to make certain that members of surgical and anesthesia teams were aware of anticipated problems or special procedures was observed. Direct communications between the surgical and anesthetic teams occurred infrequently. In one case, the failure to notify anesthesiologists of the use of a local anesthetic containing a vasoconstrictor led to a considerable increase in blood pressure. In another case, when surgeons began to dissect tissue in close proximity to the carotid bodies, they did not inform the anesthesiologist, although he could not directly observe the surgical site.

Situation awareness in the OR was generally a function of what team members could observe over the cloth barrier erected between the surgical and anesthesia groups. This barrier supposedly maintains a sterile surgical field.[2] However, functionally it shields the anesthetic team from the surgical team and prevents awareness of team activities on the other side of the fence. Surgeons cannot observe the effect of their activities on the patient's hemodynamic and respiratory variables by watching monitors, and the anesthesiologists cannot follow surgical manipulations that may impact patient status. During one operation, the patient was not sufficiently relaxed and coughed, creating concern among surgeons cutting in a critical area. The single resident present at the time was seriously overloaded and faced multiple tasks under strong time pressure. In this case, the screen prevented the surgeons from observing and appreciating the stress induced in the resident, who was new to the job. Instead of trying to establish a supportive

---

[2]Inquiries of staff regarding the use of the barrier indicated that there were no clear guidelines. In practice, the height and placement of the barrier seems to be a prerogative of the surgical nursing team.

climate in a difficult situation, the anger of surgeons hidden behind the screen prevailed.

It was observed that patient monitoring equipment produces many spurious alarms and fails to provide accurate readings on many occasions. As a result, teams frequently ignored alarms or failed to show any concern with obtaining accurate readings. On one occasion, four persons on the anesthesia team failed to react to indicated low blood-oxygen saturation for almost 15 minutes and did not remark on the intermittent failure of the monitor to show readings of the patient's status.

Observations during the maintenance phase of anesthesia, after the patient had been induced, showed wide variations in the level of vigilance and the consistency of patient status monitoring. Some team members constantly attended to monitors and attempted to follow surgeons' procedures, whereas others indulged in alternative activities including professional reading. In many cases, a single nurse or resident from the anesthesia team was present during this phase. In the event of an emergency requiring concerted action, there was not a formal procedure, other than asking help from the surgical team, to obtain assistance or to call for return of other anesthetic team members. In several cases, one of the observers was drafted to assist in high workload or abnormal situations.

Workload allocation was sometimes made without communication and consultation among team members. For example, an anesthesia consultant

- Communications/decisions
    - Surgeon's failure to inform anesthetist of drug with effects on blood pressure
    - Consultant leaving resident in OR with work overload
    - Consultant scheduling patient without informing resident or nursing staff
- Preparation/Planning/Vigilance
    - Failure to complete checklist (anesthesia machine set wrong)
    - Failure to react to blood pressure and blood oxygen alarms
    - Failure to monitor patient status during operation
    - Failure to anticipate events during a complex procedure (coming off coronary bypass)
- Workload distribution/Distraction avoidance
    - Resident reading technical manual distracted from discovering patient not sufficiently relaxed
    - Consultant distracted from making a decision to place a pulmonary artery catheter by problems reported from another operating theater

FIG. 12.7.   Classification of errors observed in the OR by behavioral category.

• Briefings (conduct and quality)
- Failures to brief own team and other teams as to plans for operation
• Inquiry/Assertion/Advocacy
- Failures to speak up to inform team of work overload or patient problems
- Failure to discuss alternative procedure and advocate own position
• Team self-critique
- Failures to debrief operation to learn from situation
• Leadership/Followership/Concern for tasks
- Failures to establish leadership for OR team
• Interpersonal relationships/Group climate
- Hostility and frustrations due to poor team coordination
• Preparation/Planning/Vigilance
- Failures to plan patient preparation
- Failures to monitor other team activities
- Failures to anticipate surgeon actions
• Teaching function
- Failures of consultants to provide training for residents
• Conflict resolution
- Unresolved conflicts between surgical team and anesthestists

FIG. 12.8.   Observed instances of inadequate group process.

wanting to give experience to a nurse in training assigned her to perform the intubation of a patient. This left the anesthesia resident in attendance with nothing to do and created an atmosphere of hostility in the OR. Failing to brief team members before the start of an operation was also observed to create false expectations and misunderstandings. One resident complained to the observer—not to the consultant in charge—that he was angry about being left alone immediately after the endotracheal tube was fixed, causing him to have to insert additional venous and arterial lines without assistance. In reality, the anesthesia consultant left to go into another operating room to induce a second patient, but he did not communicate to the resident where he was going or why he was leaving.

Despite the significant training role of this hospital, few instances were seen of effective on-the-job instruction and debriefing, after either successful or problematic operations. The negative impact of the absence of norms for the instruction of nurses and residents was apparent in a lack of awareness of effective behaviors, including completing checklists prior to placing a patient under anesthesia and planning for likely contingencies during surgery.

The observations discussed here included several instances of observable error that could have resulted in harm to the patient. Observed errors are summarized in Fig. 12.7 in the context of the behavioral categories specified in the earlier list. These errors reflect procedural deficiencies in monitoring or patient treatment. None of the observed errors indicated a lack of technical competence on the part of staff. Their roots appear to be in the interpersonal aspects of OR functioning. This can be seen by classifying observed problems in team coordination as shown in Fig. 12.8. Problems were noted in briefing others as to what to anticipate during the operation, failures to alert others to problems, communicating the status of the patient and the surgery, hostile group climate, failures in vigilance, and unresolved conflicts. Each of these reflects deficiencies in the interpersonal domain and can set the stage for technical errors.

Overall, the observations suggest that many opportunities exist for improving the coordination and efficiency of operating room teams. The inefficiencies of the present system waste much time that could be productively spent in training for improved communication and coordination. The results also indicated that quantification of team behavior is an achievable goal.

OUTCOME FACTORS

The most significant outcome factors are patient safety and quality of treatment. If team members practice effective team coordination and communication, decision making, vigilance, and monitoring along with technical task performance, the resultant processes should be greater than the sum of individual capabilities and should enhance the probability of successful outcomes. Team processes that are perceived by participants as having been effective in maintaining patient safety should impact morale and job satisfaction favorably. If these are positive, they should enhance input factors and subsequent group processes. Interventions that have a positive impact on either input or group process factors should initiate a feedback loop leading to improvement in outcomes (Helmreich & Foushee, 1993).

Understanding the implications of the input and process factors defined in Fig. 12.2 depends on accurate assessment of outcomes in the operating room. Although quality control systems are being implemented in anesthesia and surgery, most do not generate the kind of data needed to understand causal factors. It is essential that the human-factors elements be documented (e.g., breakdowns in communications, flawed decision making, inadequate leadership), especially those that center on the interface between the surgical and anesthesia groups. The focus should not be solely on instances of error or negative outcomes. It is equally critical to look at factors such as staff morale and attitudes that can indirectly influence patient safety. Special efforts should also be made to document instances of

superior performance where effective teamwork resulted in positive outcomes.

APPLYING RESEARCH TO OPTIMIZE TEAM PERFORMANCE

Research findings from the psychological literature should be used to plan
interventions to increase system effectiveness. Attitude research can specify
areas where organizational actions can improve communications, coordination, and leadership. Systematic study of team performance and the operating room environment can define maladaptive behaviors that should be
changed and organizational practices that reduce effectiveness. The following describes actions that might be taken based on data from the research
reported in this chapter.

Technical Solutions

In high-technology endeavors, the immediate response to problems involving human error with catastrophic outcomes has historically been to develop
new equipment to monitor or control the situation. In aviation, this has
resulted in automatic flight guidance computers and various warning systems. In anesthesia, the emphasis on hardware solutions is exemplified by

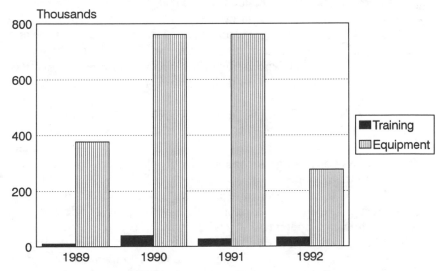

FIG. 12.9.   Relative expenditures for anesthesia equipment and staff training (Swiss
francs).

large expenditures directed at the development and installation of increasingly sophisticated patient-monitoring equipment and displays. This reaction was triggered by recognized instances of anesthesiologists failing to detect critical changes in vital signs. In short, the history of response to awareness of error has been to develop technical solutions to human problems. Figure 12.9 shows the disproportionate amounts spent on professional training and equipment enhancements over a 4-year period in the European hospital. It is the authors' thesis that although new technology is useful, *human* solutions to human problems may be more effective and less costly (Helmreich & Foushee, 1993).

Human-Oriented Solutions

There are two broad areas of intervention to increase team effectiveness—organizational change and training. Each approach should be based on empirical data and analysis of current practices and participant attitudes.

Organizational Change

Attacking team performance issues at the organizational level involves analysis of the organizational culture and its norms to determine whether there are, in Reason's (1990) terms, *resident pathogens* in the system that can interfere with team performance. For example, in the hospital studied, issues surrounding patient scheduling and preparation frequently result in conflict and can be defined as resident pathogens.

Because many of the factors interfering with effective performance involve the interface between groups, interventions need the cooperation and commitment of all components of the organization. This will usually involve compromises in preferred procedures to achieve greater overall effectiveness. Resolution of role ambiguity and leadership problems requires organizational change.

A second organizational approach is to establish norms of performance evaluation so that team members are aware of their own behaviors that are more or less effective. In parallel with evaluation is reinforcement by leaders of effective behavior through meaningful feedback on performance at the individual and team level.

A third approach is to concentrate on the selection of individuals who will function most effectively in a team environment. Individuals selected need not only technical competence but also personalities and motivation that will allow them to work effectively with fellow team members. Empirical findings suggest that personality characteristics of individuals may be a limiting factor in programs to improve team effectiveness (Chidester, Helmreich, Gregorich, & Geis, 1991). Optimizing selection involves a careful analysis of the characteristics required for successful performance and validation of selection procedures.

Training

The findings from survey and observational research in the European hospital suggest that many of the problems observed can be improved through training. Recognition of problems in team performance in aviation led to the development of training programs in team coordination and communication, known generically as *Crew Resource Management* (*CRM;* Helmreich & Foushee, 1993). Programs in CRM concentrate on changing input factors (crew member attitudes) as well as process factors such as team building, communications, leadership practices, and decision-making strategies. It is recognized that effective programs utilize multiple approaches to maximize leverage on team performance.

David Gaba and his colleagues in the United States have made a direct application of CRM training from aviation to the operating room, including the use of realistic simulation, in what they call *Crisis Resource Management* (Howard et al., 1992). Their approach involves the use of an operating room simulator to reproduce the work environment and allow participants to practice new techniques of communications and coordination. More sophisticated OR simulators are currently under development in the United States and Europe and promise to provide a powerful training tool. The optimal approach to OR simulation should involve all four teams enacting realistic scenarios that demand coordination and interface between surgery and anesthesia. It should be as important for enhancing overall team effectiveness to provide training and feedback on how to conduct routine procedures as it is to practice coping with OR crisis situations. Effective training programs include formal evaluation and continuing rather than one-shot training, focus on system issues as they affect human behavior, and concentrate on teaching specific behaviors associated with effective team performance (Helmreich & Foushee, 1993).

Validating the Impact of Interventions

Organizational changes are not without cost. Thus, it is essential for organizations to know if interventions such as training programs and policy changes reduce human error and improve team performance. Empirical demonstration of increased patient safety, enhanced efficiency, and improved job satisfaction can provide strong justification for committing resources to these efforts. In the same vein, increased safety could be reflected in decreased costs of insurance and malpractice litigation. The European hospital has instituted a quality control program that can potentially provide needed data on the effects of these efforts. To obtain needed data, multiple measures of impact should be employed to provide a range of data from attitudinal to behavioral. Useful data should include:

1. Participant reports on the value of programs and recommendations for improvement. The enthusiastic response of personnel in the European hospital to the research effort suggests that medical personnel would welcome programs that could increase their job effectiveness.
2. Measures of changes in attitudes regarding operating room management.
3. Ratings of behavior in the operating theater and in simulations by staff trained in formal evaluation of team performance.
4. Trend data on the occurrence of incidents in actual practice.

CONCLUSION

Survey data and observation of factors influencing team performance in the operating room of a European teaching hospital supported the hypothesis that interpersonal and communications issues are responsible for many inefficiencies, errors, and frustrations in this psychologically and organizationally complex environment. It is noteworthy that efforts at organizational improvement have primarily involved acquisition of new, more sophisticated hardware, whereas support for training and human interventions represents only a fraction of the funds spent on technical solutions. Organizational interventions and the provision of formal training in human-factors aspects of team performance can produce tangible improvements in performance and substantial reduction in human error.

ACKNOWLEDGMENT

Robert L. Helmreich conducted the research reported here while he was Visiting Professor in the Department of Anesthesia at the University of Basel. Special thanks are due Dr. Mark G. Ewell and John A. Wilhelm for assistance with the project.

REFERENCES

Barberat, A. L., Goldschmid, M-L, Neirynck-Carton-de-Wiart, I., Nughes, M-A., & Ricci, J-L. (1990). *La reussite professionnelle des economistes, juristes, medecins, et psychologues* [The professional success of economists, lawyers, doctors and psychologists]. Lausanne: Chaire de Pedagogie et Didactique, Ecole Polytechnique Federale de Lausanne/Suisse.

Caplan, R. A. (1989). In-depth analysis of anesthetic mishaps: Tools and techniques. *Anesthesiology, 27,* 153–160.

Chidester, T. R., Helmreich, R. L., Gregorich, S. E., & Geis, C. (1991). Pilot personality and crew coordination: Implications for training and selection. *International Journal of Aviation Psychology, 1,* 23–42.

Chopra, V., Bovill, J. G., Spierdijk, J., & Koornneef, F. (1992). Reported significant observations during anaesthesia: A prospective analysis over an 18-month period. *British Journal of Anaesthesia, 68,* 13–17.

Denison, D. R., & Sutton, R. I. (1990). Operating room nurses. In J. R. Hackman (Ed.), *Groups that work (and those that don't)* (pp. 293–308). San Francisco, CA: Jossey-Bass.

Ewell, M. G., & Adams, R. J. (1993). Aviation psychology, group dynamics, and human performance issues in anesthesiology. In *Proceedings of the Seventh International Symposium on Aviation Psychology.* Columbus: Ohio State University.

Gaba, D. M. (1989). Human error in anesthetic mishaps. *International Anesthesia Clinics, 27,* 137–147.

Gaba, D. M. (1992). Dynamic decision-making in anesthesiology: Cognitive models and training approaches. In D. A. Evans & V. L. Patel (Eds.), *Advanced models of cognition for medical training and practice* (pp. 123–147). Berlin: Springer Verlag.

Gaba, D. M., Maxwell, M., & DeAnda, A. (1987). Anesthetic mishaps: Breaking the chain of accident evolution. *Anesthesiology, 66,* 670–676.

Gough, H. B. (1987). *The California Personality Inventory.* Palo Alto, CA: Consulting Psychologists Press.

Graeber, R. C. (1988). Aircrew fatigue and circadian rhythmicity. In E. L. Wiener & D. C. Nagel (Eds.), *Human factors in aviation* (pp. 305–345). San Diego, CA: Academic Press.

Gravenstein, J. S., Cooper, J. B., & Orkin, F. K. (1990). Work and rest cycles in anesthesia practice. *Anesthesiology, 72,* 737–742.

Helmreich, R. L. (1984). Cockpit management attitudes. *Human Factors, 26,* 583–589.

Helmreich, R. L. (1992). Human factors aspects of the Air Ontario crash at Dryden, Ontario: Analysis and recommendations. In V. P. Moshansky (Commissioner), *Commission of Inquiry into the Air Ontario Accident at Dryden, Ontario: Final Report. Technical Appendices.* Ottawa, ON: Minister of Supply and Services, Canada.

Helmreich, R. L. (in press). Anatomy of a system accident: Avianca Flight 052. *International Journal of Aviation Psychology.*

Helmreich, R. L., & Foushee, H. C. (1993). Why crew resource management?: The history and status of human factors training programs in aviation. In E. Wiener, B. Kanki, & R. Helmreich (Eds.), *Cockpit resource management* (pp. 3–45). New York: Academic Press.

Helmreich, R. L., Foushee, H. C., Benson, R., & Russini, R. (1986). Cockpit management attitudes: Exploring the attitude-performance linkage. *Aviation, Space and Environmental Medicine, 57,* 1198–1200.

Helmreich, R. L., Merritt, A., Sherman, P., Gregorich, S. G., & Wiener, E. L. (1993).

The Flight Management Attitudes Questionnaire. *NASA/University of Texas/ FAA Technical Report 93-5.*

Helmreich, R. L., Wiener, E. L., & Kanki, B. G. (1993). The future of crew resource management in the cockpit and elsewhere. In E. L. Wiener, B. G. Kanki, & R. L. Helmreich (Eds.), *Cockpit resource management* (pp. 479–501). San Diego, CA: Academic Press.

Helmreich, R. L., Spence, J. T., Beane, W. E., Lucker, G. W., & Matthews, K. A. (1980). Making it in academic psychology: Demographic and personality correlates of attainment. *Journal of Personality and Social Psychology, 39,* 896–908.

Helmreich, R. L., & Wilhelm, J. A. (1991). Outcomes of crew resource management training. *International Journal of Aviation Psychology, 1,* 287–300.

Hofstede, G. (1980). *Culture's consequences: International differences in work-related values.* Beverly Hills, CA: Sage.

Howard, S. K., Gaba, D. M., Fish, K. J., Yang, G., & Sarnquist, F. H. (1992). Anesthesia crisis resource management: Teaching anesthsiologists to handle critical incidents. *Aviation, Space, and Environmental Medicine, 63,* 763–770.

Kumar, V., Barcellos, W. A., Mehta, M. P., & Carter, J. G. (1988). An analysis of critical incidents in a teaching department for quality assurance: A survey of mishaps during anaesthesia. *Anaesthesia, 43,* 879–883.

McDonald, J. S., & Peterson, S. (1985). Lethal errors in anesthesiology. *Anesthesiology, 63,* A497.

McGrath, J. E. (1964). *Social psychology: A brief introduction.* New York: Holt, Rinehart, & Winston.

Merritt, A. E., & Helmreich, R. L. (in press). Human factors on the flight deck: The influence of national culture. *Journal of Cross Cultural Psychology.*

Moshansky, V. P. (1992). *Commission of Inquiry into the Air Ontario Accident at Dryden, Ontario: Final report. (Vols. 1–4).* Ottawa, ON: Minister of Supply and Services, Canada.

National Research Council (1993). *Workload transition.* Washington, DC: National Academy Press.

Perrow, C. (1984). *Normal accidents: Living with high risk technologies.* New York: Basic Books.

Rasmussen, J., Duncan, K., & Leplat, J. (Eds.). (1987). *New technology and human error.* Chichester, England: Wiley.

Reason, J. T. (1990). *Human error.* Cambridge, England: Cambridge University Press.

Rose, R. M., Helmreich, R. L., Fogg, L., & McFadden, T. (1993). Assessments of astronaut effectiveness. *Aviation, Space, and Environmental Medicine, 64,* 789–794.

Schaefer, H.-G., & Helmreich, R. L. (1993). *The Operating Room Management Attitudes Questionnaire (ORMAQ).* NASA/University of Texas Technical Report 93-8.

Schaefer, H.-G., Scheidegger, D., & Helmreich, R. L. (1993). Human factors im Operationssaal [Human factors in the operating room]. *Schweizeriche Artztezeitschrift, 74,* 1882–1885.

Schisler, J. Q. (1992). Implementing continuous quality improvement: A private practice's experience. *International Anesthesia Clinics, 30,* 45–56.

Spence, J. T., & Helmreich, R. L. (1983). Achievement-related motives and behavior. In J. T. Spence (Ed.), *Achievement and achievement motives: Psychological and sociological approaches* (pp. 10–74). San Francisco, CA: Freeman.

Steers, R. M., & Porter, L. W. (1991). *Motivation and work behavior.* New York: McGraw-Hill.

Weinger, M. B., & Englund, C. E. (1990). Ergonomic and human factors affecting anesthetic vigilance and monitoring performance in the operating room environment. *Anesthesiology, 73,* 995–1021.

13

Operating at the Sharp End: The Complexity of Human Error

Richard I. Cook and David D. Woods
The Ohio State University

Studies of incidents in medicine and other fields attribute most bad outcomes to a category of human performance labeled *human error*. For example, surveys of anesthetic incidents in the operating room have attributed between 70% and 82% of the incidents surveyed to the human element (Chopra, Bovill, Spierdijk, & Koornneef, 1992; Cooper, Newbower, Long, & McPeek, 1978). Similar surveys in aviation have attributed more than 70% of incidents to crew error (Boeing Product Safety Organization, 1993). In general, incident surveys in a variety of industries attribute similar percentages of critical events to human error (for example, see Hollnagel, 1993, Table 1). The result is the perception, in both professional and lay communities, of a "human error problem" in medicine, aviation, nuclear power generation, and similar domains. To cope with this perceived unreliability of people, it is conventional to try to reduce or regiment the human's role in a risky system by enforcing standard practices and work rules and by using automation to shift activity away from people.

Generally, the "human" referred to when an incident is ascribed to human error is some individual or team of practitioners who work at what Reason calls the "sharp end" of the system (Reason, 1990; Fig. 13.1). Practitioners at the sharp end actually interact with the hazardous process in their roles as pilots, physicians, spacecraft controllers, or power plant operators. In medicine, these practitioners are anesthesiologists, surgeons, nurses, and some technicians who are physically and temporally close to the patient. Those at the "blunt end" of the system, to continue Reason's analogy, affect safety through their effect on the constraints and resources acting on the practitioners at the sharp end. The blunt end includes the managers, system

255

architects, designers, and suppliers of technology. In medicine, the blunt end includes government regulators, hospital administrators, nursing managers, and insurance companies. In order to understand the sources of expertise and error at the sharp end, one must also examine this larger system to see how resources and constraints at the blunt end shape the behavior of sharp-end practitioners (Reason, 1990). This chapter examines issues surrounding human performance at the sharp end, including those described as errors and those considered expert.

Most people use the term *human error* to delineate one category of potential causes for unsatisfactory activities or outcomes. Human error as a cause of bad outcomes is used in engineering approaches to the reliability of complex systems (probabilistic risk assessment) and is widely used in inci- dent-reporting systems in a variety of industries. For these investigators, human error is a specific variety of human performance that is, in retrospect, so clearly and significantly substandard and flawed that there is no doubt that the practitioner should have viewed it as substandard *at the time the act was committed*. The judgment that an outcome was due to human error is an attribution that (a) the human performance immediately preceding the incident was unambiguously flawed, and (b) the human performance led directly to the outcome.

But the term "human error" is controversial (e.g., Hollnagel, 1993). Attribution of error is a *judgment* about human performance. These judg- ments are rarely applied except when an accident or series of events have occurred that could have or nearly did end with a bad outcome. Thus, these judgments are made ex post facto, with the benefit of *hindsight* about the outcome or near miss. These factors make it difficult to attribute specific incidents and outcomes to "human error" in a consistent way. Fundamental questions arise. When precisely does an act or omission constitute an "er- ror"? How does labeling some act as a human error advance our under- standing of why and how complex systems fail? How should we respond to incidents and errors to improve the performance of complex systems? These are not academic or theoretical questions. They are close to the heart of tremendous bureaucratic, professional, and legal conflicts and tied directly to issues of safety and responsibility. Much hinges on being able to deter- mine how complex systems have failed and on the human contribution to such outcome failures. Even more depends on judgments about what means will prove effective for increasing system reliability, improving human per- formance, and reducing or eliminating human errors.

Studies in a variety of fields show that the label "human error" is prejudi- cial and unspecific. It retards rather than advances our understanding of how complex systems fail and the role of human practitioners in both successful and unsuccessful system operations. The investigation of the cognition and behavior of individuals and groups of people, not the attribution of error in itself, points to useful changes for reducing the potential for disaster in large, complex systems. Labeling actions and assessments as "errors" identifies a symptom, not a cause; the symptom should call forth a more in-depth investigation of how a system of people, organizations, and technologies functions and malfunctions (Hollnagel, 1993; Rasmussen, Duncan, & Leplat, 1987; Reason, 1990; Woods, Johannesen, Cook, & Sarter, 1994).

Recent research into the evolution of system failures finds that the story of "human error" is markedly complex (Hollnagel, 1993; Rasmussen et al., 1987; Reason, 1990; Woods et al., 1994). For example:

- The context in which incidents evolve plays a major role in human performance at the sharp end.
- Technology can shape human performance, creating the potential for new forms of error and failure.
- The human performance in question usually involves a set of interacting people.
- People at the blunt end create dilemmas and shape trade-offs among competing goals for those at the sharp end.
- The attribution of error after the fact is a process of social judgment rather than a scientific conclusion.

The goal of this chapter is to provide an introduction to the complexity of system failures and the term *human error*. It may seem simpler merely to attribute poor outcomes to human error and stop there. If one looks beyond the label, the swirl of factors and issues seems very complex. But it is in the examination of these deeper issues that one can learn how to improve the performance of large, complex systems.

We begin with an introduction to the complexity of error through several exemplar incidents taken from anesthesiology. Each of these incidents may be considered by some to contain one or more human errors. Careful examination of the incidents, however, reveals a more complicated story about human performance. The incidents provide a way to introduce some of the research results about the factors that affect human performance in complex settings such as medicine. Because the incidents are drawn from anesthesiology, most of the discussion is about human performance in the conduct of anesthesia, but the conclusions apply to other medical specialties and even to other domains.

The second part of the chapter deals more generally with the failures of large, complex systems and the sorts of problems those who would analyze human performance in such systems must encounter. It is significant that the results from studies in medicine and other domains such as aviation and nuclear power plant operation are parallel and strongly reinforcing. The processes of cognition are not fundamentally different between practitioners in these domains, and the problems that practitioners are forced to deal with are quite similar. We should not be surprised that the underlying features of breakdowns in these large, complex systems are quite similar.

Grappling with the complexity of human error and system failure has strong implications for the many proposals to improve safety by restructuring the training of people, introducing new rules and regulations, and adding technology. The third part of the chapter explores the consequences of these ideas for attempts to eliminate "human error" as a cause of large, complex system failures.

HUMAN PERFORMANCE AT THE SHARP END

What factors affect the performance of practitioners in complex settings like medicine? Figure 13.1 provides a schematic overview. For practitioners at the sharp end of the system, there are three classes of cognitive factors that govern how people form intentions to act:

1. Knowledge factors—factors related to the knowledge that can be drawn on when solving problems in context.
2. Attentional dynamics—factors that govern the control of attention and the management of mental workload as situations evolve and change over time.
3. Strategic factors—the trade-offs between goals that conflict, especially when the practitioners must act under uncertainty, risk, and the pressure of limited resources (e.g., time pressure; opportunity costs).

These three classes are depicted as interlocking rings at the sharp end of the operational system because these functions overlap. Effective system operation depends on their smooth integration within a single practitioner and across teams of practitioners. The figure does not show a single individual because these categories are not assigned to individuals in a one-to-one fashion. Rather, they are distributed and coordinated across multiple people and across the artifacts they use.

These factors govern the expression of error and expertise together with two other classes of factors. First are the demands placed on practitioners by characteristics of the incidents and problems that occur. These demands vary in type and complexity. One incident may present itself as a textbook version of a problem for which a well-practiced plan is available and appropriate. A different incident may appear embedded in a complicated background of interacting factors, creating a substantial cognitive challenge for

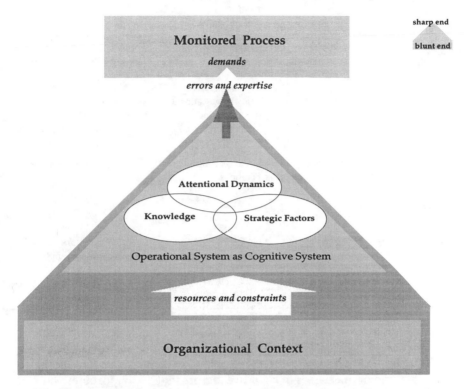

FIG. 13.1. The sharp and blunt ends of a large complex system. The interplay of problem demands and the resources of practitioners at the sharp end govern the expression of expertise and error. The resources available to meet problem demands are shaped and constrained in large part by the organizational context at the blunt end of the system (from Woods et al., 1994, reprinted by permission).

practitioners. The problem demands shape the cognitive activities of those confronting the incident at the sharp end.

The second broad class of factors arises from the blunt end of the system and includes the resources and constraints under which the practitioners function. Recent work on human error has recognized the importance of the *organizational context* in system failures (Reason, 1990, chap. 7). This context influences both the physical and cognitive resources available to practitioners as they deal with the system. For example, the knowledge available during system operations is, in part, the result of the organization's investments in training and practice. Similarly, the organizational context influences how easy it is to bring more specialized knowledge to bear as an incident evolves and escalates. Finally, organizational context tends to set up or sharpen the strategic dilemmas practitioners face. Thus, organizational (blunt end) factors provide the context in which the practitioners'

TABLE 13.1
Categories of Cognitive Factors

Category	Exemplar Incident	Cognitive Issues	Examples of Conflicts Present
Knowledge Factors	Myocardial infarction in a vascular surgery patient	• Buggy knowledge • Mental models • Knowledge calibration • Inert knowledge • Simplifications and heuristics • Imprecise knowledge	Imperfect, contradictory, incomplete domain knowledge
Attentional Dynamics	Hypotension during cardiac surgery	• Situation awareness • Fixations	Limited attentional resource demanded by multiple attractors
Strategic Factors	Busy weekend operating schedule	• Goal tradeoffs and decision choice	Goal trade-offs Procedural rules that do not apply to all cases Organizational double binds

(sharp end) knowledge factors, attentional dynamics, and strategic factors function.

This chapter begins with three exemplar incidents. Each was chosen to highlight one of the classes of cognitive factors that are important in human performance as indicated in Table 13.1. Each incident could be judged to contain one or more human errors; this judgment is usually the end point for most investigators who then tabulate the incident frequency in some sort of reporting scheme. Here, however, we take the analysis much further and reveal the complex interplay of the multiple factors sketched in Fig. 13.1 that contributed to the evolution of each incident.

Knowledge Factors: Incident #1—Myocardial Infarction

An elderly patient presented with a painful, pulseless, blue arm indicating a blood clot (embolus) in one of the major arteries that threatened loss of that limb. Emergency surgery to perform removal of the clot (embolectomy) was clearly indicated. The patient had a complex medical and surgical history with high blood pressure, diabetes requiring regular insulin treatment, a prior heart attack, and previous coronary artery bypass surgery. The patient also had evidence of recently worsening congestive heart failure, that is, shortness of breath, dyspnea on exertion and leg swelling (pedal edema). Electrocardiogram changes included inverted T waves. Chest X-ray suggested pulmonary edema. The arterial blood gas showed markedly low oxygen in the arterial blood (p_aO_2 of 56 on unknown F_iO_2). The blood glucose was high (800). The patient received furosemide (a diuretic) and 12 units of insulin in the emergency room. The patient was taken to the operating room for removal of the clot under local anesthesia with sedation provided by the anesthetist. In the operating room the patient's blood pressure was high, 210/120; a nitroglycerin drip was started and increased in an effort to reduce the blood pressure. The arterial oxygen saturation (S_aO_2) was 88% on nasal cannula and did not improve with a rebreathing mask, but rose to the high 90s when the anesthesia machine circuit was used to supply 100% oxygen by mask. The patient did not complain of chest pain but did complain of abdominal pain and received morphine. Urine output was high in the operating room. The blood pressure continued about 200/100. Nifedipine was given sublingually and the pressure fell over 10 minutes to 90 systolic. The nitroglycerin infusion rate was decreased and the pressure rose to 140. The embolectomy was successful. Postoperative cardiac enzyme studies showed a peak about 12 hours after the surgical procedure, indicating that the patient had suffered a myocardial infarction (heart attack) sometime in the period including the time in the emergency room and the operating room. The patient survived.[1]

This incident raises a host of issues regarding the nature of knowledge and its use during the evolution of the incident. Knowledge factors include those related to the knowledge available for solving problems. Especially important are those factors that conditionalize knowledge toward its use, that is, those that "call knowledge to mind." In Incident #1, it is clear that the participant was employing a great deal of knowledge. In fact, the description of just a few of the relevant aspects of knowledge important to the incident occupies several pages.

There is evidence that the participant was missing or misunderstanding important, but less obvious features of the case. It seems (and seemed to peer experts who evaluated the incident at the time; cf., Cook, Woods, & McDonald, 1991) that the practitioner misunderstood the nature of the patient's intravascular volume, believing the volume was high rather than low. This increased volume is often present in patients with the signs of congestive heart failure. In this case, however, other factors (including the high blood glucose and the prior treatment with a diuretic) were present that indicated that the patient should be treated differently. In retrospect, other practitioners argued that the patient probably should have received more intravenous fluid to replenish the low intravascular volume. They also felt that the patient should have been monitored invasively to allow precise determination of when enough fluid had been given (e.g., a catheter that goes through the heart and into the pulmonary artery).

It is also apparent that many of the practitioner's actions were appropri-

---

[1]This incident comes from Cook, Woods and McDonald, 1991 which examined a corpus of cases in anesthesiology and the associated human performance issues.

ate in the context of the case as it evolved. For example, the level of oxygen in the blood was low and the anesthetist pursued several different means of increasing the blood oxygen level, including the use of oxygen by mask. Similarly, the blood pressure was high, and this too was treated, first with nitroglycerin (which may lower the blood pressure but also can protect the heart by increasing its blood flow) and then with nifedipine. The fact that the blood pressure fell much further than intended was probably the result of depleted intravascular volume, which was, in turn, the result of the high urinary output provoked by the previous diuretic and the high serum glucose level. It is this last point that appears to have been unappreciated, at first by the physicians who saw the patient initially, and then by the anesthetist.

In the opinion of anesthesiologist reviewers of this incident shortly after it occurred, the circumstances of this case should have brought to mind a series of questions about the nature of the patient's intravascular volume. The inability to answer those questions would then have prompted the use of particular monitoring techniques before and during the surgical procedure.

Bringing knowledge to bear effectively in problem solving is a process that involves issues of knowledge *content*, knowledge *organization*, and knowledge *activation*. Research in this area has emphasized that mere possession of knowledge is not enough for expertise. It is also critical for knowledge to be organized so that it can be activated and used in different contexts (Bransford, Sherwood, Vye, & Rieser, 1986). Thus, Feltovich, Spiro, and Coulson (1989) and others emphasize that one component of human expertise is the flexible application of knowledge in new situations.

There are at least four lines of overlapping research related to knowledge use by humans in complex systems. These include (a) the role of mental models and of knowledge flaws (sometimes called "buggy" knowledge); (b) the issue of knowledge calibration; (c) the problem of inert knowledge; and (d) the use of heuristics, simplifications, and approximations. In many incidents, going behind the label "human error" demands investigating how knowledge was or could have been brought to bear in the evolving incident. Any of the previously mentioned factors could contribute to failures to activate relevant knowledge in context.

Mental Models and Buggy Knowledge.   Knowledge of the world and its operation may be complete or incomplete and accurate or inaccurate. Practitioners may act based on inaccurate knowledge or on incomplete knowledge about some aspect of the complex system or its operation. The term *mental model* has been used to describe the collection of knowledge used by a practitioner. When the mental model is inaccurate or incomplete, its use can give rise to inappropriate actions. These mental models are described as "buggy" (see Chi, Glaser, & Farr, 1988; Gentner & Stevens, 1983; and Rouse

& Morris, 1986, for some of the basic results on mental models). Studies of practitioners' mental models have examined the models that people use for understanding technological, physical, and physiological processes.

For example, Sarter and Woods (1992, 1994) found that buggy mental models contributed to the problems pilots experienced in using cockpit automation. Airplane cockpit automation has various modes of automatic flight control, ranging between the extremes of automatic and manual. The modes interact with each other in different flight contexts. Having a detailed and complete understanding of how the various modes of automation interact and the consequences of transitions between modes in various flight contexts is a demanding new knowledge requirement for the pilot in highly automated cockpits. They also found that buggy mental models played a role in automation surprises, cases where pilots are "surprised" by the automation's behavior. The buggy knowledge contributed to difficulties in monitoring and understanding automatic system behavior (What is it doing? Why did it do that?) and to projecting or anticipating future states (What will it do next?). This is a common finding in complex systems and has also been described in anesthesiologists using microcomputer-based devices (Cook, Potter, Woods, & McDonald, 1991). Significantly, the design of devices, particularly the interface between the device and human practitioners, can either aid or impede the development of useful mental models by practitioners. The presence of a buggy mental model of a device is more likely to indicate poor device design than it is some inadequacy of the user's mental machinery (Norman, 1988).

It is possible to design experiments that reveal specific bugs in practitioners' mental models. By forcing pilots to deal with various nonnormal situations, it was possible to reveal gaps or errors in their understanding of how the automation works in various situations. Although pilots were able to make the automation work in typical flight contexts, they did not fully exploit the range of the system's capabilities. Pilots tend to adopt and stay with a small repertoire of strategies, in part because their knowledge about the advantages and disadvantages of the various options for different flight contexts is incomplete. In unusual or novel situations, however, it may be essential to have a thorough understanding of the functional structure of the automated systems and to be able to use this knowledge in operationally effective ways.

Novel or unusual situations can reveal the presence of a "buggy" mental model, and many incidents are associated with situations that are unusual to some degree. It can be quite difficult to determine whether a buggy mental model was, indeed, involved in an incident. In the exemplar incident, for example, the combination of congestive heart failure (normally improved by reducing the amount of fluid in the circulation) with high urine output from high blood glucose and a diuretic drug (furosemide) was unusual. It is not clear whether the practitioner had a buggy mental model of the relationship between these factors or if the demands of attention to the low oxygen saturation and blood pressure prevented him from examining the model closely enough to discover the relationship. Alternatively, the mental model and associated knowledge may simply have been inert (see the section on inert knowledge). The inability to distinguish between these alternatives is due, in large part, to the limitations of the data about the incident.

Knowledge Calibration.   Results from several studies (Cook, Potter, Woods, & McDonald, 1991; Moll van Charante, Cook, Woods, Yue, & Howie, 1993; Sarter & Woods, 1994) indicate that practitioners may be unaware of gaps or bugs in their model of a device or system. This raises the question of knowledge calibration (Wagenaar & Keren, 1986). Everyone has some areas where their knowledge is more complete and accurate than others. Individuals are well calibrated if they are aware of how well they know what they know. People are miscalibrated if they are overconfident and believe that they understand areas where in fact their knowledge is incomplete or buggy.[2]

There are several factors that could contribute to miscalibration of practitioners' awareness about their knowledge of the domain and the technology with which they work. First, areas of incomplete or buggy knowledge can remain hidden from practitioners because they have the capability to work around these areas by sticking with a few well-practiced and well-understood methods. Second, situations that challenge practitioner mental models or force them to confront areas where their knowledge is limited and miscalibrated may arise infrequently. Third, studies of calibration have indicated that the availability of feedback, the form of feedback, and the attentional demands of processing feedback can effect knowledge calibration (e.g., Wagenaar & Keren, 1986).

Problems with knowledge calibration can be severe, especially when information technology is involved in practice. For example, many computerized devices fail to provide adequate feedback to users to allow them to learn about (to calibrate) the internal relationships of the device. A relationship between poor feedback and miscalibrated practitioners was found in studies of pilot–automation interaction (Sarter & Woods, 1994) and of physician–automation interaction (Cook, Potter, Woods, & McDonald, 1991). For example, some of the participants in the former study made comments in the postscenario debriefings such as: "I never knew that I did not know this. I just never thought about this situation." Although this is phenomenon is most easily demonstrated when practitioners attempt to use computerized devices, it is probably ubiquitous.

---

[2]One physician was recently heard to describe another as being "often wrong but never in doubt," an indication that practitioners may recognize the presence of a calibration problem.

Activating Relevant Knowledge in Context—The Problem of Inert Knowledge.   Lack of knowledge or buggy knowledge may be one part of the puzzle, but the more critical question may be factors that affect whether relevant knowledge is activated for use in the actual problem-solving context (e.g., Bransford et al, 1986). The question is not just whether the problem solver knows some particular piece of domain knowledge, but whether he or she calls it to mind when it is relevant to the problem at hand and whether he or she knows how to use this knowledge in problem solving. We tend to assume that if a person can be shown to possess a piece of knowledge in any circumstance, then this knowledge should be accessible under all conditions where it might be useful. In contrast, a variety of research has revealed dissociation effects where knowledge accessed in one context remains inert in another (Gentner & Stevens, 1983; Perkins & Martin, 1986). This situation may well have been the case in the first incident: The practitioner knew about the relationships determining the urine output in the sense that he was able to explain the relationships after the incident, but this knowledge was inert, that is, it was not summoned up during the incident.

The fact that people possess relevant knowledge does not guarantee that this knowledge will be activated when needed. The critical question is not to show that the problem solver possesses domain knowledge as might be determined by standardized tests. Rather, the more stringent criterion is that situation-relevant knowledge is accessible under the conditions in which the task is actually performed. Thus, *inert knowledge* is knowledge accessible only in a restricted set of contexts, which may not include contents of relevance to actual practice. Inert knowledge may be related to cases that are difficult not because problem solvers do not know the individual pieces of knowledge needed to build a solution, but because they have not previously confronted the need to join the pieces together. Thus, the practitioner in the first incident could be said to *know* about the relationship between blood glucose, furosemide, urine output, and intravascular volume but also *not to know* about that relationship in the sense that the knowledge was not activated at the time when it would have been useful. Studies of practitioner interaction with computerized systems show that the same pattern can occur with computer aids and automation. Sarter and Woods (1994) found that some pilots clearly possessed knowledge because they were able to recite the relevant facts in debriefing, but they were unable to apply the same knowledge successfully in an actual flight context; that is, their knowledge was inert.

Results from accident investigations often show that the people involved did not call to mind all the relevant knowledge during the incident although they "knew" and recognized the significance of the knowledge afterwards. The triggering of a knowledge item X may depend on subtle pattern recognition factors that are not present in every case where X is relevant. Alternatively, that triggering may depend critically on having sufficient time to process all the available stimuli in order to extract the pattern. This may explain the difficulty practitioners have in "seeing" the relevant details in a certain case where the pace of activity is high and there are multiple demands on the practitioner. These circumstances were present in Incident #1 and are typical of systems "at the edge of the performance envelope."

Heuristics, Simplifications and the Imprecision of Knowledge.   During the past decade, there has been much written about medical decision making, and a large portion of it is highly critical of the decision processes of practitioners. People tend to cope with complexity through simplifying heuristics, that is, through rules of thumb and simplifications. Heuristics are useful because they are usually relatively easy to apply and minimize the cognitive effort required to produce decisions. Heuristics can readily be shown to be incorrect under some circumstances (Tversky & Kahneman, 1974)[3] and, in theory, are less desirable as decision rules than precise computations, at least if the decision maker is considered to have infinite mental resources for computation. However, these simplifications may also be useful approximations that allow limited-resource practitioners to function robustly over a variety of problem demand factors (Woods, 1988).

At issue is whether a simplification is (a) generally useful because it reduces mental workload without sacrificing accuracy, or (b) a distortion or misconception that appears to work satisfactorily under some conditions but leads to error in others. The latter class is described by Feltovich et al. (1989) as an *oversimplification*. In studying the acquisition and representation of complex concepts in biomedicine, Feltovich et al. found that various oversimplifications were held by some medical students and even by some practicing physicians. They found that "bits and pieces of knowledge, in themselves sometimes correct, sometimes partly wrong in aspects, or sometimes absent in critical places, interact with each other to create large-scale and robust misconceptions" (Feltovich et al., 1989, p. 162). Examples of kinds of oversimplification include:

1. Seeing different entities as more similar than they actually are.
2. Treating dynamic phenomena statically.
3. Assuming that some general principle accounts for all of a phenomenon.
4. Treating multidimensional phenomena as unidimensional or according to a subset of the dimensions.
5. Treating continuous variables as discrete.

---

[3]Indeed, if a rule of thumb is not inaccurate in some circumstance then it is a robust rule and not a heuristic at all.

6. Treating highly interconnected concepts as separable.
7. Treating the whole as the sum of its parts (see Feltovich, Spiro, & Coulson, 1993).

These oversimplifying tendencies may occur because of requirements for cognitive effort in demanding circumstances.

> It is easier to think that all instances of the same nominal concept . . . are the same or bear considerable similarity. It is easier to represent continuities in terms of components and steps. It is easier to deal with a single principle from which an entire complex phenomenon "spins out" than to deal with numerous, more localized principles and their interactions. (Feltovich et al., 1989, p. 131)

Criticisms of practitioner decision making based on simplified or over-simplified knowledge are often used to show that practitioners make bad decisions and that their decision making would be improved by adopting a more mathematically rigorous, probabilistic reasoning approach. It can be shown mathematically, for example, that a particular strategy for contingent choice using strict criteria would be preferable to many other strategies. Such demonstrations are usually sterile exercises, however, for several reasons. First, the effort required to perform such calculations may be so large that it would keep practitioners from acting with the speed demanded in actual environments. This has been shown elegantly by Payne and colleagues (Payne, Bettman, & Johnson, 1988; Payne, Johnson, Bettman, & Coupey, 1990) who demonstrated that simplified methods will produce a higher proportion of correct choices between multiple alternatives under conditions of time pressure. Put simply, if the time and effort required to arrive at a decision is important, it may be possible to have an overall higher quality performance using heuristics than using a "mathematically ideal" approach.

The second reason that it is difficult to rely on formal decision making methods is that medical knowledge is so heterogeneous and imprecise. Much medical research data are drawn from small or only marginally representative samples; drug tests rarely include pregnant women, for example, so the effects of many drugs on pregnant women and fetuses are unknown. Much patient data are derived from coarse measurements at widely spaced intervals, whereas others (for example, the effects of exposure to anesthetic agents) are known precisely but only for a limited period of time. Thus it is possible to have quite precise knowledge about the effect of a disease or a treatment on a specific subset of patients and also to have a great deal of uncertainty about the extent to which that knowledge is useful for a given patient both because the knowledge is derived from a specific subgroup and because the patient is poorly characterized. Many important physiologic variables can be measured only indirectly with poor precision and are known to fluctuate widely even in the healthy population. Physicians often must rely on comparatively remote or indirect measures of critical variables. The precise effect of a therapy is usually only predictable for a group of patients; for example, a preoperative antibiotic will reduce the risk of postoperative infection by a small amount, but the actual benefit to an individual patient coming to the operating room for a specific procedure is extraordinarily difficult to define. All these factors tend to lead medical practitioners toward an empirically based approach to diagnosis and therapy in which successive treatments are applied until the desired result is achieved.

There are also inherent conflicts in the knowledge base that need to be resolved in each individual case by the practitioner. In Incident #1, for example, there are conflicts between the need to keep the blood pressure high and the need to keep the blood pressure low (Fig. 13.2). The heart depends on blood pressure for its own blood supply, but increasing the blood pressure also increases the work it is required to perform. The practitioner must decide what blood pressure is acceptable. Many factors enter into this decision process: What is the patient's normal blood pressure? How labile is the blood pressure now? How will attempts to reduce blood pressure affect other physiological variables? How is the pressure likely to

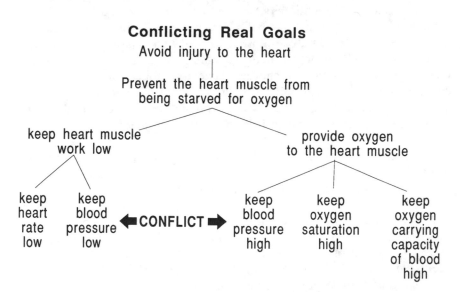

FIG. 13.2. Conflicting domain knowledge. For a cardiac surgery patient the blood pressure should be kept low to minimize the work of the heart, but the blood pressure should be kept high to maximize the blood flow to heart muscle. How practitioners at the sharp end resolve this conflict depends on several factors (from Cook, Woods, & McDonald, 1991, reprinted by permission).

change without therapy? How long will the surgery last? What is the level of surgical skill being employed? As is often the case in this and similar domains, the locus of conflict may vary from case to case and from moment to moment. It is impossible to create algorithms that adequately capture the variety of patient characteristics and risks in a highly uncertain world. These conflicts are a normal part of the medical domain and practitioners are so comfortable with them that it is hard to get the participants in an incident to be explicit about the trade-offs involved in the decisions they made.

In summary, heuristics may represent effective and necessary adaptations to the demands of real workplaces (Rasmussen, 1986). The problem, if there is one, may not always be the shortcut or simplification itself, but whether practitioners know the limits of the shortcuts, can recognize situations where the simplification is no longer relevant, and have the ability to use more complex concepts, methods, or models (or the ability to integrate help from specialist knowledge sources) when the situation they face demands it. Interestingly, practitioners are acutely aware of how deficient their rules of thumb may be and how certain situations may require abandoning the cognitively easy method in favor of more cognitively demanding "deep thinking." For example, senior anesthetists commenting on the practitioner's behavior in the first incident were critical of his performance:

> This man was in major sort of hyperglycemia and with popping in extra Lasix [furosemide] you have a risk of hypovolemia from that situation. I don't understand why that was quietly passed over, I mean that was a major emergency in itself. . . . This is a complete garbage amount of treatment coming in from each side, responding from the gut to each little bit of stuff [but it] adds up to no logic whatsoever. . . . The thing is that this patient [had] an enormous number of medical problems going on which have been simply reported [but] haven't really been addressed.

This critique is not quite correct. In fact, each problem was addressed in some way at some time. But the comment about "coming in from each side" identifies what the practitioner was missing in the incident, that is, the interactions between normally separate factors that here were closely linked. Being able to discover that link and appreciate its implications is intimately bound up with knowledge factors including mental models, heuristics, and inert knowledge.

Attentional Dynamics: Incident #2—Hypotension

During a coronary artery bypass graft procedure, an infusion controller device delivered a large volume of a potent drug to the patient at a time when no drug should have been flowing. Five of these microprocessor-based devices were set up in the usual fashion at the beginning of the day, prior to the beginning of the case. The initial sequence of events associated with the case was unremarkable. Elevated systolic blood pressure (>160 torr) at the time of sternotomy prompted the practitioner to begin an infusion of sodium nitroprusside via one of the devices. After this device was started at a drop rate of 10/min, the device began to sound an alarm. The tubing connecting the device to the patient was checked and a stopcock (valve) was found to be closed. The operator opened the stopcock and restarted the device. Shortly after restart, the device alarmed again. The blood pressure was falling by this time, and the operator turned the device off. Over a short period, hypertension gave way to hypotension (systolic pressure <60 torr). The hypotension was unresponsive to fluid challenge but did respond to repeated boluses of neosynephrine and epinephrine. The patient was placed on bypass rapidly. Later, the container of nitroprusside was found to be empty; a full bag of 50 mg in 250 ml was set up before the case.

The physicians involved in the incident were comparatively experienced device users. Reconstructing the events after the incident led to the conclusion that the device was assembled in a way that would allow free flow of drug. Drug delivery was blocked, however, by a closed downstream stopcock. The device was started, but the machine did not detect any flow of drug (the stopcock was closed), triggering visual and auditory alarms. When the stopcock was opened, free flow of fluid containing drug began. The controller was restarted, but the machine again detected no drops because the flow was wide open and no individual drops were formed. The controller alarmed again, with the same message, which appeared to indicate that no flow had occurred. Between the opening of the stopcock and the generation of the error message, sufficient drug was delivered to substantially reduce the blood pressure. The operator saw the reduced blood pressure, concluded that the sodium nitroprusside drip was not required, and pushed the button marked "off." This powered down the device, but the flow of drug continued. The blood pressure fell even further, prompting a diagnostic search for sources of low blood pressure. The sodium nitroprusside controller was seen to be off. Treatment of the low blood pressure itself commenced and was successful. The patient suffered no sequelae.[4]

This incident is used as an exemplar for the discussion of attentional dynamics, although it also involves a number of issues relevant to knowledge factors. *Attentional dynamics* refers to those factors affecting cognitive function in dynamic evolving situations, especially those involving the management of workload in time and the control of attention when there are multiple signals and tasks competing for a limited attentional focus. In many ways, this is the least explored frontier in cognitive science, especially with

---

[4]This case is described more fully in Cook, Woods, and Howie (1992), and weaknesses in the infusion device from the point of view of human-computer cooperation are covered in Moll van Charante et al. (1993).

respect to error. In dynamic, event-driven environments like the operating room, attentional factors are often crucial in the evolution of incidents (cf. Gopher, 1991; Hollister, 1986; Woods, 1992).

In Incident #2, the data are strong enough to support a reconstruction of some of the actual changes in focus of attention of the participants during the incident. A collection of infusion devices like those involved in the incident are shown in Fig. 13.3. The free flow of the drug began when one of the physicians opened the stopcock downstream of the affected device, but this source of the hypotension was not identified until the bag of fluid was nearly empty. There are a number of factors in the environment that contributed to the failure to observe (i.e., attend to) the unintended flow of drug via the infusion device, including: (a) the drip chamber being obscured by the machine's sensor, making visual inspection difficult, (b) presence of an aluminum shield around the fluid bag, hiding its decreasing size, (c) misleading alarm messages from the device, and (d) presence of multiple devices, making it difficult to trace the tubing pathways.

There are also extra-environmental factors that contributed to the failure to observe the free flow. Most importantly, the practitioners reported that they turned the device off as soon as the pressure fell and the device alarmed a second time. In their view of the external world, the device was off, therefore not delivering any drug, and therefore not a plausible source of the hypotension. When they looked at the device, the displays and alarm messages indicated that the device was not delivering drug or later that it had been turned off. The issue of whether *off* might have meant something else (e.g., that the device was powered down but a path for fluid flow remained open) might have been revisited had the situation been less demanding, but the fall in blood pressure was a critical threat to the patient and demanded the limited resource of attention. Remarkably, the practitioners intervened in precisely the right way for the condition they were facing. The choice of drug to increase the blood pressure was ideal to counteract the large dose of sodium nitroprusside that the patient was receiving. Attention did not focus on the fluid bags on the infusion support tree until the decision was made to start an infusion of the antagonist drug and a bag for that drug was being placed on the support tree.

This incident is remarkable, in part for the way in which it shows both the fragility and robustness of human performance. The inability to diagnose the cause of hypotension is in contrast to the ability to manage successfully the complications of the inadvertent drug delivery. There are a number of potential causes of hypotension in the cardiac surgery patient. In this case, successful diagnosis of the cause was less important than successful treatment of the consequences of the problem. The practitioners were quick to correct the physiologic, systemic threat even though they were unable to diagnose its source. This shift from diagnosis to what Woods (1988, 1994)

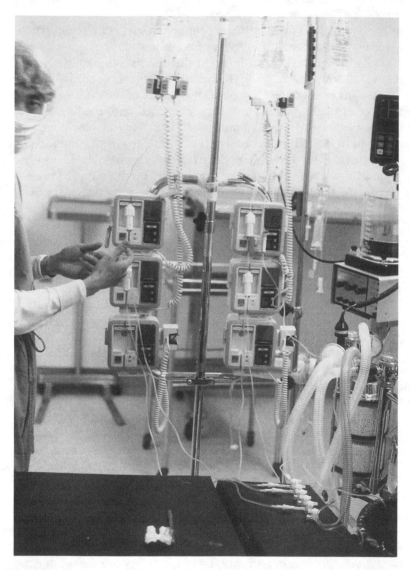

FIG. 13.3. A set-up of multiple drug infusion devices in the heart room. Drugs to raise and lower blood pressure and other cardiovascular system parameters are in the fluid bags above. The controller boxes regulate flow through the tubing based on the detection of fluid drops in drip chambers connected to the bags. The individual flows are joined together by a series of stopcocks to a single piece of tubing, which is then connected to the patient. (See Moll van Charante et al., 1993, for additional details.)

272

calls *disturbance management* is crucial in the operating room and in other domains to maintaining the system in a stable configuration to permit later diagnosis and correction of the underlying faults.

The control of attention is an important issue for those trying to understand human performance, especially in event-rich domains such as the operating room. Attention is a limited resource. One cannot attend to more than one thing at a time, and so shifts of attention are necessary to be able to "take in" the ways in which the world is changing. When something in the world is found that is anomalous (what is sensed in the world is not consistent with what is expected by the observer), attention focuses on that thing, and a process of investigation begins that involves other shifts of attention. This process is ongoing and has been described by Neisser as the *cognitive cycle* (Neisser, 1976; Tenney, Jager Adams, Pew, Huggins, & Rogers, 1992). It is a crucial concept for those trying to understand human performance because it is the basis for all diagnosis and action. Nothing can be discovered in the world without attention; no intended change in the world can be effected without shifting attention to the thing being acted upon. At least two major human performance problems can arise from alterations in attentional dynamics. The first is a loss of situation awareness, and the second is psychological fixation.

Loss of Situation Awareness. Situation awareness is a label that is often used to refer to many of the cognitive processes involved in attentional dynamics (Sarter & Woods, 1991; Tenney et al., 1992). Just a few of the cognitive processes that may pass under the label of situation awareness are: *control of attention* (Gopher, 1991), *mental simulation* (Klein & Crandall, in press), *directed attention* (Woods, 1992), and *contingency planning* (Orasanu, 1990). Because the concept involves tracking processes in time, it has also been described as *mental bookkeeping* to track multiple threads of different but interacting subproblems (Cook, Woods, & McDonald, 1991; Dorner, 1983). These terms refer to tracking the shifting pattern of interactions in the system under control. For example, the state of chemical paralysis of the patient and the "depth" of anesthesia are two different threads. Normally these may be treated independently, but under some circumstances they may interact in ways that have implications for the future course of the patient.

Maintaining situation awareness necessarily requires shifts of attention between the various threads. It also requires more than attention alone, for the object of the shifts of attention is to inform and modify a coherent picture or model of the system as a whole. Building and maintaining that picture requires cognitive effort.

Breakdowns in these cognitive processes can lead to operational difficulties in handling the demands of dynamic, event-driven incidents. In aviation circles, this is known as "falling behind the plane," and in aircraft carrier flight operations it has been described as "losing the bubble" (Roberts & Rousseau, 1989). In each case what is being lost is some of the operator's internal representation of the state of the world at that moment and the direction in which the forces active in the world are taking the system that the operator is trying to control.

Obtaining a clear, empirically testable model for situation awareness is difficult. For example, Hollister (1986) presented an overview of a model of divided attention operations—tasks where attention must be divided across a number of different input channels and where the focus of attention changes as new events signal new priorities. This model then defines an approach to breakdowns in attentional dynamics (what has been called a divided attention theory of error) based on human divided attention capabilities balanced against task demands and adjusted by fatigue and other performance-shaping factors. Situation awareness is clearly most in jeopardy during periods of rapid change and where a confluence of forces makes an already complex situation critically so. This condition is extraordinarily difficult to reproduce convincingly in a laboratory setting. Practitioners are, however, particularly sensitive to the importance of situation awareness, even though researchers find that a clear definition remains elusive (Sarter & Woods, 1991).

Failures to Revise Situation Assessments: Fixation or Cognitive Lockup. The results of several studies (Cook, McDonald, & Smallhout, 1989; De Keyser & Woods, 1990; Gaba & DeAnda, 1989; Johnson, Moen, & Thompson, 1988; Johnson & Thompson, 1981; Woods, O'Brien, & Hanes, 1987) strongly suggest that one source of error in dynamic domains is a *failure to revise* situation assessment as new evidence comes in. Evidence discrepant from the agent's or team's current assessment is missed or discounted or rationalized as not really being discrepant with the current assessment. In addition, it seems that several major accidents involved a similar pattern of behavior from the operational teams involved; examples include the Three Mile Island accident (Kemeny et al., 1979) and the Chernobyl accident.

Many critical real-world human problem solving situations take place in dynamic, event-driven environments where the evidence arrives over time and situations can change rapidly. Incidents rarely spring, full blown and complete; incidents *evolve*. In these situations, people must amass and integrate uncertain, incomplete, and changing evidence; there is no single well-formulated diagnosis of the situation. Rather, practitioners make provisional assessments based on partial and uncertain data. These assessments are incrementally updated and revised as more evidence comes in. Furthermore, situation assessment and plan formulation are not distinct sequential stages, but rather they are closely interwoven processes with partial and provisional plan development and feedback leading to revised situation assessments (Klein, Orasanu, Calderwood, & Zsambok, 1993; Woods & Roth, 1988).

In psychological fixations, the initial situation assessment tends to be appropriate, in the sense of being consistent with the partial information available at that early stage of the incident. As the incident evolves, however, people fail to revise their assessments in response to new evidence, evidence that indicates an evolution away from the expected path. The practitioners become fixated on an old assessment and fail to revise their situation assessment and plans in a manner appropriate to the data now present in their world. A *fixation* occurs when practitioners fail to revise their situation assessment or course of action and maintain an inappropriate judgment or action *in the face of opportunities to revise.*

Several criteria are necessary to describe an event as a fixation. One critical feature is that there is some form of *persistence* over time in the behavior of the fixated person or team. Second, opportunities to revise are cues, available or potentially available to the practitioners, that could have started the revision process if observed and interpreted properly. In part, this feature distinguishes fixations from simple cases of lack of knowledge or other problems that impair error detection and recovery (Cook et al., 1989).[5] The basic defining characteristic of fixations is that the immediate problem-solving context has biased the practitioners in some direction. In naturally occurring problems, the context in which the incident occurs and the way the incident evolves activate certain kinds of knowledge as relevant to the evolving incident. This knowledge in turn affects how new incoming information is interpreted. After the fact or after the correct diagnosis has been pointed out, the solution seems obvious, even to the fixated person or team.

De Keyser and Woods (1990) describe several patterns of behavior that have been observed in cases of practitioner fixation. In the first one, "everything but that," the practitioners seem to have many hypotheses in mind, but never entertain the correct one. The external behavior looks incoherent because they are jumping from one action to another without any success. The second pattern of behavior is the opposite: "this and nothing else." The practitioners are stuck on one strategy, one goal, and they seem unable to shift or to consider other possibilities. The persistence in practitioner behavior can be remarkable. For example, practitioners may repeat the same action or recheck the same data channels several times. This pattern is easily identified because of the unusual number of repetitions despite an absence

---

[5]Of course, the interpretation problem is to define a standard to use to determine what cue or when a cue should alert the practitioners to the discrepancy between the perceived state of the world and the actual state of the world. There is a great danger of falling into the hindsight bias when evaluating after the fact whether a cue "should" have alerted the problem solvers to the discrepancy.

of results. The practitioners often detect the absence of results themselves but without any change in strategy. A third pattern is "everything is OK" (Perrow, 1984). Here the practitioners do not react to the change in their environment. Even if there are a lot of cues and evidence that something is going wrong, they do not seem to pay much attention to them. The practitioners seem to discount or rationalize away indications that are discrepant with their model of the situation.

There are certain types of problems that may encourage fixations by mimicking other situations. This, in effect, leads practitioners down a *garden path*. In garden path problems, "early cues strongly suggest [plausible but] incorrect answers, and later, usually weaker cues suggest answers that are correct" (Johnson et al., 1988). It is important to point out that the erroneous assessments resulting from being led down the garden path are not due to knowledge factors. Rather, they seem to occur because "a problem solving process that works most of the time is applied to a class of problems for which it is not well suited" (Johnson et al., 1988). This notion of garden path situations is important because it identifies a task genotype in which people become susceptible to fixations. The problems that occur are best attributed to the interaction of particular environmental (task) features and the heuristics people apply (locally rational strategies given difficult problems and limited resources), rather than to the any particular bias or problem in the strategies used. Simply put, going down a garden path is not an "error" per se. It is how the problem presents to practitioners that makes it easy to entertain plausible but erroneous possibilities. Anesthesiology and similar domains have inherent uncertainties in diagnostic problems, and it may be necessary for practitioners to entertain and evaluate what turn out to be erroneous assessments. Problems arise when the revision process breaks down and the practitioner becomes fixated on an erroneous assessment, missing, discounting or reinterpreting discrepant evidence (see Johnson et al., 1988; Roth, Woods, & Pople, 1992, for analyses of performance in garden path incidents). What is important is the process of "error" detection and recovery, which fundamentally involves searching out and evaluating discrepant evidence in order to keep up with a changing incident.

Fixation is a characteristic of practitioners in an incident. There are several cognitive processes involved in attentional dynamics that may give rise to fixation:

1. Breakdowns in shifting or scheduling attention as the incident unfolds.
2. Factors of knowledge organization and access that make critical knowledge inert.
3. Difficulties calling to mind alternative hypotheses that could account

for observed anomalies—problems in the processes underlying hypothesis generation.

4. Problems in strategies for situation assessment (diagnosis) given the probability of *multiple* factors, for example, how to value parsimony (single-factor assessments) versus multifactor interpretations.

Fixation may represent the down side of normally efficient and reliable cognitive processes involved in diagnosis and disturbance management in dynamic contexts. Although fixation is fundamentally about problems in attentional dynamics, it may also involve inert knowledge (calling to mind potentially relevant knowledge such as alternative hypotheses) or strategic factors (trade-offs about what kinds of explanations to prefer).

It is clear that in demanding situations where the condition of the patient and the operating room system is changing rapidly, there is a potential conflict between the need to revise the situation assessment and the need to maintain coherence. Not every change is important; not every signal is meaningful. The practitioner whose attention is constantly shifting from one item to another may not be able to formulate a complete and coherent picture of the state of the system. For example, the practitioner in Incident #1 was criticized for failing to build a complete picture of the patient's changing physiological state. Conversely, the practitioner whose attention does not shift may miss cues and data that are critical to updating the situation assessment. This latter condition may lead to fixation. How practitioners manage this conflict is largely unstudied.

Strategic Factors: Incident #3—Busy Weekend Operating Schedule

On a weekend in a large tertiary care hospital, the anesthesiology team (consisting of four physicians of whom three are residents in training) was called on to perform anesthetics for an in vitro fertilization, a perforated viscus, reconstruction of an artery of the leg, and an appendectomy, in one building, and one exploratory laparotomy in another building. Each of these cases was an emergency, that is, a case that cannot be delayed for the regular daily operating room schedule. The exact sequence in which the cases were done depended on multiple factors. The situation was complicated by a demanding nurse who insisted that the exploratory laparotomy be done ahead of other cases. The nurse was only responsible for that single case; the operating room nurses and technicians for that case could not leave the hospital until the case had been completed. The surgeons complained that they were being delayed and their cases were increasing in urgency because of the passage of time. There were also some delays in preoperative preparation of some of the patients for surgery. In the primary operating room suites, the staff of nurses and technicians were only able to run two operating rooms simultaneously. The anesthesiologist in charge was under pressure to attempt to overlap portions of procedures by starting one case as another was finishing so as to use the available resources maximally. The hospital also served as a major trauma center, which means that the team needed to be able to start a large emergency case with minimal (less than 10 minutes) notice. In committing all of the residents to doing the waiting cases, the anesthesiologist in charge produced a situation in which there were no anesthetists available to start a major trauma case. There were no trauma cases, and all the surgeries were accomplished. Remarkably, the situation was so common in the institution that it was regarded by many as typical rather than exceptional.

The third incident is remarkable in part because it is regarded as unremarkable by the participants. These kinds of scheduling issues recur and are considered by many to be simply part of the job. In the institution where the incident occurred, the role of being anesthetist in charge during evening and weekend duty is to determine which cases will start and which ones will wait. Being in charge also entails handling a variety of emergent situations in the hospital, including calls to intubate patients on the floor, requests for pain control, and emergency room trauma cases. The in-charge person also serves as a backup resource for the operations in progress. In this incident, the anesthetist in charge committed all of her available resources, including herself, to doing anesthesia. This effectively eliminated the in-charge person's ability to act as a buffer or extra resource for handling an additional trauma case or a request from the floor. There were strong incentives to commit the resources, but also a simultaneous incentive to avoid that commitment. Trauma severe enough to demand immediate surgery occurs in this institution once or twice a week.

Factors that played a role in the anesthetist's decision to commit all available resources included the relatively high urgency of the cases, the absence of a trauma alert (indication that a trauma patient was in route to the hospital), the time of day (fairly early; most trauma is seen in the late evening or early morning hours), and the pressure from surgeons and nurses. Another seemingly paradoxical reason for committing the resources was the desire to free up the resources by getting the cases completed before the late evening when trauma operations were more likely. These factors are not severe or even unusual. Rather, they represent the normal functioning of a large urban hospital as well as the nature of the conflicts and double binds that occur are part of the normal playing field of the specialty.

The conflicts and their resolution presented in Incident #3 and the trade-offs between highly unlikely but highly undesirable events and highly likely but less catastrophic ones are examples of strategic factors. People have to make trade-offs between different but interacting or conflicting goals, between values or costs placed on different possible outcomes or courses of action, and between the risks of different errors. People make these trade-offs when acting under uncertainty, risk, and the pressure of limited resources (e.g., time pressure, opportunity costs). One may think of these trade-offs in terms of simplistic global examples like safety versus economy. Trade-offs also occur on other dimensions. In dynamic fault management, for example, there is a trade-off with respect to when to commit to a course of action. Practitioners have to decide whether to take corrective action early in the course of an incident with limited information, or to delay the response to wait for more data to come in, to search for additional findings, or to ponder additional alternative hypotheses. Practitioners also trade-off between following operational rules or taking action based on reasoning about the case itself (cf. Woods et al., 1987). Do the standard rules apply to this particular situation when some additional factor is present that complicates the textbook scenario? Should we adapt the standard plans, or should we stick with them regardless of the special circumstances? Strategic trade-offs can also involve coordination among agents in the distributed human-machine cognitive system (Roth, Bennett, & Woods, 1987). A machine expert recommends a particular diagnosis or action, but your own evaluation is different. What is enough evidence that the machine is wrong to justify disregarding the machine expert's evaluation and proceeding on your own evaluation of the situation? The pulse oximeter may provide an unreliable reading, especially when perfusion is poor and the oxygen saturation is low. Is the current reading of 80% indicative of an artifact or an accurate representation of the patient's oxygen saturation?

Criterion setting on these different trade-offs may not be a conscious process or a decision made by individuals. More likely, it may be an emergent property of systems of people, either of small groups or larger organizations. The criteria may be fairly labile and susceptible to influence, or they may be relatively stable and difficult to change. The trade-offs may create explicit choice points for practitioners embedded in an evolving situation, or they may cast a shadow of influence over the attentional dynamics relating intertwined events, tasks, and lines of reasoning.

In hindsight, practitioners' choices or actions can often look to be simple blunders. Indeed, most of the media reports of "human error in medicine" focus on such cases. But a more careful assessment of the distributed system including the patient, physicians, and the larger institutions comprising the hospital may reveal strategic factors at work. Behavior in the specific incident derives from how the practitioners set their trade-off criteria across different kinds of risks from different kinds of incidents that could occur. Because incidents are evaluated as isolated events, such trade-offs can appear in hindsight to be unwise or even bizarre. This is because the individual incident is used as the basis for examining the larger system (see later discussion of hindsight). There are many strategic factors that can be elaborated; two forms are discussed here. The first is the presence of goal conflicts, and the other is the responsibility–authority double bind.

Goal Conflicts. Multiple goals are simultaneously relevant in actual fields of practice. Depending on the particular circumstances, the means to influence these multiple goals will interact, potentially producing conflicts between different goals. To perform an adequate analysis of the human performance in an evolving incident requires an explicit description of the strategic factors acting in the incident, including the interacting goals, the trade-offs being made, and the pressures present that shift the operating points for these trade-offs.

The impact of potential conflicts may be quite difficult to assess. Consider the anesthetist. Practitioners' highest level goal (and the one most often explicitly acknowledged) is to protect patient safety. But that is not the only goal. There are other goals, some of which are less explicitly articulated. These goals include reducing costs, avoiding actions that would increase the likelihood of being sued, maintaining good relations with the surgical service, maintaining resource elasticity to allow for handling unexpected emergencies, and others (Fig. 13.4).

In a given circumstance, the relationships between these goals can produce conflicts. In the daily routine, for example, maximizing patient safety and avoiding lawsuits creates the need to maximize information about the

Conflicting Real Goals

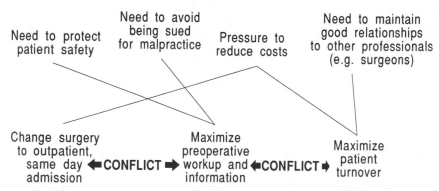

FIG. 13.4.   Conflicting goals in anesthesiology. Maximizing patient safety and avoiding lawsuits creates the need to maximize information about the patient through preoperative workup. The cost-reduction goal provides an incentive for the use of same-day surgery and limits preoperative workup. The anesthetist may be squeezed in this conflict (from Cook, Woods, & McDonald, 1991, reprinted by permission). Compare this conflict with the one shown in Fig. 13.2.

patient through preoperative workup. The anesthetist may find some hint of a potentially problematic condition and consider further tests that may incur costs, risks to the patient, and a delay of surgery. The cost reduction goal provides an incentive for a minimal preoperative workup and the use of same-day surgery. This conflicts with the other goals (Fig. 13.4). The anesthetist may be squeezed in this conflict—gathering the additional information, which in the end may not reveal anything important, will cause a delay of surgery and decrease throughput. The delay will affect the day's surgical schedule, the hospital and the surgeons' economic goals, and the anesthesiologists' relationship with the surgeons. The external pressures for highly efficient performance are strongly and increasingly in favor of limiting the preoperative workup of patients and omitting tests that are unlikely to yield important findings. But failing to acquire the information may reduce the ill-defined margin of safety that exists for this patient and contribute to the evolution toward disaster if other factors are present. Increasing external economic pressure, in particular, can generate sharp conflicts in anesthesiology and in other areas of medicine (Eddy, 1993a, 1993b).

For an example from outside of medicine, consider the task of en route flight planning in commercial aviation. Pilots sometimes need to modify their flight plans en route when conditions change (e.g., weather). Some of the goals that need to be considered are avoiding passenger discomfort (i.e., avoiding turbulence), minimizing fuel expenditure, and minimizing the difference between the target arrival time and actual arrival time. Depending on the particulars of the actual situation where the crew and dispatchers have to consider modifying the plan, these goals can interact, requiring prioritization and trade-offs. Layton, Smith, and McCoy (in press) created simulated flight situations where goal conflicts arose and studied how the distributed system of dispatchers, pilots, and computer-based advisors attempted to handle these situations.

In another aviation example, an aircraft is deiced and then enters the queue for takeoff. After the aircraft has been deiced, the effectiveness of the deicing agent degrades with time. Delays in the queue may raise the risk of ice accumulation. However, leaving the queue to go back to an area where the plane can be deiced again will cause additional delays, plus the aircraft will have to re-enter the takeoff queue again. Thus, the organization of activities (where deicing occurs relative to queuing in the system) can create conflicts that the practitioners must resolve because they are at the sharp end of the system. The dilemmas may be resolved through conscious effort by specific teams to find ways to balance the competing demands, or practitioners may simply apply standard routines without deliberating on the nature of the conflict. In either case, they may follow strategies that are robust (but still do not guarantee a successful outcome), strategies that are brittle (work well under some conditions but are vulnerable given other circumstances), or strategies that are very vulnerable to breakdown. Analyses of past disasters frequently find that goal conflicts played a role in the accident evolution. For example, there have been several crashes where, in hindsight, crews accepted delays of too great a duration and ice did contribute to a failed takeoff (Moshansky, 1992; National Transportation Safety Board, 1993).

Goal conflicts can involve economic pressures but also intrinsic characteristics of the field of activity. An example from anesthesiology is the conflict between the desirability of a high blood pressure to improve cardiac perfusion (oxygen supply to the heart muscle) and a low one to reduce cardiac work (Fig. 13.2). Specific actions will depend on details of the context. The appropriate blood pressure target adopted by the anesthetist depends in part on the individual's strategy, the nature of the patient, kind of surgical procedure, circumstances within the case that may change (e.g., the risk of major bleeding), and negotiation between different people in the operating room team (e.g., the surgeon who would like the blood pressure kept low to limit the blood loss at the surgical site).

Constraints imposed by the organizational or social context represent another source of goal competition. Some of the organizational factors producing goals include management policies, legal liability, regulatory guidelines, and economic factors. Competition between goals generated at the organizational level was an important factor in the breakdown of safety barriers in the system for transporting oil through Prince William Sound that preceded the Exxon *Valdez* disaster (National Transportation Safety Board, 1990). Finally, some of the goals that play a role in practitioner decision making relate to the personal or professional interests of the people in the operational system (e.g., career advancement, avoiding conflicts with other groups).

It should not be thought that the organizational goals are necessarily simply the written policies and procedures of the institution. Indeed, the messages received by practitioners about the nature of the institution's goals may be quite different from those that management acknowledges. Many goals are indirect and implicit. Some of the organizational influences on how practitioners will negotiate their way through conflicting goals may not be explicitly stated or written anywhere. These covert factors are especially insidious because they affect behavior and yet are unacknowledged. For example, the Navy sent a clear message to its commanders by the differential treatment it accorded to the commander of the *Stark* following that incident (U.S. House of Representatives Committee on Armed Services, 1987) as opposed to the *Vincennes* following that incident (Rochlin, 1991; U.S. Department of Defense, 1988).

In Incident #3, economic factors, intrinsic characteristics of the domain of practice, and organizational factors all contributed to the goal conflicts the practitioner faced.

Expertise consists, in part, of being able to negotiate among interacting goals by selecting or constructing the means to satisfy all sufficiently. But practitioners may fail to deal with goal conflicts adequately. Some practitioners will not follow up hints about some aspect of the patient's history because to do so would impact the usual practices relative to throughput and economic goals. In a specific case, that omission may turn out to be important to the evolution of the incident. Other practitioners will adopt a defensive stance and order tests for minor indications, even though the yield is low, in order to be on the safe side. This generates increased costs and incurs the wrath of their surgical colleagues for the delays thus generated. In either case, the nature of the goals and pressures on the practitioner are seldom made explicit and rarely examined critically.

In postincident analysis, in hindsight, the consequences will be apparent. It should be clear, however, that the external pressures for highly efficient performance are strongly in favor of limiting the preoperative workup of patients and omitting tests that are unlikely to yield important findings. Assessments after the incident will always identify factors that if changed would have produced a more favorable result; large, complex systems always have many such factors available for scrutiny. Thus, if those practitioner actions that are shaped by the goal conflict contribute to a bad outcome in a specific case, then it is easy for postincident evaluations to say that a human error occurred—the practitioners should have delayed the surgical procedure in order to investigate the hint. The role of the goal conflict may never be noted.

To evaluate the behavior of the practitioners involved in an incident, it is important to elucidate the relevant goals, the interactions between these goals and the factors that influenced criterion setting on how to make tradeoffs in particular situations. The role of these factors is often missed in evaluations of the behavior of practitioners. As a result, it is easy for organizations to produce what appear to be solutions that in fact exacerbate conflict between goals rather than helping practitioners handle goal conflicts in context. In part, this occurs because it is difficult for many organizations (particularly in regulated industries) to admit that goal conflicts and trade-off decisions arise. However distasteful to admit or whatever public relations problems it creates, denying the existence of goal interactions does not make such conflicts disappear and is likely to make them even tougher to handle when they are relevant to a particular incident. As Feynman remarked regarding the Challenger disaster, "For a successful technology, reality must take precedence over public relations, for nature cannot be fooled" (Rogers et al., 1986, Appendix F, p. 5). The difference is that, in medical practice, one can sweep the consequences of attempting to fool nature under the rug by labeling the outcome as the consequence of "human error."

Responsibility–Authority Double Binds.

Another strategic factor that plays a role in incidents and especially in medical practice is responsibility–authority double binds. These are situations in which practitioners have the responsibility for the outcome but lack the authority to take the actions they see as necessary. Regardless of how the practitioners resolve a trade-off, from hindsight they are vulnerable to charges of and penalties for error. In particular, control via regimentation and bureaucratically derived policies (just follow the procedures) or the introduction of machine-cognitive agents that automatically diagnose and plan responses, can undermine the effective authority of the practitioners on the scene. However, these same people may still be responsible and held accountable both formally and informally for bad outcomes. The results of research on the role of responsibility and authority are limited but consistent—splitting authority and responsibility appears to have bad consequences for the ability of operational systems to handle variability and surprises that go beyond preplanned routines (Hirschhorn, 1993; Roth et al., 1987).

There is one important investigation of the effects of responsibility–authority double binds in the industrial literature. Hirschhorn (1993) examined an organization's (i.e., the managers) attempts to balance the need to adapt on line to complicating factors (relative to throughput and other goals) with the goal of adhering absolutely strictly to written procedures. In part this is the result of the regulatory climate that believes that absolute adherence to procedures is the means to achieve safe operations and avoid "human error." This creates conflicts in some situations and generates dilemmas for the people involved. If they follow the standard procedures strictly, the job will not be accomplished adequately; if they always wait for formal permission to deviate from standard procedures, throughput and productivity will degrade substantially. If they deviate and it later turns out that there is a problem with what they did (e.g., they did not adapt adequately), they may create safety or economic problems. The double bind arises because they are held responsible for the outcome (the bad outcome, the lost productivity, the erroneous adaptation) but do not have authority for the work practices because they are expected to comply exactly with the written procedures. Notice the similarity to the evolving nature of medical practice today, with the introduction of increasing regulation and so- called "practice parameters" (Arens, 1993).

After the Three Mile Island accident, utility managers were encouraged by the Nuclear Regulatory Commission to develop detailed and comprehensive work procedures to reduce the likelihood of another major disaster. The management at a particular nuclear power plant instituted a policy of verbatim compliance with the procedures developed at the blunt end of the system. However, for the people at the sharp end of the system, who actually did things, strictly following the procedures posed great difficulties because (a) the procedures were inevitably incomplete, contradictory, and buggy, and (b) novel circumstances arose that were not anticipated in the written procedures. The policy created a double bind because the people would be wrong if they violated a procedure even though it could turn out to be an inadequate procedure, and they would be wrong if they followed a procedure that turned out to be inadequate. As Hirschhorn (1993) said:

> They had much responsibility, indeed as licensed professionals many could be personally fined for errors, but were uncertain of their authority. What freedom of action did they have, what were they responsible for? This gap between responsibility and authority meant that operators and their supervisors felt accountable for events and actions they could neither influence nor control.

Workers coped with the double bind by developing a covert work system that involved, as one worker put it, "doing what the boss wanted, not what he said" (Hirschhorn, 1993). There were channels for requesting changes to the procedures, but the process was cumbersome and time-consuming. This is not surprising: If modifications are easy and liberally granted, then it may be seen as undermining the policy of strict procedure following. The increasingly complex and bureaucratic policies and procedures of U.S. hospitals seems likely to generate a situation similar to that described by Hirschhorn.

The *n*-Tuple Bind

The three incidents that have been described are exemplars for the different cognitive demands encountered by practitioners who work at the sharp end of large, complex systems, including anesthetists, aircraft pilots, nuclear power plant operators, and others. Each category of cognitive issue (knowledge factors, attentional dynamics, and strategic factors) plays a role in the conduct of anesthesia and hence plays a role in the genesis and evolution of incidents. The division of cognitive issues into these categories provides a tool for analysis of human performance in complex domains. The categories are united, however, in their emphasis on the conflicts present in the domain. The conflicts exist at different levels and have different implications, but the analysis of incidents depends in large part on developing an explicit description of the conflicts and the way in which the practitioners deal with them (Table 13.1).

Together the conflicts produce a situation for the practitioner that appears to be a maze of potential pitfalls. This combination of pressures and goals that produce a conflicted environment for work is what we call *the n-tuple bind*.[6] The practitioner is confronted with the need to follow a single course of action from a myriad of possible courses. The choice of how to proceed is constrained by both the technical characteristics of the domain and the need to satisfy the "correct" set of goals at a given moment chosen from the many potentially relevant ones. This is an example of an overconstrained problem, one in which it is impossible to maximize the function or work product on all dimensions simultaneously. Unlike simple laboratory worlds with a "best" choice, real complex systems intrinsically contain conflicts that must be resolved by the practitioners at the sharp end. Retrospective critiques of the choices made in system operation will always be informed by hindsight. For example, if the choice is between obtaining more information about cardiac function or proceeding directly to surgery with a patient who has soft signs of cardiac disease, the outcome will be a potent determinant of the "correctness" of the decision. Proceeding with undetected cardiac disease may lead to a bad outcome (although this is by no means certain), but obtaining the data may yield normal results, cost money, "waste" time, and incur the ire of the surgeon. Possessing knowledge of the outcome, because of the hindsight bias, trivializes the situation confronting the practitioner and makes the "correct" choice seem crystal clear.

This *n-tuple* bind is most easily seen in Incident #3, where strategic factors dominate. The practitioner has limited resources and multiple demands for them. There are many sources of uncertainty. How long will the in vitro fertilization take? It should be a short case, but it may not be. The exploratory laparotomy may be either simple or complex. With anesthetists of different skill levels, whom should she send to the remote location where that case will take place? Arterial reconstruction patients usually have associated heart disease, and the case can be demanding. Should she commit the most senior anesthetist to that case? Such cases are also usually long, and committing the most experienced anesthetist will tie up that resource for a long time. What is the likelihood that a trauma case will come during the time when all the cases will be going on simultaneously (about an hour)? There are demands from several surgeons for their case to be the next to start. Which case is the most medically important one? The general rule is that an anesthetist has to be available for a trauma; she is herself an anesthetist and could step in, but this would leave no qualified individual to

---

[6]This term derives from the mathematical concept of a series of numbers required to define an arbitrary point in an n-dimensional space. The metaphor here is one of a collection of factors that occur simultaneously within a large range of dimensions, an extension of the notion of a *double bind*.

go to cardiac arrests in the hospital or to the emergency room. Is it desirable to commit all the resources now and get all of the pending cases completed so as to free up the people for other cases that are likely to follow?

It is not possible to measure accurately the likelihood of the various possible events that she considers. As in many such situations in medicine and elsewhere, she is attempting to strike a balance between common but lower consequence problems and rare but higher consequence ones. Ex post facto observers may view her actions as either positive or negative. On the one hand, her actions are decisive and result in rapid completion of the urgent cases. On the other hand, she has produced a situation where emergent cases may be delayed. The outcome influences how the situation is viewed in retrospect.

A critique often advanced in such situations is that the patient's "safety" should outweigh all other factors and be used to differentiate between options. Such a critique is usually made by naive individuals or administrative personnel not involved in the scene. Safety is not a concrete entity, and the argument that one should always choose the safest path (in the sense of the path that minimizes risk to the patient) misrepresents the dilemmas that confront the practitioner. The safest anesthetic is the one that is not conducted, just as the safest airplane is the one that never leaves the ground. All large, complex systems have intrinsic risks and hazards that must be incurred in order to perform their functions, and all such systems have had failures. The investigation of such failures and the attribution of cause and effect by retrospective reviewers is discussed next.

SYSTEM FAILURES AND HUMAN PERFORMANCE

Large, Complex System Failures: The Latent Failure Model

The spectacular failures of large, semantically complex, time-pressured, tightly coupled, high consequence, high-reliability systems[7] have prompted the study of how such systems fail and the role of human operators in successful and unsuccessful operation. The complexity of these systems arises in large part from the need to make them reliable. All such complex systems include potentially disastrous failure modes and are carefully crafted to reduce the risk of such failures. Significantly, these systems usually have multiple redundant mechanisms, "safety" systems, and elaborate policies

---

[7]These failures include the explosion of *Apollo 13*, the destruction of the space shuttle *Challenger*, the *Herald of Free Enterprise* ferry capsizing, the Clapham Junction railroad disaster, the grounding of the tanker *Exxon Valdez*, a number of airplane crashes, the reactor explosion at Chernobyl, and a host of other nuclear power incidents, most particularly the destruction of the reactor at Three Mile Island. Some of these are reviewed in Perrow (1984) and Reason (1990).

and procedures to keep them from failing in ways that produce bad outcomes.

The results of combined operational and technical measures make systems relatively safe from single-point failures; that is, they are protected against the failure of a single component or procedure. For example, the routine oxygen and nitrous oxide supply for anesthesia machines is derived from a hospital-wide pipeline. Each machine, however, has its own supply tanks available as a backup should the hospital supply fail, as well as elaborate valving mechanisms to insure automatic switch over to the cylinder supply. There are even special backups designed to shut off the flow of nitrous oxide (which will not support life) if the oxygen pressure falls below a preset level. In addition, the machines are gas powered and will operate even if external electrical supplies are lost. Of course, there are components and procedures that cannot be protected through redundancy. An example of such a component is the nuclear power plant's reactor containment building. The building is critical to plant safety and there is only one, but it is lavishly constructed to withstand extreme events. Similarly, the anesthesia machine has internal piping and mechanisms that make the machine vulnerable to single-point failures, although these failures are few and the components are conservatively designed (Andrews, 1990).

When large system failures do occur, they are the result of multiple, apparently innocuous faults that occur together (Perrow, 1984; Reason, 1990; Turner, 1978). All complex systems have many such apparently minor faults. These can include such simple items as a burned out indicator bulb on a seldom-used control panel, a checklist that is out of date because the corresponding equipment has been modified, or an apparently minor failure of some backup system (for example, an emergency generator). For the most part, the minor faults are inactive, play no role in system operation, and are therefore described as *latent failures* (Reason, 1990). These latent failures may be found at any level within an organization from the corporate boardroom to the individual physical components of the system. System failures occur when a particular collection of latent failures are present together in a combination that allows the system to fail. Rather than being derived from the massive failure of a single component, system failures arise from the insidious accumulation of individual faults, each of which seems too small or unimportant to threaten the larger system. Thus *Challenger* failed because of the combination of the brittle O-ring seals *and* the unexpectedly cold weather *and* the reliance on the seals in the design of the boosters *and* the change in roles between the contractor and the NASA engineers *and* other factors. None of these conditions was individually able to provoke a system failure, but together they were able to disrupt an extraordinarily safety-oriented system in a catastrophic way. In the field of aviation, a combination of factors were responsible for the simultaneous failure of all three engines of an L-1011 jumbo jet (Norman, 1992). In medicine, a similar case can be found in the failure of the Therac-25 radiation therapy machine. This device would, under certain highly unusual circumstances, deliver huge doses of radiation to patients. These circumstances, although unlikely, did arise and injure several patients. Review of the design and use of the Therac-25 showed that multiple small failures in software, in testing, and in operator procedures were required to generate the system failure (Leveson & Turner, 1993). It is important to note that the latent failures can involve technology, operational practices, and even organizational elements: Latent failures can occur in any part of the larger system (Reason, 1990).

These large system failures have several notable characteristics. First, they are *complex*. Large system failures are comprised of multiple failed components or procedures. Predicting this combination is likely to be difficult or impossible for human operators; the failure mode is hard to foresee and prevent. Second, failures are likely to be *catastrophic* rather than minor. The multiple redundancies and robust design characteristics of a large system tend to limit small-scale failures and to minimize their consequences. In addition, the cost of the so-called safety systems and redundancies generally encourages the development of ever larger and more economically efficient systems in order to reduce the average cost of each unit of performance. Thus, it is not an oil tanker accident but supertanker accident, not a plane crash but a jumbo jet crash, not an overdose of radiation but a massive overdose.

Third, the potential for catastrophic failure encourages the employment of *human skill* and expertise at the final few links in the causal chain of events. The more delicate the system, the more important its function, the more often a person will be charged with protecting the system's integrity or accomplishing some critical goal. Moreover, the systems are so complex and are operated under such variable conditions that only human operators can be expected to have both the flexibility and judgment necessary to control them. Fourth, because disasters are composed of a collection of latent failures occurring together, large system failures appear in retrospect to be unique. After-accident reviews will show that the failure depended on having a particular pattern of small faults. As the number of latent failures required to produce the system failure increases (i.e., as the system becomes, in some sense, "safer"), the odds against repeating a precise pattern of failure become astronomical. Paradoxically, this will make *any* future system failure seem extremely unlikely, even though the accumulation of latent failures actually makes the system more failure prone.

Gaba's group (Gaba, Maxwell, DeAnda, 1987) at Stanford suggested that this model of large system behavior and failure might apply to anesthesia practice in particular and, by extension, to medical practice in general. He noted that anesthesia practice includes many of the characteristics of complex systems and that the infrequent anesthesia mishaps appeared to be similar to the disasters studied in other complex systems. The anesthesiologist works in a highly complex, technologically intensive environment. The conduct of an anesthetic is a critical process that is severely time pressured, and the elements of the system are tightly coupled together in ways that do not provide much slack.[8] Cooper's group (Cooper et al., 1978; Cooper, Newbower, & Kitz, 1984) at Massachusetts General Hospital noted that anesthesia incidents appeared to be unique and were difficult to analyze, exactly as one would expect in a system that had been refined to eliminate single-point failures. Significantly, the loci of single-point failures in the conduct of the anesthetic have been studied and largely removed or buffered by redundant components or safety procedures. Although their study predated the latent failure model of large system failure, Cooper's data also indicated that critical incidents that progressed toward bad outcomes required multiple, simultaneous failures. Thus, there are reasons to consider that anesthesia practice and, by extension, modern medical practice, has the characteristics of a large, complex system and may be expected to fail in similar ways. A recent study supports this view (Cook, Woods, & McDonald, 1991).

One consequence of the latent failure model of large system failures is that efforts to improve the overall system performance by "fixing" particular latent failures that contributed to a past mishap are unlikely to markedly reduce the accident rate. Because that particular pattern of contributors is so unlikely to recur and because there are many unrecognized latent failures that remain in the system, the correction of one set of specific flaws is of limited value. The usual response to a system failure is to attempt to make certain that it doesn't happen again by producing new rules and regulations, new equipment, and new training of key personnel. Because the exact set of flaws is unlikely to recur, these attempts will add more cost and make the system even more complex and brittle than it was before the accident. Because the system is already highly reliable, some time will pass between instituting these changes and the next accident. This leads those who promulgated the changes to believe that they have significantly improved the system safety ("after all, it hasn't failed like that since we instituted our program X"). After some time, however, another accident occurs, but this time with a different sequence of events derived from a different collection of latent failures. This apparently unique accident is seen in isolation and the cycle is repeated (see Fig. 13.5).

---

[8]For a detailed discussion of coupling, see Perrow, 1984, chap. 3 and especially Table 3.2. It is interesting to note that Perrow does not include the operating room and anesthesia in his Fig. 3.1, although based on our studies, it would lie somewhere between aircraft and nuclear plants.

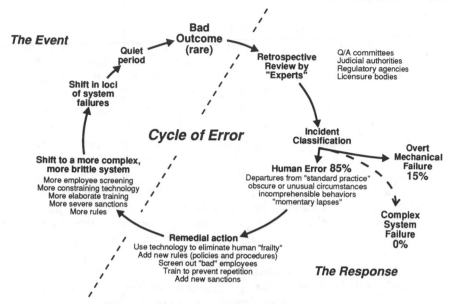

FIG. 13.5. The cycle of error. Attributing system failures to human operators generates demands for more rules, automation, and policing. But these actions do not significantly reduce the number of latent failures in the system. Because overt failures are rare, a quiet period follows institution of these new policies, convincing administrators that the changes have been effective. When a new overt failure occurs, it seems to be unique and unconnected to prior failures (except in the label human error), and the cycle repeats. With each pass through the cycle, more rules, policies, and sanctions make the system more complicated, conflicted, and brittle, increasing the opportunities for latent failures to contribute to disasters. (© 1993, R.I. Cook, reprinted by permission)

Retrospective Evaluations of Human Performance in System Transients

Attributing System Failures to Practitioners.   System failures, near failures, and critical incidents are the usual source for investigations of human performance. When critical incidents do occur, operator failure or human error will almost always be indicted as a major cause of any bad outcome. In fact, large, complex systems can be readily identified by the percentage of critical incidents that are considered to have been caused by "human error": The rate for these systems is typically about 70% or 75%. Incident rates attributed to human error are the same in several domains including aviation, nuclear power, shipping, and, most recently, in anesthesia and medicine (cf., Hollnagel, 1993). Cooper et al. (1978) found that anesthesiologists were contributors in 82% of critical incidents. Wright, Mackenzie, Buchan, Cairns, and Price (1991) and Chopra et al. (1992) found similar rates in the operating room and intensive care unit, respectively. The repeated finding of about three fourths of incidents arising from human error has built confidence in the notion that there is a problem with human error in these domains. Indeed, it is the belief that fallible humans are responsible for large system failures that has led many system designers to use more and more technology to try to eliminate the human operator from the system or to reduce the operator's possible actions so as to forestall these errors.

Attributing system failure to the human operators nearest temporally and spatially to the outcome ultimately depends on the judgment by someone that the processes in which the operator engaged were faulty and led to the bad outcome. Deciding which of the many factors surrounding an incident are important and what level or grain of analysis to apply to those factors is the product of *human* processes (social and psychological processes) of causal attribution. What we identify as the cause of an incident depends on with whom we communicate, on the assumed contrast cases or causal background for that exchange, and on the purposes of the inquiry (Woods et al., 1994).

For at least four reasons, it is actually not surprising that human operators are blamed for bad outcomes. First, operators are available to blame. These large and intrinsically dangerous systems have a few well-identified humans at the sharp end. Those humans are closely identified with the system function, and so it is unlikely that a bad outcome will occur without having them present. Moreover, these individuals are charged, often formally and institutionally, with maintaining the system's safe operation as well as the efficient functioning of the system. For any large system failure, there will be a human in close temporal and physical relationship to the outcome (e.g., a ship captain, pilot, air traffic controller, physician, nurse) and available to blame.

The second reason that human error is often the verdict after accidents is that it is so difficult to trace backward through the causal chain that led to the system failure (Rasmussen, 1986). It is particularly difficult to construct a sequence that passes back through humans in the chain. To construct such a sequence requires the ability to reconstruct, in detail, the cognitive processing of operators during the events that preceded the bad outcome. There are few tools for doing this in any but the most simple laboratory settings. The environment of the large system makes these sorts of reconstructions extremely difficult. Indeed, a major area of research is the development of tools to help investigators trace the cognitive processing of operators as they deal with normal situations, situations at the edges of normality, and system faults and failures. The incidents described in the first part of this chapter are unusual in that substantial detail about what happened and what the participants saw and did was available to researchers. In general, most traces of causality will begin with the outcome and work backward in time until they encounter a human whose actions seem to be, in hindsight, inappropriate or suboptimal. Because so little is known about how human operators process a multitude of conflicting demands of large, complex systems (e.g., avoid delays in the train schedule but also keep the trains from colliding), incident analyses rarely demonstrate the ways in which the actions of the operator made sense at the time and from their perspective.

The third reason that human error is often the verdict is paradoxical. Human error is the attributed cause of large system accidents because human performance in these complex systems is so good. Failures of these systems are, by almost any measure, rare and unusual events. Most of the system operations go smoothly; incidents that occur do not usually lead to bad outcomes. These systems have come to be regarded as "safe" by *design* rather than by *control*. Those closely studying human operations in these complex systems are usually impressed by the fact that the opportunity for large-scale system failures is present all the time and that expert human performance is able to prevent these failures. As the performance of human operators improves and failure rates fall, there is a tendency to regard system performance as a marked improvement in some underlying quality of the system itself, rather than the honing of the operator skills and expertise to a fine edge. The studies of aircraft carrier flight operations by Rochlin, La Porte, and Roberts (1987) point out that the qualities of human operators are crucial to maintaining system performance goals and that, by most measures, failures should be occurring much more often than they do. As consumers of these systems' products (health care, transportation, defense) society is lulled by success into the belief that these systems are intrinsically low risk and that the expected failure rate should be zero. Only catastrophic failures receive public attention and scrutiny. The remainder of the system operation is generally regarded as unflawed because of the low overt failure rate, even though there are many incidents that could become overt failures. Thorough after-accident analyses often indicate that there were numerous incidents or "dress rehearsals" that preceded an accident, as has been reported for the mode error at the heart of the crash of an advanced commercial aircraft at Strasbourg (Woods et al., 1994).

This ability to trace backward with the advantage of hindsight is the fourth major reason that human error is so often the verdict after accidents. Hindsight bias, as Fischhoff (1975) puts it, is the tendency for people to "consistently exaggerate what could have been anticipated in foresight." Studies have shown consistently that people have a tendency to judge the quality of a process by its outcome. The information about outcome biases their evaluation of the process that was followed. After a system failure, knowledge of the outcome biases the reviewer toward attributing failures to system operators. During postevent reviews, knowledge of the outcome will give reviewers the sense that participants ignored presumably obvious or important factors and that the participants therefore erred. Indeed, this effect is present even when those making the judgments have been warned about the phenomenon and been advised to guard against it (Fischhoff, 1975, 1982). Fischhoff (1982) wrote:

> It appears that when we receive outcome knowledge, we immediately make sense out of it by integrating it into what we already know about the subject. Having made this reinterpretation, the reported outcome now seems a more or less inevitable outgrowth of the reinterpreted situation. "Making sense" out of what we are told about the past is, in turn, so natural that we may be unaware that outcome knowledge has had any effect on us. . . . In trying to reconstruct our foresightful state of mind, we will remain anchored in our hindsightful perspective, leaving the reported outcome too likely looking. (p. 343)

In effect, reviewers will tend to *simplify* the problem-solving situation that was actually faced by the practitioner. The dilemmas facing the practitioner in situ, the uncertainties, the trade-offs, the attentional demands and double binds, all may be underemphasized when an incident is viewed in hindsight. In complex, uncertain, highly conflicted settings, such as anesthesia practice and the other similar disciplines such as military situations (Lipshitz, 1989), critics will be unable to disconnect their knowledge of the outcome in order to be able to make unbiased judgments about the performance of human operators during the incident (Baron & Hershey, 1988).

Interestingly, although the phenomenon of *hindsight bias* is well known in psychology, medical practice has had to rediscover it *de novo*. More than a decade after Fischhoff's seminal papers, a study demonstrated the phenomenon in physician judgment. Caplan, Posner, and Cheney (1990) asked two groups of anesthesiologists to evaluate human performance in sets of cases with the same descriptive facts but with the outcomes randomly assigned to be either bad or neutral. The professionals consistently rated the care in cases with bad outcomes as substandard, whereas they viewed the same behaviors with neutral outcomes as being up to standard even though the care (i.e., the preceding human acts) were identical. Typically, hindsight bias in evaluations makes it seem that participants failed to account for information or conditions that "should have been obvious"[9] or behaved in ways that were inconsistent with the (now known to be) significant information. Thus, the judgment of whether or not a human error occurred is critically dependent on knowledge of the outcome, something that is impossible before the fact. Indeed, *it is clear from the studies of large system failures*

---

[9]When someone claims that something "should have been obvious," hindsight bias is virtually always present.

that hindsight bias is the greatest obstacle to evaluating the performance of humans in complex systems after bad outcomes.

Outcome Failures and Process Defects.   It is reasonable to ask if there are means for any evaluation of human performance in complex systems. Indeed, the preceding argument seems a little disingenuous. On the one hand, it is claimed that human performance is critical to the operations of complex systems, and on the other hand, it is argued that there is no scientific way to describe something as a human error and therefore that it is necessarily impossible to distinguish between expert and inexpert performance.

One resolution of this apparent paradox is to distinguish between outcome failures and process defects. Outcome failures are defined in terms of a categorical shift in consequences on some performance dimension. Note that outcome failures are necessarily defined in terms of the language of the domain, for example, sequelae such as neurological deficit, reintubation, myocardial infarction within 48 hours, or an unplanned ICU admission. Process defects are departures from a standard about *how* problems should be solved. Generally, the process defect, if uncorrected, would lead to or increase the risk of some type of outcome failure. Process defects can be defined in domain terms—for example, insufficient intravenous access, insufficient monitoring, regional versus general anesthetic, decisions about canceling a case, or problematic plans or actions with regard to the anesthetic agent of choice. They may also be defined psychologically in terms of deficiencies in some cognitive or information-processing function—for example, activation of knowledge in context, situation awareness, diagnostic search, goal trade-offs.

The distinction between outcome and process is important because the relationship between them is not fixed. Not all process defects are associated with bad outcomes. The defect may be insufficient to create the bad outcome by itself. In addition, as Edwards (1984) said, "a good decision cannot guarantee a good outcome," that is, bad outcomes may result even if there are no defects in process. This is especially true for domains such as anesthesiology where bad outcomes can occur despite the exercise of nearly flawless expertise by the medical personnel involved (cf. Keats, 1979, 1990).

The rate of process defects may be frequent when compared with the incidence of overt system failures. This is so because the redundant nature of complex systems protects against many defects. It is also because the systems employ human operators whose function is, in part, to detect such process flaws and adjust for them before they produce bad outcomes (a process of error detection and recovery). Just such a situation can be seen in Incident #2. Evaluating human performance by examining the process of human problem solving in a complex system depends on specifying a standard about how problems should be handled. There are several categories of standards that can be used to evaluate defects in the process of solving a problem.

One standard is a *normative model of task performance*. This method requires detailed knowledge about precisely how problems should be solved, that is, nearly complete and exhaustive knowledge of the way the system works. Such knowledge is, in practice, rare. At best, some few components of the larger system can be characterized in this exhaustive way. Unfortunately, normative models rarely exist or are not applicable to complex situations like anesthesia practice. Those models are largely limited to mathematically precise situations such as games or artificial tasks in bounded worlds.

Another standard is the comparison of actual behavior to *standard operating practices* (e.g., standards of care, policies, and procedures). These practices are mostly compilations of rules and procedures that are acceptable behaviors for a variety of situations. They include various protocols (e.g., the Advanced Cardiac Life Support protocol for cardiac arrest, the guidelines for management of the difficult airway), policies (e.g., it is the policy of the hospital to have informed consent from all patients prior to beginning an anesthetic), and procedures (e.g., the chief resident calls the attending anesthesiologist to the room before beginning the anesthetic but after all necessary preparations have been made). These standards may be of limited value because they are either codified in ways that ignore the real nature of the domain[10] or because the coding is too vague to use for evaluation. For example, one senior anesthesiologist, when asked about the policy of the institution regarding the care for emergent Cesarean sections replied, "Our policy is to do the right thing." This seemingly curious phrase in fact sums up the problem confronting those at the sharp end of large, complex systems. It recognizes that it is impossible to comprehensively list all possible situations and appropriate responses because the world is too complex and fluid. Thus the person in the situation is required to account for the many factors that are unique to that situation. What sounds like a nonsense phrase is, in fact, an expression of the limitations that apply to all structures of rules, regulations, and policies (cf. e.g., Roth et al., 1987; Suchman, 1987). The set of rules is necessarily incomplete and sometimes

---

[10]It is not unusual, for example, to have a large body of rules and procedures that are not followed because to do so would make the system intolerably inefficient. The "work to rule" method used by unions to produce an unacceptable slowdown of operations is an example of the way in which reference to standards is unrealistic. In this technique, the workers perform their tasks to an exact standard of the existing rules and the system performance is so degraded by the extra steps required to conform to all the rules that it becomes nonfunctional (e.g., Hirschhorn, 1993).

contradictory. It is the role of the human at the sharp end to resolve the apparent contradictions and conflicts in order to satisfy the goals of the system.

In general, procedural rules are too vague to be used for evaluation if they are not specific enough to determine the adequacy of performance before the fact. Thus, a procedural rule such as "the anesthetic shall not begin until the patient has been properly prepared for surgery" is imprecise, whereas another such as "flammable anesthetic agents shall not be used" is specific. When the rules are codified as written policies, imprecise rules usually function simply to provide administrative hierarchies the opportunity to assign blame to operators after accidents and to finesse the larger institutional responsibility for creating the circumstances that lead to accidents (see the report on the aircraft accident at Dryden, in Moshansky, 1992). Significantly, the value of both the normative and standard practices methods of evaluating the problem-solving process of human operators is limited to the most simple systems and generally fails as system size and complexity increase.

A third approach is called the *neutral observer criterion* by De Keyser and Woods (1990). The neutral observer criterion is an empirical approach that compares practitioner behavior during the incident in question to the behavior of similar practitioners at various points in the evolving incident. In practice, the comparison is usually accomplished by using the judgment of similar practitioners about how they would behave under similar circumstances. Neutral observers make judgments or interpretations about the state of the world (in this domain, the patient and related monitors and equipment), relevant possible future event sequences, and relevant courses of action. The question is whether the path taken by the actual problem solver is one that is plausible to the neutral observers. One key is to avoid contamination by hindsight bias; knowledge about the later outcome may alter the neutral observers' judgment about the propriety of earlier responses. The function of the neutral observer is to help define the envelope of appropriate responses given the information available to the practitioner at each point in the incident.

The writers' research, and that of others, is based on the development of neutral observer criteria for actions in complex systems. This method involves comparing actions that were taken by individuals to those of other experts placed in the same situation. Note that this is a strong criterion for comparison and it necessarily requires that the evaluators possess the same sort of expertise and experience as was employed during the incident. It does not rely on comparing practitioner behaviors with theory, rules, or policies. It is particularly effective for situations where the real demands of the system are poorly understood and where the pace of system activity is high (i.e., in large, complex systems). The writers have used this technique in examining human performance in incidents from several different sources in anesthesia and in other domains. The technique is complex, as the descriptions and discussions of the three exemplar incidents indicate, but the complexity simply matches that of the domain and the human behaviors being evaluated.

Did the Practitioners Commit Errors?

The three exemplar incidents in this chapter are not remarkable or unusual; rather they reflect the normal, day-to-day operations that characterize busy, urban tertiary care hospitals. In each incident, human performance is closely tied to system performance and to eventual outcome, although the performance of the practitioners is not the sole determinant of outcome. The myocardial infarction following the events of Incident #1 may well have happened irrespective of any actions taken by practitioners. That patient was likely to have an infarction, and it is not possible to say whether the anesthetist's actions caused the infarction. The incidents and the analysis of human performance that they prompt (including the role of latent failures in system transients) may make us change our notion of what constitutes a human error.

Arguably, the performance in each exemplar incident is flawed. In retrospect, things can be identified that might have been done differently and that would have forestalled or minimized the incident or its effect. In the myocardial infarction incident, intravascular volume was misassessed, and treatment for several simultaneous problems was poorly coordinated. In the hypotension incident (#2), the device set-up by practitioners probably contributed to the initial fault. The practitioners were also unable to diagnose the fault until well after its effects had cascaded into a near crisis. In the scheduling incident (#3), a practitioner violated policy. She chose one path in order to meet certain demands, but simultaneously exposed the larger system to a rare but important variety of failure. In some sense, each of the exemplar incidents constitutes an example of human error. Note, however, that each incident also demonstrates the complexity of the situations confronting practitioners and the way in which practitioners adjust their behavior to adapt to the unusual, difficult, and novel aspects of individual situations.

Especially in the hypotension incident (#2), the resiliency of human performance in an evolving incident is demonstrated. The practitioners were willing to abandon their efforts at diagnosis and shift to a *disturbance management* mode of response in order to preserve the patient's life pending resolution of the disturbance. The practitioner was also busy during the myocardial infarction incident, although in this instance the focus was primarily on producing better oxygenation of the blood and control of the blood pressure and not on correcting the intravascular volume. These efforts were significant and, in part, successful. In both Incidents #1 and #2, attention is drawn to the practitioner performance by the outcome.

In retrospect, some would describe aspects of these incidents as human error. The high urine output with high blood glucose and prior administration of furosemide *should* have prompted the consideration of low (rather than high) intravascular volume. The infusion devices *should* have been set up correctly, despite the complicated set of steps involved. The diagnosis of hypotension *should* have included a closer examination of the infusion devices and their associated bags of fluid, despite the extremely poor device feedback. Each of these conclusions, however, depends on knowledge of the outcome; each conclusion suffers from hindsight bias. To say that something *should* have been obvious, when it manifestly was not, may reveal more about our ignorance of the demands and activities of this complex world than it does about the performance of its practitioners. It is possible to generate an infinite list of shoulds for practitioners in anesthesiology and other large systems, but these lists quickly become unwieldy and, in any case, focus only on the most salient failures from the most recent disaster. It is easy to slip into the "cycle of error" (Fig. 13.5), focusing on error out of context, increasing the complexity of the larger system, exacerbating conflicts, and creating more opportunities for latent failures to accumulate and come together in an accident.

The scheduling incident (#3) is somewhat different. In that incident, it is clear how knowledge of the outcome biases the evaluations of practitioner performance. As Abraham Lincoln said, "If the end brings me out all right what is said against me won't amount to anything. If the end brings me out wrong, ten angels swearing I was right will make no difference." Is there a human error in Incident #3? If a trauma case had occurred in this interval where all the resources had been committed to other cases, would her decision then be considered an error? On the other hand, if she had delayed the start of some other case in order to be prepared for a possible trauma case that never happened and the delay contributed to some complication for that patient, would her decision then be considered an error?

From this discussion, we are being forced to conclude that the human error is a judgment made in hindsight. In a real sense, then, for scientists and investigators, *there is no such thing as human error* (cf. Hollnagel, 1993). Human error does not comprise a distinct category of human performance. As the incidents suggest, human performance is not simply either adequate or inadequate. Neither is it either faulty or fault-free. Rather, human performance is as complex and varied as the domain in which it is exercised. Credible evaluations of human performance must be able to account for all of the complexity that confronts the practitioner. This is precisely what most evaluations of human performance do not do: They simplify the situations and demands confronting practitioners until it is obvious that the practitioners have erred. By stripping away the complexities and contradictions inherent in operating these large systems, the evaluators eliminate the richness of detail that might help to show how the activities of the practitioners were locally rational and miss the bottlenecks and dilemmas that challenge practitioner expertise and skill. The term *human error* should not represent the concluding point but rather the starting point for studies of accident evolution in large systems.

The schema of knowledge factors, attentional dynamics, and strategic factors provide one means of categorizing the activities of teams of practitioners.[11] The model of large system failure arising from the concatenation of multiple small latent failures provides an explanation for the mysteriously unique appearance of failures. That model also explains the limited success achieved by the pursuit of first causes in the cycle of error. It also suggests that the human practitioner's role in large systems may be in part to uncouple elements of the system in order to minimize the propagation of the latent failures resident in the system (Perrow, 1984).

Together, the exemplar incidents and their analyses imply that many of the changes occurring in medical practice may make the system more brittle and increase the apparent contribution of human error. In response to incidents, organizations generate more rules, regulations, policies, and procedures that make it more likely that medical practitioners will be found to have erred by postincident analyses (Fig. 13.5). Emphasis on cutting costs and increasing efficiency generates more pressure on practitioners, making scenarios like that of the scheduling incident more likely. Increased use of technology such as the computer-based infusion devices in the hypotension incident (#2) raises the complexity of incidents and creates new modes of failure. Even the burgeoning volume of medical knowledge plays a role, making the likelihood of the sort of inert knowledge problems of the myocardial infarction incident more probable (Feltovich et al., 1989). In the face of these pressures, a quality management system that steadfastly maintains that human error is the root cause of system failures can be relied on to generate a huge volume of error statistics that, in turn, become part of the cycle of error and its consequences.

---

[11]The practitioners need not be human; the same schema may be used for evaluating the performance of machine "expert systems" and the performance of teams of human and machine cognitive agents.

ENHANCING HUMAN PERFORMANCE

Training

If human performance is critical to system performance, then it seems reasonable to try to enhance it. One method of improving human performance is retraining. Unfortunately, most retraining is predicated on the presence of a human error problem, that is, that flawed human performance is the root cause of system failures and that eliminating this failure mode is the key to success. Under this assumption, many training programs consist merely of routinization of tasks according to a rote method. This approach is sometimes called *blame and train,* because it begins with the concept that human error is the source of the problem and that this error arises from capriciousness or inattentiveness by practitioners. This was exactly what happened following the Three Mile Island accident in 1979. The regulatory agencies and organizations responded in part with an emphasis in training on rote performance of compiled procedures, and the result was that operational personnel confronted a variety of dilemmas about whether to depart from the standard procedures in more complicated incidents (Woods et al., 1987).

There are several methods in use in the aviation and anesthesia domains that represent contrasting approaches to training. Cockpit resource management (CRM) is a tool used by several major air carriers in an effort to improve crew performance (Wiener, Kanki, & Helmreich, 1993). CRM acknowledges that air crews are a resource for solving problems and attempts to give the crews more experience in working together in crisis situations where coordination is critical. Unlike blame and train methods that seek to regiment human performance to eliminate human error, CRM implicitly views human performance as the critical resource in dealing with novel, threatening situations and focuses on developing in pilots and engineers the ability to work together as a coordinated team. In anesthesia, Gaba's group at Stanford developed a similar tool called Crisis Resource Management (CRM) that provides anesthetists with opportunities to see themselves act under the pressure of simulated disasters (Gaba & DeAnda, 1989). Gaba's group uses material from the aviation CRM as well. Again, the implicit view of this training method is that the human practitioner is a resource and the only likely source of system resilience in the event of an incident. The anesthesia CRM concentrates on infrequently experienced but quite realistic scenarios (e.g., a complete power outage) as a test bed for improving human performance. Both CRMs are qualitatively different from the majority of training approaches generally in use. Both make extensive use of elaborate simulators and large bodies of domain knowledge and can be quite expensive.

Technology

All of the large systems with which we are concerned (anesthesia, nuclear power operations, aviation) are intensely technological, so much so that they do not exist apart from their technology. During the past decade, each of these domains has seen the introduction of automation, the purpose of which is to eliminate human activity as the source of errors. The record of these systems is mixed and controversial (Woods et al., 1994).

Much technological innovation is supposed to reduce human error by reducing human workload. The introduction of microprocessor-controlled infusion pumps, for example, can eliminate the cumbersome process of adjusting a manual valve to regulate drips. However, these same devices create other demands, including set-up, operation, and fault diagnosis as seen in the hypotension incident. Similar equipment in the cockpit and the operating room are actually examples of *clumsy automation* (Cook et al., 1990; Wiener, 1989), where the workload reduction is limited to the times when the operator is not busy at any rate (e.g., mid-flight), and the cost is a substantial increase in workload at peak workload times (e.g., takeoff and landing). Such systems are poor amplifiers of human performance and are likely to degrade performance at critical times.[12] Clumsy automation also includes technologies that provide great increases in precision but demand equally precise operation, such as the newer generation of drug infusion pumps. One such device has a library of drug concentration data and is able, given suitable input parameters, to compute and deliver precisely metered doses of potent drugs based on patient weight. The set-up procedure for this device is more complicated and time consuming than its predecessors and increases the potential for large (i.e., order of magnitude) errors in dosing. Practitioners may also encounter black box systems whose behavior is extraordinarily difficult to understand even when the function performed is relatively simple (Cook, Potter, Woods, & McDonald 1991).

Another use of technology is to eliminate human decision making as a source of error by eliminating the decision making entirely. This trend is most advanced in the commercial aircraft cockpit, but it has also been demonstrated in the operating room and the nuclear plant control center. One effect of attempting to automate decision making can be to increase the intensity of the responsibility–authority double bind. In the larger context, practitioners faced with such devices confront a double problem: Not only do they have to understand the situation confronting them, but they must also understand how the machine sees that situation and be able to evaluate

---

[12]This may be one reason practitioners are sometimes reluctant to embrace such technologies.

the machine's proposed responses to the situation (Roth et al., 1987; Sarter & Woods, in press; Woods et al., 1994).

Eliminating Human Error Versus Aiding Human Performance

Clearly, those strategies that derive from a desire to minimize human error are different from those that seek to aid human performance. Rules, regulations, sanctions, policies, and procedures are largely predicated on the belief that human error is at the heart of large system failures and that a combination of restrictions and punishments will transform human behavior from error to an error-free state. The same basis exists for some training and technology programs, for example, blame and train and automated decision systems, whereas others (notably CRM) regard human performance as the primary means for dealing with system transients and look for ways to produce more effective human performance. The distinction is an important one and not simply a matter of degree; the choice of path depends critically on the validity of the whole notion of human error.

CONCLUSION

Human operator performance in large systems and the failures of these systems are closely linked. The demands that large, complex systems operations place on human performance are mostly cognitive. The difference between expert and inexpert human performance depends on the timely and appropriate action that in turn is shaped by knowledge factors, attentional dynamics, and strategic factors. A brief examination of a few incidents occurring in anesthesia practice has demonstrated that human performance is complex in proportion to the complexity of the domain itself. Analyses of the human role, especially those that take place after an incident or accident, must provide a satisfactory account of that complexity and its impact on human decision making and activity. The schema of knowledge factors, attentional dynamics, and strategic factors can provide a framework for laying out the issues confronting practitioners at the sharp end.

There are at least two different ways of interpreting human performance in complex systems. The conventional way views human performance as the source of errors that can be eliminated by restricting the range of human activity or eliminating the performer from the system. According to this view, *human error* is seen as a distinct category that can be counted and tabulated.

This chapter has presented a second approach, one that views human performance as the means for resolving the uncertainties, conflicts, and competing demands inherent in large, complex systems (Hollnagel, 1993). This view acknowledges the presence of both blunt and sharp ends of the system. The blunt end, including regulatory bodies, administrative entities, economic policies, and technology development organizations, can affect sharp-end practitioners by creating and amplifying conflicts and by determining the resources available for resolving those conflicts. The analyses guided by this approach explicitly avoid the term *human error* because it obscures more than it reveals.

Human error is not a distinct category of human performance. After the outcome is clear, any attribution of error is a social and psychological judgment process, not a narrow, purely technical or objective analysis. Outcome knowledge biases retrospective evaluations. Different judges with different background knowledge of the events and context or with different goals will judge the performance of human practitioners differently. Recognizing the limits of the label *human error* can lead us in new, more fruitful directions for improving the performance of complex systems (Woods et al., 1994).

So how should we view a large, complex system failure? If a bad outcome is seen as yet another incident containing one or more human errors by some practitioners, that is, we adopt the conventional view, what shall we do then? The options are few. We can try to train people to remediate the apparent deficiencies in their behavior. We can try to remove the culprits from the scene or, at least, prevent these sorts of defective people from becoming practitioners. We can try to police practitioner activities more closely.

This chapter suggests quite a different approach. It proposes that system failures are a form of information about the system in which people are embedded. They do not point to a single independent (and human) component (a culprit) as the source of failure. Instead, system failures indicate the need for analysis of the decisions and actions of individuals and groups embedded in the larger system that provides resources and constraints. To study human performance and system failure requires studying the function of the system in which practitioners are embedded. Failures tell us about situations where knowledge is not brought to bear effectively, where the attentional demands are extreme, where the *n-tuple* bind is created. Knowledge of these systemic features allows us to see how human behavior is shaped and to examine alternatives for shaping it differently.

In this view, the behavior that people, in hindsight, call "human error" is the end result of a large number of factors coming to bear at the sharp end of practice. Social and psychological processes of causal attribution lead us to label some practitioner actions as "human error" and to regard other actions as acceptable performance. Hindsight bias leads us to see only those forks in the road that practitioners decided to take—we see "the view from one side of a fork in the road, looking back" (Lubar, 1993, p. 1168). This view is fundamentally flawed because it does not reflect the situation confronting the practitioners at the scene. The challenge we face as evaluators of human performance is to reconstruct what the view was like or would have been like had we stood on the same road.

The few examples in this chapter give the flavor of what operating at the sharp end really demands of practitioners. It is not surprising that human operators occasionally should be unable to extract good outcomes from the conflicted and contradictory circumstances in which they work. The surprise is that they are able to produce good outcomes as often as they do.

ACKNOWLEDGMENTS

This preparation of this chapter was supported in part by grants from the Anesthesia Patient Safety Foundation and from the National Aeronautics and Space Administration.

REFERENCES

Andrews, J. J. (1990). Inhaled anesthesia delivery systems. In R. D. Miller (Ed.), *Anesthesia* (3rd ed., pp. 171–223). New York: Churchill Livingstone.

Arens, J. F. (1993). A practice parameters overview. *Anesthesiology, 78,* 229–230.

Baron, J., & Hershey, J. (1988). Outcome bias in decision evaluation. *Journal of Personality and Social Psychology, 54*(4), 569–579

Boeing Product Safety Organization. (1993). *Statistical summary of commercial jet aircraft accidents: Worldwide operations, 1959–1992.* Seattle, WA: Boeing Commercial Airplanes.

Bransford, J., Sherwood, R., Vye, N., & Rieser J. (1986). Teaching and problem solving: Research foundations. *American Psychologist, 41,* 1078–1089.

Caplan, R. A., Posner, K. L., Cheney, F. W. (1991). Effect of outcome on physician judgments of appropriateness of care. *Journal of the American Medical Association, 265,* 1957–1960.

Chi, M., Glaser, R., & Farr, M. (1988). *The nature of expertise.* Hillsdale, NJ: Lawrence Erlbaum Associates.

Chopra, V., Bovill, J. G. , Spierdijk, J., & Koornneef, F. (1992). Reported significant observations during anaesthesia: A prospective analysis over an 18-month period. *British Journal of Anaesthesia, 68,* 13–17.

Cook, R. I., McDonald, J. S., & Smalhout, R. (1989). *Human error in the operating room: Identifying cognitive lock up.* (Cognitive Systems Engineering Laboratory Tech. Rep. 89-TR-07). Columbus, OH: The Ohio State University.

Cook, R. I., Potter, S. S., Woods, D. D., & McDonald, J. S. (1991). Evaluating the human engineering of microprocessor controlled operating room devices. *Journal of Clinical Monitoring, 7,* 217–226.

Cook, R. I., Woods, D. D., & Howie, M. B. (1990). The natural history of introducing new information technology into a dynamic high-risk environment. *Proceedings of the 34th Annual Human Factors Society Meeting* (pp. 429–433).

Cook, R. I., Woods, D. D., & Howie, M. B. (1992). Unintentional delivery of vasoactive drugs with an electromechanical infusion device. *Journal of Cardiothoracic and Vascular Anesthesia, 6,* 1–7.

Cook, R. I., Woods, D. D., & McDonald, J. S. (1991). *Human performance in anesthesia: A corpus of cases. Report to the Anesthesia Patient Safety Foundation.* (Cognitive Systems Engineering Laboratory Technical Report 91-TR-03). Columbus, OH: The Ohio State University.

Cooper, J. B., Newbower, R. S., & Kitz, R. J. (1984). An analysis of major errors and equipment failures in anesthesia management: Conditions for prevention and detection. *Anesthesiology, 60,* 43–42.

Cooper, J. B., Newbower, R. S., Long, C. D., & McPeek, B. (1978). Preventable anesthesia mishaps: A study of human factors. *Anesthesiology, 49,* 399–406.

DeKeyser, V., & Woods, D. D. (1990). Fixation errors: Failures to revise situation assessment in dynamic and risky systems. In A. G. Colombo & A. Saiz de Bustamante (Eds.), *System reliability assessment* (pp. 231–251). Dordercht, The Netherlands: Kluwer.

Dorner, D. (1983). Heuristics and cognition in complex systems. In R. Groner, M. Groner, & W. F. Bischof (Eds.), *Methods of heuristics* (pp. 89–107). Hillsdale, NJ: Lawrence Erlbaum Associates.

Eddy, D. M. (1993a). Clinical decision making: Broadening the responsibilities of practitioners [Letters]. *Journal of the American Medical Association, 270,* 708–710.

Eddy, D. M. (1993b). Redefining the role of practitioners: The team approach. *Journal of the American Medical Association, 269,* 1849–1855.

Edwards, W. (1984). How to make good decisions. *Acta Psycholgica, 56,* 5–27.

Feltovich, P. J., Spiro, R. J., & Coulson, R. (1989). The nature of conceptual understanding in biomedicine: The deep structure of complex ideas and the development of misconceptions. In D. Evans & V. Patel (Eds.), *Cognitive science in medicine: Biomedical modeling.* Cambridge, MA: MIT Press.

Feltovich, P. J., Spiro, R. J., & Coulson, R. (1993). Learning, teaching and testing for complex conceptual understanding. In N. Fredericksen, R. Mislevy, & I. Bejar (Eds.), *Test theory for a new generation of tests.* Hillsdale, NJ: Lawrence Erlbaum Associates.

Fischhoff, B. (1975). Hindsight does not equal foresight: The effect of outcome knowledge on judgment under uncertainty. *Journal of Experimental Psychology: Human Perception and Performance, 1,* 288–299.

Fischhoff, B. (1982). For those condemned to study the past: Heuristics and biases in hindsight. In D. Kahneman, P. Slovic, & A. Tversky (Eds.), *Judgment under uncertainty: Heuristics and biases* (pp. 335–354). New York: Cambridge University Press.

Gaba, D. M., & DeAnda, A. (1989). The response of anesthesia trainees to simulated critical incidents. *Anesthesia and Analgesia, 68,* 444–451.

Gaba, D. M., Maxwell, M., & DeAnda, A. (1987). Anesthetic mishaps: Breaking the chain of accident evolution. *Anesthesiology, 66,* 670–676.

Gentner, D., & Stevens, A. L. (Eds.). (1983). *Mental models.* Hillsdale, NJ: Lawrence Erlbaum Associates.

Gopher, D. (1991). The skill of attention control: Acquisition and execution of attention strategies. In *Attention and performance XIV* (pp. 299–322) Hillsdale, NJ: Lawrence Erlbaum Associates.

Hirschhorn, L. (1993). Hierarchy versus bureaucracy: The case of a nuclear reactor. In K. H. Roberts (Ed.), *New challenges to understanding organizations,* (pp. 137–149). New York: Macmillan.

Hollister, W. M. (Ed.). (1986). Improved guidance and control automation at the man–machine interface (AGARD Advisory Report No. 228). AGARD-AR-228.

Hollnagel, E. (1993). *Reliability of cognition: Foundations of human reliability analysis.* London: Academic Press.

Johnson, P. E., Moen, J. B., & Thompson, W. B. (1988). Garden path errors in diagnostic reasoning. In L. Bolec & M. J. Coombs (Eds.), *Expert system applications* (pp. 395–428) New York: Springer-Verlag.

Johnson, P. E., & Thompson, W. B. (1981). Strolling down the garden path: Detection and recovery from error in expert problem solving. Paper presented at *the Seventh International Joint Conference on Artificial Intelligence.* Vancouver, British Columbia.

Keats, A. S. (1979). Anesthesia mortality—A new mechanism. *Anesthesiology, 68,* 2–4.

Keats, A. S. (1990). Anesthesia mortality in perspective. *Anesthesia and Analgesia, 71,* 113–119.

Kemeny, J. G., et al. (1979) *Report of the President's Commission on the accident at Three Mile Island.* New York: Pergamon.

Klein, G. A., & Crandall, B. (in press). The role of mental simulation in problem solving and decision making. In J. Flach, P. Hancock, J. Caird, & K. Vicente (Eds.), *The ecology of human-machine systems.* Hillsdale, NJ: Lawrence Erlbaum Associates.

Klein, G. A., Orasanu, J., Calderwood, R., & Zsambok, C. (Eds.). (1993). *Decision making in action: Models and methods.* Norwood, NJ: Ablex.

Layton, C., Smith, P. J., & McCoy, E. (in press). Design of a cooperative problem solving system for enroute flight planning: An empirical evaluation. *Human Factors.*

Leveson, N. G., & Turner, C. S. (1993, July). *An investigation of the Therac-25 accidents. IEEE Computer,* 18–41.

Lipshitz, R. (1989). "Either a medal or a corporal": The effects of success and failure on the evaluation of decision making and decision makers. *Organizational Behavior and Human Decision Processes, 44,* 380–395.

Lubar, S. (1993, May 21). Review of "The evolution of useful things" by Henry Petroski. *Science, 260,* 1166–1168.

Moll van Charante, E., Cook, R. I., Woods, D. D., Yue, Y., & Howie, M. B. (1993). Human–computer interaction in context: Physician interaction with automated intravenous controllers in the heart room. In H. G. Stassen (Ed.), *Analysis, design and evaluation of man–machine systems 1992* (pp. 263–274). London: Pergamon.

Moshansky, V. P. (1992). *Final report of the Commission of Inquiry into the Air Ontario crash at Dryden, Ontario.* Ottawa: Minister of Supply and Services, Canada.

National Transportation Safety Board (1990). *Marine accident report: Grounding of the U.S. Tankship Exxon Valdez on Bligh Reef, Prince William Sound, Near Valdez, Alaska, March 24, 1989* (Report No. NTSB/MAT-90/04). Springfield, VA: National Technical Information Service.

National Transportation Safety Board (1993). *US Air Flight 405 LaGuardia Airport March 22, 1992* (Report No. AAR-93/02). Springfield, VA: National Technical Information Service.

Neisser, U. (1976). *Cognition and reality: Principles and implications of cognitive psychology.* San Francisco, CA: Freeman.

Norman, D. A. (1988). *The psychology of everyday things.* New York: Basic Books.

Norman, D. A. (1992). *Turn signals are the facial expressions of automobiles.* New York: Addison-Wesley.

Orasanu, J. (1990). *Shared mental models and crew decision making* (Tech. Rep. No. 46). Princeton NJ: Princeton University, Cognitive Science Laboratory.

Payne, J. W., Bettman, J. R., & Johnson, E. J. (1988). Adaptive strategy selection in decision making. *Journal of Experimental Psychology: Learning, Memory, and Cognition, 14,* 534–552.

Payne, J. W., Johnson, E. J., Bettman, J. R., & Coupey, E. (1990). Understanding contingent choice: A computer simulation approach. *IEEE Transactions on Systems, Man, and Cybernetics, 20,* 296–309.

Perrow, C. (1984). *Normal accidents. Living with high-risk technologies.* New York: Basic Books.

Perkins, D., & Martin, F. (1986). Fragile knowledge and neglected strategies in novice programmers. In E. Soloway & S. Iyengar (Eds.), *Empirical studies of programmers* (Vol. 1, pp. 213–279). Norwood, NJ: Ablex.

Rasmussen, J. (1986). *Information processing and human–machine interaction: An approach to cognitive engineering.* New York: North-Holland.

Rasmussen, J., Duncan, K., & Leplat, J. (Eds.). (1987). *New technology and human error.* New York: Wiley.

Reason, J. (1990). *Human error.* Cambridge, England: Cambridge University Press.

Roberts, K. H., & Rousseau, D. M. (1989). Research in nearly failure-free, high-reliability organizations: Having the bubble. *IEEE Transactions in Engineering Management, 36,* 132–139.

Rochlin, G. I., La Porte, T. R., & Roberts, K. H. (1987, Autumn). The Self-designing high-reliability organization: Aircraft carrier flight operations at sea. *Naval War College Review,* pp. 76–90.

Rochlin, G. I. (1991). Iran Air Flight 655 and the USS Vincennes. In T. LaPorte (Ed.), *Social responses to large technical systems* (pp. 99–126). Dordrecht, Netherlands: Kluwer Academic.

Rogers, W. P., et al. (1986). *Report of the Presidential Commission on the Space Shuttle Challenger Accident.* Washington, DC: U.S. Government Printing Agency.

Roth, E. M., Bennett, K. B., & Woods, D. D. (1987). Human interaction with an "intelligent" machine. *International Journal of Man–Machine Studies, 27,* 479–525.

Roth, E. M., Woods, D. D., & Pople, H. E., Jr. (1992). Cognitive simulation as a tool for cognitive task analysis. *Ergonomics, 35,* 1163–1198.

Rouse, W. B., & Morris, N. M. (1986). On looking into the black box: Prospects and limits in the search for mental models. *Psychological Bulletin, 100,* 359–363.

Sarter, N. B., & Woods, D. D. (1991). Situation awareness: A critical but ill-defined phenomenon. *International Journal of Aviation Psychology, 1,* 43–55.

Sarter, N. B., & Woods, D. D. (1992). Pilot interaction with cockpit automation I: Operational experiences with the Flight Management System. *International Journal of Aviation Psychology, 2,* 303–321.

Sarter, N. B., & Woods, D. D. (1994). Pilot interaction with cockpit automation II: An experimental study of pilots' mental model and awareness of the Flight Management System (FMS). *International Journal of Aviation Psychology, 4,* 1–28.

Sarter, N. B., & Woods, D. D. (in press). "How in the world did we get into that mode?" Mode error and awareness in supervisory control. *Human Factors (Special Issue on Situation Awareness).*

Suchman, L. (1987). *Plans and situated actions. The problem of human machine communication.* Cambridge, England: Cambridge University Press.

Tenney, Y. J., Jager Adams, M., Pew, R. W., Huggins, A. W. F., & Rogers, W. H. (1992). *A principled approach to the measurement of situation awareness in commercial aviation* (NASA Contractor Report No. NAS1-18788). Hampton, VA: NASA Langley Research Center.

Turner, B. A. (1978). *Man-made disasters.* London: Wykeham.

Tversky, A, & Kahneman, D. (1974). Judgment under uncertainty: Heuristics and biases. *Science, 185,* 1124–1131.

U.S. Department of Defense. (1988). *Report of the formal investigation into the circumstances surrounding the downing of Iran Air Flight 655 on 3 July 1988.* Washington, DC: Author.

U.S. House of Representatives Committee on Armed Services. (1987). *Report on the staff investigation into the Iraqi attack on the USS Stark.* Springfield, VA: National Technical Information Service.

Wagenaar, W., & Keren, G. (1986). Does the expert know? The reliability of predictions and confidence ratings of experts. Decision making in complex systems. In E. Hollnagel, G. Mancini, & D. D. Woods (Eds.), *Intelligent decision making in process control environments* (pp. 87–103). Berlin: Springer-Verlag.

Wiener, E. L. (1989). *Human factors of advanced technology ('glass cockpit') transport aircraft* (Tech. Rep. 117528). Moffett Field, CA: NASA Ames Research Center.

Wiener, E., Kanki, B., & Helmreich, R. (Eds.). (1993). Cockpit resource management. New York: Academic Press.

Woods, D. D. (1988). Coping with complexity: The psychology of human behavior in complex systems. In L. P. Goodstein, H. B. Andersen, & S. E. Olsen, (Eds.), *Tasks, errors, and mental models* (pp. 128–148). New York: Taylor & Francis.

Woods, D. D. (1992). *The alarm problem and directed attention.* (Cognitive Systems

Engineering Laboratory Technical Report 92-TR-01). Columbus, OH: The Ohio State University.

Woods, D. D. (1994). Cognitive demands and activities in dynamic fault management: Abductive reasoning and disturbance management. In N. Stanton (Ed.), *The human factors of alarm design* (pp. 63–92). London: Taylor & Francis.

Woods, D. D., O'Brien, J., & Hanes, L. F. (1987). Human factors challenges in process control: The case of nuclear power plants. In G. Salvendy (Ed.), *Handbook of human factors/ergonomics.* New York: Wiley.

Woods, D. D., & Roth, E. M. (1988). Cognitive systems engineering. In M. Helander (Ed.), *Handbook of human–computer interaction.* New York: Elsevier.

Woods, D. D., Johannesen, L., Cook, R. I., & Sarter, N. (1994). *Behind human error: Cognitive system, computers and hindsight.* Wright-Patterson AFB, OH: Crew Systems Ergonomics Information Analysis Center.

Wright, D., Mackenzie, S. J., Buchan, I., Cairns, C. S., & Price, L. E. (1991, September 14). Critical incidents in the intensive therapy unit. *Lancet, 338,* 676–678.

14 Fatigue, Performance, and Medical Error

Gerald P. Krueger
U.S. Army Research Institute of
Environmental Medicine

The provision of quality health care requires medical personnel to give careful attention to important life-sustaining details, such as monitoring critical vital signs of patients in intensive care, administering correct levels of anesthetic gases to prepare a patient for surgery, or dispensing proper levels of prescribed medications. Like other workers, health-care providers are affected by many physiological, psychological, and behavioral variables associated with their jobs and their particular lifestyles. For example, those who work much longer than a normal 8-hour workday, and those who have frequent shiftwork duty cycle changes often experience fatigue on the job, which can lead to lapses in attention or performance, and contribute to medical error and health-care mishaps.

The focus of this chapter is on effects of shiftwork, disruption of bodily circadian rhythms, sleep loss, fatigue, and performance expectations of health-care providers as they pertain to sustained vigilant monitoring of patients, the likelihood of fatigue-related error, and other safety aspects of providing institutional health-care services.

BACKGROUND

According to a recent U.S. Congress Office of Technology Assessment report (Liskowsky, 1991) approximately 20 million U.S. workers are on some form of nonstandard work schedule (i.e., different from standard 8 a.m. to 5 p.m. jobs). The OTA report states that nonstandard work schedules (e.g., night shiftwork—11 p.m. to 7 a.m.), when most people would normally be sleeping, disrupt workers' circadian biological rhythms. Extending working hours longer than the conventionally expected 8 hours per day on a frequent basis and the accompanying sleep disruption usually associated with changing shiftwork schedules can adversely affect workers' health, safety, and performance.

In this country, many institutional health-care providers work through the night and on weekends and holidays. This pattern raises concerns about worker fatigue, the possibility of medical errors, and their impact on the quality of health care. This is particularly true for medical residents or interns in training who typically work extensively long hours, and often through the night. Traditionally, medical residents are expected to watch over patients nearly continuously for days and nights to observe the progressive sequelae of disease. Residents regularly participate in long, intensive work hours over a 1- to 2-year period, sometimes working more than 130 hours per week in shifts of 12 to 60 hours in duration. Such health-care providers obtain only limited, frequently interrupted, nonrestful sleep. They have recurrent problems with circadian bodily rhythm disruption and are not always maximally alert when administering life-sustaining health care. Those individuals are at risk for making mistakes and becoming involved in costly medical errors.

It is easy to conceptualize the difficulties of hospital workers who work through the night. They include loss of sleep, difficulties performing certain tasks in the middle of the night, disrupted social life, and so on. Similar problems can be apparent for health-care providers who work extensively long duty shifts, that is, longer than a normal 8-hour duty day. Surgical teams of nurses, anesthesiologists, surgeons, and other specialists frequently conduct arduously long medical procedures lasting 20 or more hours. Teams often schedule back-to-back surgeries that result in very long work periods. Due to shortages of critical professional personnel, for scheduling convenience, or simply to amass overtime hours with incentive pay, some health-care workers often work two 8-hour shifts in succession. This is common among nurses, but young salaried physicians also work extra shifts in emergency rooms to earn additional money.

Provision of institutional health care is a 24-hour-a-day business. Hospitals and nursing centers must be staffed around the clock with professional medical care providers to provide continuous health care. This necessitates the use of standard shiftwork and some nonstandard work schedules. When evaluating issues of medical error, hospital management and personnel supervisors should carefully consider the safety and quality of care implications associated with predictable performance decrements attributable to worker fatigue. Among such decrements are slowed reaction time, lapses of attention to critical details, errors of omission, compromised problem-solving initiative, reduced motivation, and decreased vigor for successful completion of important medical care tasks.

Management understanding of shiftwork, circadian rhythms, sustained performance requirements, and fatigue effects on workers is essential for designing work schedules with the least negative impact on health-care providers and their performance (Liskowsky, 1991).

WORK SCHEDULE IMPLICATIONS

In addressing the likelihood of fatigue-related error in medical institutions such as hospitals or nursing homes, three work-scheduling concepts pertain: Shiftwork scheduling, continuous operations, and sustained performance.

Shiftwork schedules generally have been configured by societal convention into three 8-hour work shifts per day to emulate factory production schedules (Colquhoun & Rutenfranz, 1980). In hospital settings, these shift duty times generally are (a) 6:45 a.m. to 3:15 p.m., (b) 2:45 p.m. to 11:15 p.m., and (c) 10:45 p.m. to 7:15 a.m. These schedules allow for 8 hours of work, a half-hour lunch break, and some overlap among workers at the hand-off time for shift changeover transitions. Any continued care situation could be configured into numerous different work-shift combinations (Tepas & Monk, 1987). Many large hospitals in the United States are presently converting to 12-hour work shifts. Hospital care facilities operate on so many different shiftwork systems, and within a hospital, many workers adhere to such different assigned work schedules, that their description would be too lengthy here.

Use of three duty shifts per day permits *continuous operations (CONOPS),* which involve scheduling rotating teams of workers to provide uninterrupted service or performance around the clock for days, weeks, or even months in succession (Englund & Krueger, 1985; Krueger & Englund, 1985). Such team scheduling occurs in never-close manufacturing production factories; in providing 24-hour police, fire, and communication coverage; and also in providing hospital or nursing home in-patient health care 24 hours per day. From an institutional or organizational perspective, because work never stops, continuous operations are the epitome of human productivity. Teams of workers replace one another, rotating into the job on regular shift-change schedules. At the end of a work shift, individual workers physically leave the workplace, and presumably go home where they have opportunities for rest and sleep. However, in some on-call hospital schedules, workers must sleep at the hospital itself. For example, interns and residents on duty through the night usually sleep down the hall from the actual workplace. Sleep in that situation is often brief, fragmented or scattered, and not very restful or restorative of performance alertness.

Individuals who perform nonstop without relief on shifts of 12 or more hours participate in *sustained operations (SUSOPS;* Krueger, 1989). They usually get little opportunity for rest and no opportunity for sleep. Staffing busy emergency rooms, intensive care units, lengthy intern work schedules, overtime nursing ward shifts, exceptionally lengthy surgeries, or back-to-back surgical procedures constitute SUSOPS for medical care workers.

WORK SCHEDULE EFFECTS ON MEDICAL CARE SPECIALISTS

Fatigue, sleep loss, patient surveillance, variation in worker performance, and disrupted circadian rhythms can all contribute to the likelihood of error in medical care settings. The following definitions should aid in understanding the effects of nonstandard work schedules on medical personnel.

Fatigue

Everyone has experienced varying degrees of fatigue and probably remembers how it affected performance of important tasks, such as driving a car late at night. Fatigue, which results from continued physical or mental activity, is characterized by diminished ability to do work and is accompanied by subjective feelings of tiredness. Fatigue also may result from inadequate rest, sleep loss, or from displaced or disrupted biological rhythms.

With *physical fatigue,* the sensory receptors or bodily muscles exhibit a temporary loss of power to respond. Physical fatigue usually is induced by continued physical stimulation or an extensive work period and is accompanied by feelings of muscular tiredness and decreased physical performance. Nurses occasionally experience muscular fatigue, strain, or back discomfort while bending over in awkward body postures to work on countless patients in hospital beds. Muscular fatigue also may occur when surgical attendants stand for extended periods during lengthy surgical procedures.

Physical fatigue is not a simple construct, because it can also have cognitive components, such as when people decide they are tired or fatigued before their muscles actually are. For example, a person may request a rest break from hard physical chores, but then immediately voluntarily play a vigorous volleyball game at lunchtime.

General or mental fatigue is characterized by subjective feelings of weariness after hours of repeated performance of predominately nonphysical tasks. Such fatigue may be a result of reduced afferent nerve impulses or reduced feedback from the cortex to the reticular activating system in the brain stem (Grandjean, 1968). A lack of novel stimuli may also bring feelings of monotony and boredom. If an individual is deprived of sleep prior to beginning a set of tasks, the tiredness or drowsiness attributable to loss of sleep accentuates feelings of general fatigue. Reaction time and performance decline, mood and motivation drop, and initiative and enthusiasm decrease.

Acute mental fatigue, resulting from excessive cognitive work, is generally short lived, as it is readily relieved by adequate rest or sleep. However, chronic mental fatigue, from overwork carried out for successive weeks or months or from cumulative stresses, is not usually alleviated by rest alone. Like job "burn-out" it requires something akin to a vacation or a change of jobs to restore full mental effectiveness.

Sleep and Sleep Loss

Individuals differ in terms of how much sleep they require to maintain health, well-being, peak mental alertness, and performance effectiveness (Monk, 1991; Scott, 1990; Webb, 1982). To permit reasonable on-the-job alertness, most adults require 6 to 8 hours of sleep per 24-hour period. Many teenagers require 8 to 10 hours. There are, of course, many individual differences in our responses to moderate amounts of sleep deprivation.

Adults generally do not function at their peak levels of mental performance with fewer than 5 hours of sleep per 24-hour period, even if it is only for 2 to 3 days in duration. However, ample sleep research data indicate that with 4 hours of sleep per night, many adults can function reasonably well, but certainly below peak levels for bursts of activity lasting 2 to 3 days (Haslam, 1985a, 1985b). Observable decrements in performance are marked in terms of slowed responses and decreased initiative. Workers with fewer than 4 hours of sleep per day generally exhibit dramatically compromised mental performance overall.

When 2 to 3 hours of needed sleep are missed each 24-hour day, an increasing accumulation of sleep loss occurs, establishing for the body a chronic *sleep debt*. That sleep debt gradually and insidiously develops to the point that such workers seldom are maximally alert. If sleep debt continues to accumulate over a number of days (e.g., 5 to 10 days), at some point general mental performance will be observably degraded. This decrement in performance is similar to that which occurs after deprivation of an entire night or two of sleep.

Physical performance, such as digging, lifting, or using large tools, does not degrade noticeably with sleep loss, in part because such physical activity is the antithesis of sleeping. However, when an individual is sleep deprived and muscle fatigue sets in after extensive muscular work, recovery of those muscle tissues takes longer for an otherwise well-rested person who has merely become muscularly tired through extensive physical activity.

On the other hand, after significant sleep loss, mental performance degrades quickly. Sleep loss leads to decrements in cognitive job performance in three ways. A steady reduced central nervous system arousal level reduces the overall speed of performance. Continued sleep loss leads to occasional brief mental lapses or microsleeps (lasting from 1 to 10 seconds). These lapses result in slowed response and sometimes errors of omission. Sleep loss also lowers mood and motivation and reduces morale and initiative. When combined with the general slower responses, this results in less work accomplished and often exhibits sloppy attention to details. Such decrements in cognitive performance can lead to mistakes, accidents, or medical care mishaps.

Missing even one night of sleep has marked negative effects on an individual's performance (Cox & Krueger, 1989; Webb, 1982). After significant sleep loss, over 2 or more days in succession, sustained cognitive performance declines noticeably. For example, after missing one night's sleep and working continuously 18 hours on the job, cognitive performance can degrade 20% to 25%. By 24 hours, it may drop to 70% of baseline, remain stable for about 18 more hours, and then drop again to about 40% effectiveness (Angus & Heslegrave, 1985). Behavioral studies using standardized cognitive tests, including those embedded in job tasks, demonstrate that several successive nights of sleep loss negatively affect decision-making ability. Initiative, information integration, planning, and plan execution generally degrade most quickly. Unfortunately, this usually occurs without the individual knowing it.

When highly trained equipment operators who place a premium on task accuracy are overly tired, they usually make trade-offs in work strategy to maintain accuracy. Fatigued operators usually take longer to accomplish tasks and consequently accomplish less work per unit time than they do when they are rested. In settings where speed of response is not so critical, sleep-deprived workers may still accomplish useful work, just less of it. In medical care settings, sleep loss is potentially detrimental to those who carry out surgical procedures or perform patient vital signs surveillance (e.g., surgical anesthetists) and for those involved in providing continuing patient care and monitoring, especially in instances where quick reaction time is important.

Accumulations of sleep debt are a particular problem for workers who change the time of their work shifts every few weeks. Sustained workloads (8 to 12 hours or longer on the job) or irregular work schedules (shiftwork changes) lead to desynchronized circadian rhythms and disrupted sleep patterns, and compromise restful sleep (Scott, 1990; Tepas & Monk, 1987).

It is such a prevalent problem in our society that it prompts questions about whether or not America is chronically sleep deprived.

Vigilance

Monitoring (patient surveillance or alert watchfulness) of predominately nonchanging or slowly changing stimuli (e.g., displays of patient vital signs information) is particularly sensitive to the influence of sustained work sessions, especially when combined with sleep loss. In carrying out a lengthy vigilant surveillance task, one can anticipate performance decrements manifested as decreases in reaction time and a drop in visual or auditory alarm detection probability (Mackie, 1976). This vigilance decrement usually occurs as early as 20 to 30 minutes into the task, especially if the worker is tired. Even if the person was well rested at the beginning, a few hours into an intense monitoring vigil, decrements in performance will be noticeable.

Sleep loss and fatigue present additional performance considerations for the accomplishment of any sustained vigilance task. Sleep-deprived individuals exhibit intermittent lapses in attention: periods of reduced arousal, akin to 1- to 10-second microsleeps, resulting in slower response times and occasional errors of omission, or very long reaction times (Williams, Lubin, & Goodnow, 1959). Such lapses are generally intermittent and therefore hard to predict. Performance between lapses is often at an acceptable but generally reduced level. With continued time on the task and especially with increasing sleep debt, microsleeps increase in frequency and duration, making omission errors likely.

The relevance of vigilance to medical care is noted by the World Congress of Anesthesiologists, whose logo is an artist's rendition of a globe of the earth emblazoned with a large banner containing a single word: "vigilance." During lengthy surgeries, vigilance decrements can be a problem for anesthesiologists or nurse anesthetists who must continuously monitor various instruments, scopes, displays of gaseous mixtures, or patient vital signs, and administer medications. If workers have been at the job for a long time, or if they have recently missed significant amounts of sleep, vigilance decrements or microsleeps could lead to slowed recognition of changes in surgical patient status or to missing important signals leading to a failure to perform a required sequence of treatment events. Thus, microsleeps or lapses in performance could contribute to the occurrence of medical error.

Knowledge of vigilance decrements prompts job designers to insist on rest pauses, breaks, or rotations in job task duties at least every 4 hours for intense surveillance and monitoring work. Naval ship command, control and communication centers, and air traffic control centers all use such safety practices. Such knowledge of vigilance behavior should be accounted for in the design of hospital and other health-care job tasks.

Circadian Rhythms and Variation in Worker Performance

People exhibit daily cyclical rhythmic variations in numerous biological, physiological, and behavioral functions. These variations, repeated every 24 hours, are called *circadian rhythms*. Circadian rhythms reflect our synchronization of consistent internal organization of bodily functions with such environmental events as sunrise, sunset, regularly scheduled meal times, and sleep periods (Comperatore & Krueger, 1990; Monk, 1991).

On the job, workers exhibit predictable daily rhythmic variations in performance. These cyclical variations in efficiency and alertness are attributable in part to the body's circadian rhythms. For those people who usually work by day and sleep at night, two predictable periods of lower performance are expected daily, and these performance decrements coincide with noticeable daily changes in body temperature.

Performance generally will be somewhat less effective from 2:30 p.m. to 5:00 p.m. when the body's daily temperature rise tends to level off in mid-afternoon. If people work extensively long schedules through the night, their performance will be noticeably lower in the circadian rhythm nadir from 1:00 a.m. to 5:00 a.m., when body temperature is at its lowest ebb of the 24-hour day. A drop in mood and motivation parallels these rhythmic performance declines, as do increased reports of fatigue, sleepiness, and decreased alertness (Coleman, 1986).

Many daytime workers exhibit periods of particularly good performance when body temperature is rising from about 7:30 a.m. to 1:00 p.m. and again from 6:30 p.m. to 10:00 p.m. However, there are wide individual differences in when people perform at their best. The practical implications of people claiming to be *larks* who work best in the morning, or *owls* who claim to work best in the late evening are debated among behavioral scientists. Nonetheless, numerous people claim that these terms describe their experience and their performance. Credence should be given to scheduling people to work at the time of the day they claim they work best.

Shiftwork Implications

Circadian rhythms are highly resistant to alteration. Many bodily rhythms are linked to light–dark cycles accompanying sunrise and sunset and to the time of day meals are eaten, or when sleep is initiated. Shiftworkers who change work schedules from a day to night shift must undergo sleep and circadian rhythm adjustments. Some individuals require up to 2 weeks on a new work schedule before appreciable adjustments in bodily rhythms are made to match the new cycle (Comperatore & Krueger, 1990; Rosa & Tepas, 1990; Scott, 1990). Shiftworkers who settle into permanent night work schedules as a way of life exhibit circadian rhythms whose expected peaks and valleys of body temperature, mood, and performance are slightly displaced or phase-shifted by a few hours from those of the daytime worker.

It is difficult to adapt biologically and psychologically to frequent shift changes. Shiftworkers often work 5 days of scheduled 8-hour night shifts, or evening shifts. This is followed by 2 days off duty (e.g., weekends), during which they attempt to match the expectations of society's clock by interacting with family and friends and living according to a different schedule on those 2 days. When they return to work on Monday, they bring about another biological schedule shift. This action consequently affects their sleep schedule about every 7 days and exacerbates their circadian rhythm desynchronosis.

Shiftworker maladies and complaints in hospitals often include sleep disturbances, shift-cycle fatigue, job fatigue, impaired job performance, and greater likelihood of injuries and accidents (Scott, 1990). There also are numerous reports of anxiety, irritability, nervousness, gastrointestinal and digestive disorders, nervousness, psychological and musculoskeletal symptoms, menstrual dysfunction, family schedule disturbances, child care needs, and disrupted social life (Liskowksy, 1991; Tepas & Monk, 1987).

In our society, it is common for shiftwork changeovers to be scheduled in a manner that adds to workers' circadian rhythm desynchronosis. For example, many workers are scheduled to work night shifts (midnight to 8 a.m.) for a period of time, followed by the evening shift (4 p.m. to midnight), and in sequence followed then by working the day shift (8 a.m. to 4 p.m.). Such a counterclockwise (backward) directional change, in which the employee reports to work 8 hours earlier on the first day of each subsequent shift schedule change, is less suited to the body's natural rhythms. That is, because the report-to-work time is earlier, the workers' bodily adjustments advance slowly and tend to "lag behind" their new clock schedule. In a resultant *shift-lag* phenomenon, the worker usually experiences feelings of general malaise, fatigue, and desynchronization of various bodily functions, much the same as jet lag affects travelers flying across time zones.

To ease circadian rhythm adjustments to shiftwork schedule changes, it is preferable that subsequent start-to-work times be changed in the *forward direction* of the clock (Comperatore & Krueger, 1990). That is, after working a week at the day shift, the next shift change should be to evenings, and after that convert to the night shift. In this arrangement, at each subsequent shift change, workers start 8 hours later in the day (clockwise, forward direction). In this situation, the bodily circadian rhythms need to adjust for an 8-hour delay from the week's previous schedule, a much easier adjustment for the body to make starting with staying awake longer to meet the new schedule on shift changeover day.

In an analogous way, our practical experiences with jet lag are that it is generally worse traveling east, whereby the sun is rising earlier and our body adjustments tend to lag behind the advanced time zone, than it is for traveling west for which the first night's solution is merely to remain awake a little longer (phase delay) until normal bedtime for the new time zone.

The backward directional change (i.e., days to nights to evenings), requiring the less easily accomplished circadian phase advance, is often encountered in manufacturing industries and police departments. It prompts worker complaints because of the difficulties in making bodily adjustments to the new clock changes, especially making adjustments to new sleep schedules, and it often results in circadian desynchronosis. In designing hospital shiftwork schedules, it is advisable to learn these known principles of circadian rhythm physiology and account for them in the workplace.

REST PAUSES AND SLEEP MANAGEMENT

Prolonged working without pauses or rest diminishes performance. Effects of fatigue are insidious. Poor concentration and inattention can lead to human error and accidents. Adequate pauses on the job, rest breaks, and off-the-job rest and sleep help maintain acceptable alertness in those who provide medical care.

Many laboratory and field studies are in consensus about the positive influence of scheduled pauses or rest breaks on the job (Alluisi & Morgan, 1982). Industrial psychologists do not agree upon optimum rest-pause break schedules for various kinds of job tasks. For jobs in which workers are able to control their own rate of productivity, occasional rest pauses on the job tend to result in enhanced productivity. Where rates of work and productivity are under less direct control of the employee, for example, machine-paced work or when the patient's status requires constant care or monitoring, work pauses should be carefully planned or agreed on with work colleagues. Rest pauses should be considered in designing important hospital medical care tasks, as they may be particularly helpful for individual workers in lengthy surgery, postoperative care, the intensive care unit, and other attention-demanding medical tasks.

Sleep management plans including sleep discipline are needed for continuous performance in hospital care settings. Such considerations are particularly relevant to health-care workers who remain at the hospital for extended periods of time. They are especially important for those who sleep down the hall. The idea of educating employees on sleep discipline may not at first seem relevant for employees who go home from work to obtain their rest. However, hospital management should educate all employees on the topics of circadian rhythms, sleep loss effects, fatigue, and resultant performance implications in the same way that safety program considerations are generally given high-level attention. Such education and adherence to a sleep discipline policy are important to prevention of medical error.

People cannot store up sleep, but generally they can prerest, or sleep before anticipated periods of sustained work. This activity staves off inevitable fatigue and loss of alertness as work continues without rest. Sleep debt can be made up only by sleep. Generally, intermittent, broken sleep is not considered to be as restorative as a long continuous period of sleep. Sleeping when the body's core temperature is low, during the circadian nadir, 1:00 a.m. to 5:00 a.m., and during the mid-afternoon, 2:30 p.m. to 5:00 p.m., is believed to be the best time to obtain restorative sleep. During those periods, a person is not acting against the body's temperature-warming tendencies. The warming period is a sign that the body expects to be awake (Naitoh & Angus, 1989).

A preferred sleep schedule can be developed based on circadian rhythm principles. All workers should strive for 6 to 8 hours of restful sleep every 24 hours. Such sleep should be obtained in a single contiguous block of time. It is best to sleep from midnight to 6:00 a.m. because the body's temperature is normally falling at that time, which presumably is when the body prefers to sleep. A good second choice is to sleep from 1:00 p.m. to 6:00 p.m. because the daily body temperature rise generally has leveled off at this time. It is interesting to note that this time coincides with the tropical siesta period. Sleeping in the evening from 6:00 p.m. to midnight is less preferred because the body's temperature is again rising in early evening. It is difficult to awaken at midnight and to be expected to be at peak alertness through the night. The least preferred time to sleep is from 6:00 a.m. to noon, because the body's temperature is rising during this time, the sun is shining, and there is activity everywhere.

If long uninterrupted sleep periods are not feasible, judicious use of naps is recommended. It can be argued that the times when the body's temperature is dropping (2:30 p.m. to 5:00 p.m. and 1:00 a.m. to 5:00 a.m.) are the best times to take short naps. Proponents say that napping when the body temperature is rising, that is, 7:30 a.m. to noon, is not easily accomplished because the person is attempting to sleep when the body wants to be awake.

In sustained military operations, opportunities to take naps of 10 minutes to 1 hour are recommended when the military situation is safe to do so. To break the adage of "it's not macho for soldiers to take a nap," the U.S. Army now refers to "power naps." Troops are encouraged to take naps whenever they safely can. Judicious use of naps of anywhere from 10 to 15 minutes to several hours in duration should also be encouraged for hospital personnel engaged in extended work schedules.

Depending on the state of tiredness and how deeply one has fallen into sleep during a nap, awakening from a nap may result in *sleep inertia*. Sleep inertia is grogginess immediately on awakening. Thus, when awakening from a nap, performance decrements of slow and inappropriate responses may occur until a person is fully alert. Sleep inertia sometimes lasts 15 to 20 minutes. There is a tendency for sleep inertia to be more pronounced after one awakens from a nap during the circadian lull from 1:00 a.m. to 5:00 a.m.

If a person usually has problems with sleep inertia on awakening from naps, the person should be aroused from sleep in sufficient time to awaken gradually for those occasions when that individual is expected to perform well immediately. Expecting daytime physicians to serve on call through the night is stacking the deck against them and their patients. Inevitable performance decrements can be anticipated in the circadian lull, and the risks of sleep inertia are apparent. Resident physicians sleeping near the hospital emergency room while on call through the night need to be awakened early to respond to on-call work and to avoid error-prone sleep inertia.

In providing continuous hospital care, viable work–rest schedules should be used to enforce good sleep discipline among hospital staff, especially surgical and intensive care unit team members. Even after establishing sleep policies, military leaders and command and control personnel are notorious for not taking the time to obtain needed sleep (Krueger, 1991). Leaders are the ones who make most of the critical decisions, and they frequently do so in a sleep-deprived state. It is therefore most important for management to ensure that leaders and supervisors personally adhere to their own sleep discipline as well as being concerned about that of others. Hospital personnel such as surgical chiefs, anesthesiologists, and ICU personnel should pay heed to these sleep management cautions as well.

A system of developing and training *night-shift teams*—surgical or emergency room teams who work at night and obtain rest and sleep during the day—should be adopted in those hospital centers with continuous provision of emergency health care. Lessons can be learned from police, manufacturing industries, and the military to optimize performance (Krueger, 1991). Adopting semipermanent shiftworkers for around-the-clock surgical wards could be effective in reducing medical errors.

NUTRITION

Shiftworkers and SUSOPS workers are notoriously lacking in good nutritional habits. They skip meals, eat coin machine food, and do not maintain proper hydration. Junk food in general does not maintain proper nutrition. They also ingest too much caffeine and engage in other less desirable dietary practices. These affect fatigue levels and, ultimately, worker health and performance. Those who must sustain performance should select appropriate body nutrients for their diet. To the extent possible, medical care workers should eat regularly scheduled nutritious meals. Good nutrition will help them maintain fitness, health, and mental alertness on the job.

PHARMACOLOGICAL INTERVENTION

Sleeping pills (hypnotics) have been suggested for potential use in promoting sleep during off-duty hours. Important biomedical research questions need to be answered concerning the quality of sleep obtained with hypnotics. It needs to be determined whether users can awaken easily and respond normally in an emergency, if there are hangover effects, how that impacts performance, and what the risks are of repeated use or abuse of hypnotics. Recent experimentation with oral administration of melatonin (a hormone from the pineal gland) demonstrates great promise to provide a nondrug sleep enhancer for daytime sleep (Comperatore & Lieberman, personal communication, 1993).

A search continues for stimulants to maintain alertness to meet extended performance demands in select circumstances. Concerns with most pharmacological enhancements include not well-understood effects on body biochemistry and physiology and on performance. There also is concern regarding perceptual distortions, safety, rebound effects, and likely addictions.

A recent survey of research on the use of stimulants to ameliorate the effects of sleep loss and their impact on performance (Babkoff & Krueger, 1992) indicates that significant progress has been made identifying characteristics of suitable stimulant compounds, but much more research and development needs to be done. Until there is a proven hypnotic or stimulant, it is best not even to consider the use of either class of pharmaceuticals in health-care settings.

CONCLUSION

Medical management and personnel supervisors should consider the implications of the foregoing treatise for safety and quality health care in their institutions. The following summary tips for coping with shiftwork, extensively long working hours, and the inherent risks of inducing costly medical errors may be of help:

1. Allocate proper staffing for around-the-clock operations.
2. Cross-train several workers to be able to accomplish the same tasks so as to be able to relieve one another to permit periodic rest breaks.
3. Train staff to perform select tasks so well (overlearning) that they are less likely to be subject to fatigue effects.

4. Honor known bodily circadian rhythm principles when designing shiftwork schedules.

5. Schedule rest breaks in long work sessions.

6. Encourage employees to eat nutritious meals.

7. Stress the importance of rest and sleep for an alert medical care staff.

As a medical specialty group, anesthesiologists are the *pace-setters* in addressing human error in medicine. In the past 5 years, anesthesiologists have regularly exchanged much formal information on this topic. Through various symposiums, workshops, applied research, training using simulators, and even through hospital management action groups, anesthesiologists are raising the medical profession's consciousness to many concerns for safety and excellence in the provision of superb medical care.

Anesthesiologists exhibiting concern and attention to performance-related issues have been analyzing how their working conditions impact patient care and safety (Howard, Gaba, Fish, Yang, & Sarnquist, 1992). They are working to improve the human engineering and ergonomic design of work stations, patient monitoring, and alarm systems so that they present less workload and make workers less prone to accidents or mishaps (Loeb, Weinger, & Englund, 1991). Anesthesiologists are working to control their shiftwork schedules and are decreasing scheduling extensive back-to-back surgeries to avoid fatigue-likely situations. They have been developing and implementing numerous ways to improve their work environment to reduce the likelihood of error.

It is recommended that *hospital management* combine the forces of responsible-minded surgeons, physicians, anesthesiologists, nurses, and other personnel to systematically address causes of medical error and address worker-related issues such as those described in this chapter.

ACKNOWLEDGMENT

This chapter is based upon Gerald P. Krueger's series of lectures presented on this topic at:

1. Anaesthesia Patient Safety Symposium, 9th World Congress of Anesthesiologists, Washington, DC, May 1988.

2. Conference on Human Error in Anesthesia, for the Anesthesia Patient Safety Foundation and U.S. Food and Drug Administration, Pacific Grove, CA, February 1991.

3. Human Error in Medicine Symposium at American Association for

the Advancement of Science (AAAS*92) Annual Meeting, Chicago, IL, February 1992.

4. Human Factors and Biomedical Devices Meeting at the Department of Defense Human Factors Engineering Technical Group (DoD HFE TG) 29th Meeting, New Orleans, LA, May 1992.

5. Human Performance and Anesthesia Technology Meeting, Society for Technology in Anesthesia, New Orleans, LA, February 1993.

REFERENCES

Alluisi, E. A., & Morgan, B. B. (1982). Temporal factors in human performance and productivity. In E. A. Alluisi, & E. A. Fleishman (Eds.), *Human performance and productivity, Vol. 3: Stress and performance effectiveness* (pp. 165–247). Hillsdale, NJ: Lawrence Erlbaum Associates.

Angus, R. G., & Heslegrave, R. J. (1985). Effects of sleep loss on sustained cognitive performance during a command and control simulation. *Behavior Research Methods, Instruments & Computers, 17*(1) 55–67.

Babkoff, H., & Krueger, G. P. (1992). Use of stimulants to ameliorate the effects of sleep loss during sustained performance. *Military Psychology, 4*, 191–205.

Coleman, R. M. (1986). *Wide awake at 3:00 A.M.: By choice or by chance?* New York: Freeman.

Colquhoun, W. P., & Rutenfranz, J. (Eds.). (1980). *Studies of shiftwork*. London: Taylor & Francis.

Comperatore, C. A., & Krueger, G. P. (1990). Circadian rhythm desynchronosis, jet lag, shift lag, and coping strategies. In A. J. Scott (Ed.), *Occupational medicine: Shiftwork. State of the art reviews* (pp. 323–341). Philadelphia, PA: Hanley & Belfus.

Cox, T., & Krueger, G. P. (Eds.). (1989). Stress and sustained performance [Special issue]. *Work & Stress, 3*(1), 1–102.

Englund, C. E., & Krueger, G. P. (1985). Introduction to 1st special section: Methodological approaches to the study of sustained work/sustained operations. *Behavior Research Methods, Instruments, & Computers, 17*(1), 3–5.

Grandjean, E. P. (1968). Fatigue: Its physiological and psychological significance. *Ergonomics, 11*, 427–436.

Haslam, D. R. (1985a). Sleep deprivation and naps. *Behavior Research Methods, Instruments, and Computers, 17*, 46–54.

Haslam, D. R. (1985b). Sustained operations and military performance. *Behavior Research Methods, Instruments, and Computers, 17*, 90–95.

Howard, S. K., Gaba, D. M., Fish, K. J., Yang, G., & Sarnquist, F. H. (1992). Anesthesia crisis resource management training: Teaching anesthesiologists to handle critical incidents. *Aviation, Space, and Environmental Medicine, 63*(9), 763–770.

Krueger, G. P. (1989). Sustained work, fatigue, sleep loss and performance: A review of the issues. *Work & Stress, 3*, 129–141.

Krueger, G. P. (1991). Sustained military performance in continuous operations: Combatant fatigue, rest and sleep needs. In R. Gal & A. D. Mangelsdorff (Eds.), *Handbook of military psychology* (pp. 255–277). Chichester, England: Wiley.

Krueger, G. P., & Englund, C. E. (1985). Introduction to 2nd special section: Methodological approaches to the study of sustained work/sustained operations. *Behavior Research Methods, Instruments, & Computers, 17*(6), 587–591.

Liskowsky, D. R. (Ed.). (1991). *Biological rhythms: Implications for the worker* (U.S. Congress, Office of Technology Assessment, Rep. No. OTA-BA-463). Washington, DC: U.S. Government Printing Office.

Loeb, R. G., Weinger, M. B., & Englund, C. E. (1991). Ergonomics of the anesthesia workspace. In J. Ehrenwerth & J. B. Eisenkraft (Eds.), *Anesthesia equipment: Principles and applications* (pp. 385–404). St. Louis, MO: Mosby.

Mackie, R. R. (Ed.). (1976). *Vigilance: Theory, operational performance, and physiological correlates*. New York: Plenum.

Monk, T. H. (Ed.). (1991). *Sleep, sleepiness, and performance*. Chichester, England: Wiley.

Naitoh, P., & Angus, R. G. (1989). Napping and human function during prolonged work. In D. F. Dinges & R. J. Broughton (Eds.), *Sleep and alertness: Chronobiological, behavioral and medical aspects of napping* (pp. 221–246). New York: Raven.

Rosa, R., & Tepas, D. I. (Eds.). (1990). Factors for promoting adjustment to night and shift work [Special Issue]. *Work & Stress, 4*(3).

Scott, A. J. (Ed.). (1990). *Occupational medicine: Shiftwork. State of the art reviews*. Philadelphia, PA: Hanley & Belfus.

Tepas, D. I., & Monk, T. H. (1987). Work schedules. In G. Salvendy (Ed.), *Handbook of human factors* (pp. 819–843). New York: Wiley.

Webb, W. B. (Ed.). (1982). *Biological rhythms, sleep, and performance*. Chichester, England: Wiley.

Williams, H. L., Lubin, A., & Goodnow, J. J. (1959). Impaired performance with acute sleep loss. *Psychological Monographs No. 484, 73*(14), 1–26.

15 Errors in the Use of Medical Equipment

William A. Hyman
Texas A&M University

Problems with the use of medical equipment or devices can be broadly divided into two types. One type is when the equipment malfunctions as a result of a technical problem that is not caused by the user. This type of malfunction may be a result of an inherent defect in the design or manufacturing of the device (e.g., software error or inadequate mechanical assembly), or it may be a result of a random failure of a component. It may occur without warning to the user, and it can be relatively innocuous or can result in direct harm to the patient.

The second type of problem is when the user causes or initiates a malfunction. This type of malfunction can be one in which a technical failure occurs secondarily to the user's actions, or it may be one in which the medical objective of the device is compromised although the condition of the equipment remains unchanged. This can present a classic human-factors question in the evaluation of user error. Is the design adequate because it does allow proper use of the equipment, or is it inadequate in that it creates situations in which a user error is predictable?

The types of medical equipment that are associated with user problems range from the relatively simple, such as catheters and syringes, to the most technologically complex, such as computer-controlled diagnostic equipment. Inclusively, they are covered under the definition of a medical device that is used by the U.S. Food and Drug Administration (FDA). Under this definition, a medical device is:

> an instrument, apparatus, implement, machine, contrivance, implant, in vitro reagent, or other similar or related article, including any component, part, or

accessory, which is intended for use in the diagnosis of disease or other conditions, or in the cure, mitigation, treatment, or prevention of disease, or is intended to effect the structure or any function of the body, and which does not achieve its primary function through chemical action or by being metabolized.

This definition is found in section 201 of the Food, Drug and Cosmetic Act (as amended, 1992).

User problems in the use of medical equipment are of concern because of their effect on the quality and cost of health care. Preventing user error is a design, user-training, and risk-management issue. In this regard, users of interest include physicians, nurses, other health-care providers, support personnel, and patients, and other lay users. User error is also an issue for the FDA in its role in regulating the marketing of medical devices. In the aftermath of patient injuries, it is also an issue for the courts as attempts are made to assign blame for the injury. From any of these perspectives, a prerequisite to understanding causation and prevention is an understanding of the nature of user error and design principles for error control.

TECHNICAL FAILURE VERSUS USER ERROR

The distinction between technical failure of medical equipment and user-initiated failure or malfunction of the equipment is an important one with respect to prevention of related patient injuries. User-initiated malfunctions can be generally characterized as ones in which the user had the opportunity to operate the device without incident but, for any of a variety of reasons, failed to do so. The user may be a physician, nurse, therapist, or other direct patient-care professional. In addition, relevant users of the equipment can include installers, clinical engineers and maintenance personnel, house-keeping staff, and other hospital employees who work with or around the equipment.

The patient can also be a user of medical equipment in the hospital and in the home. In addition, family members and visitors can be both authorized and unauthorized users of medical equipment. It is anticipated that health-care cost pressures and the identification of new potential markets will increase the number of devices proposed for home use. As home use increases, design and training issues associated with the lay user will demand increasing attention.

In the broadest sense, the hospital itself is a user of medical equipment through its corporate responsibility to select and maintain the equipment, assure the training of hospital personnel, and establish appropriate policies and procedures concerning staffing and equipment-use issues. It has recently been suggested in this regard that the term *user* should include the entire spectrum of responsible individuals, and those with hands-on responsibility should be termed *operators* (Shepherd & Brown, 1992). Under this classification, operators become a subset of users. It is not yet known if this attempt to clarify terminology will find universal acceptance.

Operator-initiated problems have been identified as the most frequent cause of medical equipment incidents (Bogner, 1993). In anesthesia, for example, operator-caused equipment problems have been cited in the past as very significant, including a high percentage that occur during the set-up of the equipment (Cooper, Newbower, & Kitz, 1984; Cooper, Newbower, Long, & McPeek, 1978).

It is operator problems that are the focus of human factors, including attention to the user (e.g., training, policies and procedures, working conditions, etc.) and to the design of the equipment (e.g., types and locations of switches and displays, durability, operating requirements, etc.). This range of the human-factors perspective is of particular interest in that one of the principles of human-factors engineering, as it relates to the design of safe systems, is that reliance on the operator is the last choice of hazard-prevention methods. This choice is always to be preceded by elimination of the hazard, providing guards to protect against the hazard, or providing automated warnings of hazardous operating conditions.

Operator-initiated problems have often been characterized with the term *user error* or *operator error* with the implication that the problem was totally caused by, and can therefore be blamed on, the operator. In using this term, it must be remembered that the human-factors perspective often rejects this assessment of cause through the concept that bad design can cause users to make mistakes, that such mistakes may be predictable and preventable, and that therefore the result can be blamed on the design. In this perspective, bad designs have been considered "accidents waiting to happen" (Cooper & Newbower, 1975) or "traps" for the user (Peterson, 1984). Thus the term *user error,* although perhaps correctly identifying what occurred, should not be interpreted as necessarily attributing any or all of the blame to the user.

The importance of understanding the causes of human error in the use of medical devices is, therefore, to reduce patient and other injuries and the burdens on medical personnel that may impair their effectiveness. Risk management efforts are thereby enhanced and the waste of materials and money reduced.

IMPACT OF USER ERROR

It is fortunate that many equipment problems in the medical setting do not cause significant, if any, harm to the patient. In some of these cases, it may be that the malfunction is easily detected and corrected or the equipment replaced without incident. In other cases, the equipment may not have been doing anything particularly important at the time of the malfunction, or the malfunction may be compensated for by other equipment. Also, the event that the equipment was meant to deal with may not have occurred. Thus, a misset monitor alarm is not of direct consequence unless the event being monitored occurs and is missed as a result of the lack of alarm function.

The frequency of relatively harmless events is hard to determine because they generally go unnoticed or unreported. This does not mean that they are unimportant. An improperly functioning or otherwise misused device that is innocuous sometimes may cause injury at other times. Physically harmless device use problems can also have other consequences, such as increased cost as a disposable must be discarded or a diagnostic procedure repeated.

Whereas many instances of device malfunction may be without serious consequence, some equipment mishaps, whether caused by technical malfunction or the user, can be of catastrophic consequence to the patient. This can occur when the equipment or procedure is highly invasive, otherwise life threatening, or life supporting. There can also be serious consequences if an important diagnostic result is erroneous or misleading as a result of user error or malfunction. In addition, equipment problems associated with one patient can have adverse effects on other patients. For example, attention and effort can be diverted toward the equipment problem and away from the patients.

Finally, it should be noted that the patient is not the only one at risk with respect to user-related equipment problems. Others in the hospital such as staff and visitors can also be at risk from medical equipment. Risks to various hospital staff include the danger of accidental needle sticks. Improved collection containers have been marketed, and point of use signage is now common. However, according to a medical device industry newsletter, relatively little has been seen in the form of new products that will automatically protect the user ("Reducing needle stick injuries," 1993). Thus, the effort to date emphasizes improvement in the behavior of the user rather than the improvement in the equipment. However, the latter is always preferred from the human-factors and safety engineering perspectives because it provides reduced risk in a way that is independent of the training of the user.

USER ERROR: REGULATION AND STANDARDS

Medical devices sold in the United States are under the regulatory purview of the FDA. The current law is codified in the Food, Drug, and Cosmetic Act (as amended, 1992) and is the result of substantial amendments in 1976,

1990, and 1992. Applicable regulations are found in Title 21 of the CFR. The legal and regulatory requirements include the submission of designated materials to the FDA before a device can be marketed.

Depending on the type of device and its history of use, these materials can range from a relatively brief overview of the device's intended use, design, and instructional materials, to extensive documentation and data from clinical trials. There are then corresponding degrees of FDA scrutiny of the submitted materials prior to market release. The FDA also conducts a variety of postmarket activities including the maintenance of mandatory manufacturer and user reporting requirements for mishaps associated with medical devices. The FDA has also worked with industry and medical professionals on the development of training materials for the user community.

Although a complete review of applicable requirements is beyond the scope of the present chapter, it is noteworthy that FDA activity includes human-factors and user-error issues. One example of such training materials is the joint FDA, industry, and user development of the 1987 videotape "Human Factors in Hemodialysis." This tape focuses on mishaps attributed to user error. Another example is the promulgation of anesthesia set-up checklists in 1987 and 1992 that are intended to guide the user through the initial critical steps that can help prevent anesthesia incidents (see later discussion of these checklists). The FDA has also been involved in quality control guidelines for radiology facilities in an effort to reduce patient exposures resulting from retakes.

These efforts appear to focus primarily on improvement of the performance of the user rather than improvement of the design of the equipment. However, a 1990 FDA study of recalls (FDA, 1990) noted that many device design problems could have been prevented if adequate care had been exercised, including consideration of the user.

Other official FDA communications are issued to inform the user community about current information relevant to particular types of medical devices. These include Medical Device Safety Alerts, which contain information on risk of substantial harm from a medical device in commercial use. Postmarket information is also issued by manufacturers and may be in the form of Medical Device Safety Alerts, notifications, or recalls. Safety alerts are voluntary, whereas notifications are issued at the request of the FDA. A recall is action taken either to remove a product from the market or to conduct a field correction.

The FDA monitors manufacturer alerts and recalls. The FDA can also request or order a recall. The reasons for an alert, notification, or recall may include user-oriented problems. Safety alerts, notifications, and recalls are reported in the weekly FDA Enforcement Report.

The FDA has invited proposals from time to time for outside investigators to study user error and other issues associated with particular equipment. For example, under the 1993 Small Business Innovation Research (SBIR) program, the FDA listed the topic "Perform human factors analysis of design and operation of one or more medical devices, such as infusion pumps, defibrillators or endoscopes" (U.S. Small Business Administration, 1993).

In the past, the FDA has sponsored studies of particular equipment problems that may include human-factors issues (e.g., a 1983 study of accidental breathing system disconnects; 1988 State studies of defibrillator use and maintenance; 1990 State studies of the use of anesthesia checkout procedures; a 1990 study of human factors in blood glucose monitoring; and a 1992 human factors evaluation of contact lens labeling for wearers of soft contact lenses.)

Perhaps most specific to the user error issue are the current Medical Device Reporting (MDR) requirements (21CFR803), under which both user facilities (e.g., hospitals) and manufacturers must report certain types of clinical events involving medical equipment directly to the FDA. Reportable events are those in which there has been a death or serious injury or illness. Also reportable are malfunctions that, if repeated, would be likely to cause death or serious injury or illness. Of particular note here is that reportable events specifically include those caused by clinical and maintenance user errors.

Relevant coded responses from the November 1991 draft MDR reporting form (FDA, 1991) included a variety of matters pertaining to the user. Included under device issues are "design—human factors," "labeling—incomplete," "labeling—difficult to understand," "labeling—inadequate instructions for use," and "labeling—incorrect instructions for use." Additional relevant responses are provided under the use of device category, including "failure to service/maintain according to manufacturer recommendations," "failure to follow instructions," "incorrect technique/procedure," "misapplication of device," "reused nonreusable device," "reused device beyond label specifications," and "unapproved use of device." Similar user error categories are included in the June 1993 57-page interim coding manual for use with a new draft form designated 3500A (FDA, 1993c). User facilities will be required to use either this or a subsequent revision following the issuance of the final mandatory reporting regulations.

In addition to the mandatory reporting under MDR, the FDA has recently reorganized its voluntary reporting programs under the overall umbrella of MEDWatch (Kessler, 1993). This program, which uses form 3500 as opposed to 3500A, is intended to add to the information flow to FDA from the field on serious adverse events and product problems. A further aim of the MEDWatch program is to make reporting part of the culture of health-care providers. Although the program seeks general product problem information, human-factors issues have not been targeted under voluntary reporting as they have been under MDR.

The user error categories under MDR and the MEDWatch program raise the concern that the reports may be viewed as an admission of responsibility for an adverse event. The objective of the FDA reporting requirements is to establish a comprehensive record of actual use histories of medical devices in the clinical environment for regulatory purposes. This is in contrast to the FDA becoming directly involved in assigning responsibility for individual events. This objective was established in the 1990 U.S. Safe Medical Devices Act following the long-standing perception that, with the exception of a few efforts at concentrated study of particular devices, there was inadequate reporting of real user experiences.

In the absence of systematic study, most of the information on user errors has come from anecdotal reports, recalls and safety alerts, and civil litigation. The limitation in available information is based in part on the negative public relations and risk management incentives associated with revealing hospital-based equipment use errors and other problems. One interesting exception is a study on the causes of problems in the first year of CT arterial portography. This study found that 26% of the procedures were technical failures. These failures were caused by several factors, including catheter misplacements and equipment operation errors (Paulson et al., 1992). No patient injuries were reported in this study.

Reporting requirements similar to those of the FDA have been instituted by some states. New York has requirements that extend beyond medical device events to all unexpected adverse outcomes. Based on such reporting, New York has acted rigorously when the reporting system indicated widespread problems with specific devices. (See later discussion of laparoscopic surgery.) The Joint Commission on Accreditation of Healthcare Organizations (JCAHO) has also taken specific interest in user error, and such data are now required to be gathered and acted on for inhouse quality monitoring and improvement programs (JCAHO, 1993).

USER ERROR: LITIGATION

Medical mishaps may result in civil litigation in which patients or their families seek monetary compensation for their injuries and financial losses. A case of this kind can include physicians, nurses, other health-care providers, the hospital, and the manufacturers of medical devices as defendants. In general, it would be alleged that the medical personnel and hospital were negligent in the provision of care and that this negligence was the direct cause of the patient's injuries.

With respect to the medical device manufacturer, it would generally be alleged that the manufacturer was negligent in the design, manufacturing, or the adequacy of the instructions associated with the medical device. Alternatively, under the theory of strict liability, it can be alleged that the medical device was in a defective, unreasonably dangerous condition regardless of the care taken by the manufacturer in designing, producing, and marketing the product. On one hand, the various defendants may be allies in resisting the claims of the patient; on the other hand, they may seek to distribute or redistribute blame among themselves.

In a typical user error mishap, there will be no precipitating technical failure of the equipment and in fact, the equipment will be in normal operating condition after the mishap. In this type of case there are issues for the user, the manufacturer, and the hospital. For the user, the question is the degree to which they should be directly blamed for the error. With respect to the manufacturer, the question is the degree to which the manufacturer should be blamed for designing equipment that is prone to error or failing to provide adequate instructions and warnings. For the hospital, the question is the degree to which the hospital is responsible for selecting and controlling the use of equipment that may be prone to error.

These issues must always be considered in the context of the real environment of use, which may include fatigue, aggressive personalities, and the stress of medical emergencies. The resolution of these issues is dependent on the recognition of the role of user issues and device design in controlling errors in the use of equipment.

EXAMPLES OF MEDICAL EQUIPMENT USER ERROR

The human factors literature has identified a variety of common types of operator error that are applicable to the selection and use of medical equipment. These principles apply across the spectrum of user sophistication, training, and experience, from the medical professional to the lay user at home.

One type of user error is the use of equipment under inappropriate circumstances. This may include use contrary to its intended purpose or reliance on the equipment to achieve a purpose that is unsuitable with respect to the condition of the patient.

The use of monitoring equipment presents many opportunities for such misapplication. For example, consider the case of the use of a heart-rate monitor to supplement the direct observation of a pediatric patient with a tracheal tube who is under respiratory distress. Typical heart-rate monitors include user-adjusted high and low alarm limits. The person setting up the alarm may, in the absence of direct physician orders, have to select appropriate heart-rate limits. Depending on the design of the monitor, the limits that have been set may not be continuously displayed. If this is the case, other clinical personnel would not be aware of the monitoring setting unless they made an active effort to determine it.

The fact that the patient is being electronically monitored may lead users to the specific or subconscious feeling that if the patient's condition worsens, the alarm will alert them. However, if the patient's condition deteriorates as a result of secretions blocking the tracheal tube, the immediate physiological response will be loss of gas exchange. Shortly thereafter, the patient's heart rate will increase in response to the decrease in available oxygen. Depending on the high limit set by the user, this increase in heart rate may not be great enough to trigger the alarm. Subsequently, the patient's heart rate will decrease as the loss of oxygen affects the heart's ability to pump. Eventually, this will cause the heart-rate monitor to respond at the selected low heart-rate limit. By the time the low heart-rate limit is reached, the patient may have been without respiration for so long that the brain is irreversibly damaged.

There are two human-factors issues in this example. One is the decision to use heart rate as the primary monitored variable on a respiratory patient. The second is allowing reliance on the monitor to the point where it replaces continuously effective bedside nursing. In addition to the monitoring question, appropriate use of the tracheostomy tube is an issue with respect to reliance on it to maintain a patent airway under the patient's specific condition. In addition to individual staff decision making, this incident is also dependent on hospital policy with respect to monitor use and nurse staffing.

In an attempt to avoid inappropriate use of a monitoring device, one manufacturer warned users that their pulse oximeter was not to be used as an apnea monitor (a device to detect cessation of breathing; "Safety alert warns," 1992).

Two at-home monitoring devices that have been controversial with respect to user issues are those for infant apnea and blood glucose. Problems with infant apnea monitors have included set-up, the effect of false alarms, and inappropriate reliance. Capillary blood glucose measurements have been reported to be sensitive to user procedural errors such that health might be endangered if inaccurate data are used as the basis for regimen adjustment. This risk has prompted the development of patient education strategies along with continued interest in a measurement method that is not technique sensitive (Marrero, 1993).

Set-up and Adjustment

A second type of error associated with the use of medical equipment concerns its set-up and periodic adjustment, given that the equipment is appropriate for the patient. As noted earlier, many monitoring devices require the user to select the values of the monitored parameters that will cause the alarm to sound or display. In addition to setting the limits, it is common for the user to be able to select whether the alarm is in use at all. The user also selects the volume of the audible portion of the alarm. Volume adjustments may include a volume so low that the alarm becomes inaudible. Thus, there are two different ways to disable the alarm: turning it off, or turning the volume so low that it is in effect off. Alarms can also be made ineffectual by deliberately or inadvertently setting the limits outside of the physiological range of the patient. Given these possibilities, it is predictable that the occasion will occur in which an alarm is inadvertently disabled and then still relied on to call attention to a deteriorating patient condition.

One brand of respiratory flow monitor, which is an accessory to a ventilator, illustrates a sequence of design approaches to preventing the inadvertently turned off alarm. The first version of the device had a simple toggle switch to turn the alarm on and off. In this design, the status of the alarm could be determined only by relatively close inspection. Therefore, the design failed to prevent the condition of the alarm being off inadvertently. Note, however, that it was theoretically possible to operate the alarm in exactly the intended manner.

A second design was introduced in which the switch to turn off the alarm was made integral to a larger mechanical flag. This flag was displayed and labeled in red "Alarm Off" when the alarm was turned off. This design increased the visibility of the alarm off condition but continued to provide the option to turn the alarm off if desired. It can be argued that this design also continued to allow the alarm to be left off inadvertently because, although easier to see, the alarm off flag could still be lost in the visual noise of the bedside environment. The subsequent third design integrated the alarm being on with the respirator being on, thereby eliminating the choice to turn the alarm off manually .

In some devices, eliminating the possibility to turn an alarm off can produce other human-factors problems such as annoying noise after an alarm sounds but before the condition causing the alarm is resolved. Unnecessary alarms can also occur during nursing procedures that imitate the conditions triggering the alarm's function. For example, a respiratory flow alarm might sound while a patient was being suctioned. These types of events have been dealt with in designs that provide a preset timed alarm silence switch that allows for a period of quiet after which the alarm will sound again if the condition causing the alarm has not been resolved.

The working environment associated with multiple audible alarms and frequent false alarms must also be considered. It has been anecdotally reported that frequently sounding alarms will be ignored or result in deliberate modification of the offending noisy device by the clinical staff.

The proper set-up and preuse checkout of anesthesia equipment has received extensive attention, including the promulgation of FDA recommended checklists in 1987 and 1992 (FDA, 1987, 1992a). The 1992 list revised the 1987 recommendations to match the components of more modern anesthesia systems. The checklist approach, if carefully prepared and scrupulously followed, can eliminate many errors of omission and commission in preparing equipment for use. This is particularly valuable when the design of the equipment allows simple errors to go unnoticed. Preuse care must be extended to all equipment prepared for a case, including backup units. This is in anticipation of having to switch equipment and rapidly bring in the new unit while maintaining system safety.

In the home environment, peritoneal dialysis has seen recent attention with respect to redesigned equipment (Rifkin, 1993). The new equipment is much easier to set up and therefore less prone to error. In approaching this design, it was noted that the home user may have secondary problems such as poor vision. These problems can compromise their ability to use equipment properly. Part of the design goal was therefore to make a machine that someone could use "blindfolded wearing hockey gloves."

Meeting User Expectations

An additional issue in the adjustment and use of medical devices concerns the way in which the knobs and switches are arranged and operated and the relationship between these arrangements and common human experience and expectations. Most rotating adjustments are expected to increase the variable they effect when the knob is rotated clockwise, and decrease the value when the knob is rotated counterclockwise. Two notable exceptions to this expectation are gas cylinder regulators, which increase flow when turned counterclockwise, and some hot water faucets (which are supposed to be on the left), which, for symmetry, increase hot water flow when turned counterclockwise. Despite these exceptions, we generally expect rotating knobs to function as described.

Given the expectation of turning clockwise to increase, consider the design of a particular older anesthesia ventilator. It had five knobs on the front to control various parameters of the ventilatory function. Some of these knobs increased their respective parameter with clockwise rotation, and others increased their parameter through counterclockwise rotation. In addition, the five knobs were arrayed in a straight line, with no obvious relationship between those that worked clockwise and those that worked counterclockwise. Each knob was accompanied by an indication of the direction of increase marked on the case of the device. However, it can be argued that this design is conducive to error because the user at some time, regardless of training, will probably turn a knob in the wrong direction with possible adverse consequences to the patient. Many other examples are available of expectations about the way in which things should work and examples of the design of medical equipment that violates these expectations.

There is a general expectation that a high-limit adjustment knob would be above or to the right of a low-limit knob. Consider, then, the design of a high–low adjustment in which the knob is pushed in to adjust the low limit and pulled out to adjust the high limit. This is clearly technically feasible and can be done correctly, but it invites confusion. Similarly, there is a high–low alarm adjustment on the market in which the outer or back ring sets the low limit, and the inner or forward ring sets the high limit. This might be more potentially confusing than two correctly arranged separate knobs.

Such confusing medical designs are reminiscent of the automobile that had the horn button on the end of the directional signal arm. Although functional, it was repeatedly demonstrated that drivers could not find it when they needed it, even if they actually knew were it was. Moreover, the occasional driver of such a vehicle would have no idea where the horn button was. This design was abandoned.

The use of knobs and buttons is often accompanied by pictographs that are intended to communicate their function to the user. The meaning of these pictographs, especially in the absence of any effort to standardize or internationalize them, can render their interpretation as challenging as the use of the device itself. Modern technology has also provided soft keys in which the same physical button has different functions, depending on what software the system is running. In some applications, the user may not realize that the system is essentially a multipurpose computer. The availability of such technology can lead to the creation of devices and designs that offer more capability than is usual or necessary in the application. This possibly may be to the detriment of human performance and safety. In this regard, concern has been expressed about increased fluoroscopy exposures resulting in part from the greater operational complexity and radiation output of some fluoroscopic systems (FDA, 1992b).

The misuse and overuse of technology has also been addressed for products in general (Norman, 1988; Sedgwick, 1993). One common example is the often expressed inability to use a VCR correctly. Another example of the creation of unreasonable technical complexity that can defeat user understanding is a commercially available calculator that can perform 2,100 functions. New technology can also bring new problems such as increased sensitivity of electronic devices to electromagnetic interference. This has occurred at the same time that the number of electronic devices including radiofrequency devices has been rapidly increasing. This increases the opportunities for interference to occur.

Another significant contemporary example of the problems caused by the discrepancy between the way things appear and the way they are is the challenge presented by the video display used in laparoscopic surgery. This technique has been introduced relatively rapidly and is sometimes used by surgeons with limited experience (Altman, 1993). In addition to lack of experience with the laparoscopic method, some surgeons using the technique may not be skilled in the conventional surgery. Therefore, they would be unprepared to revert to the conventional method when difficulties arise during the procedure.

Beyond the usual need for adequate training in a new technique, the technology of laparoscopic surgery, as currently designed, presents some unique challenges. The technique requires the use of a video display to visualize the surgical field. This display is used to guide the manual manipulation of small instruments that have been inserted along with the videocamera through surface incisions. Unlike conventional surgery, laparoscopic surgery provides only indirect observation and a limited, two-dimensional field of view. A different form and degree of hand–eye coordination is therefore required. In addition, the orientation of the image on the monitor relative to the orientation of the patient varies as the camera is manipulated inside the patient. Thus a direction on the screen (e.g., up) is not always the same direction in the patient. Overall, the technology of laparoscopic surgery requires perceptive and manual skills that are significantly different than what is usually required in surgery. The New York State mandatory hospital reporting system revealed a seemingly large number of mishaps in the use of laparoscopic equipment. This prompted New York to mandate minimum experience requirements before a surgeon could proceed with the technique unsupervised ("New York state adopts," 1992).

Durability

A different type of instance in which devices do not meet the expectations of the operator is when components are not of adequate strength to survive reasonably anticipated conditions of use. In this regard, the various parts of catheter connectors should not be so fragile that they are easily cracked or broken during tightening. The emphasis here should be on designing the unit to be strong enough to resist normal and anticipated use rather than rely on, if it was considered at all, the user to manipulate the device so gently as not to overstress it. In addition to fracture, some tubular devices can be rendered unusable or dangerous if they become kinked. One such case resulted in a safety alert for an implantable vena cava filter ("Vena cava filter," 1993).

If a device can only be made so that it is unavoidably fragile, then great care must be taken to transmit this information to the user at the point of use if possible. Another problem of inability to function under real conditions of use is represented by the 1993 medical device safety alert issued for a rotating adaptor used in cardiac catheterization. As presented in the FDA Enforcement Report (FDA, 1993a), the alert stated that this device could lead to an air embolism if the manifold were subject to back-and-forth or side loading.

A variation on the durability problem is whether devices can survive the activation of various functions at unanticipated times or out of sequence. An example is a 1983 recall of a defibrillator that would become inoperative if the charge cycle were initiated while the paddle switch push buttons were depressed (FDA, 1983). A more recent example is an infusion pump in which an incorrect infusion rate or volume would result if the internal program were altered while the device was in an alarm mode ("Incorrect infusion pump rate," 1992). In each of these cases, the operator action that initiated the malfunction was not protected against by the design of the equipment.

Device Interaction

Issues in the assembly and use of devices from varying manufacturers present special challenges. It is generally the case that none of the involved manufacturers provide adequate directions on integrating their device with a variety of other devices. Slight design or manufacturing variations may also make devices unexpectedly incompatible. The assembly of system components during initial installation, clinical set-up, or maintenance can therefore present significant challenges.

One type of anesthesia-switching valve in which each of three ports were physically identical, and therefore could be easily misconnected, was eventually recalled by the manufacturer (Morse, 1984). The recall notice included the interesting admonition not only to identify and remove the unit from service, but also to strike it with a heavy object to render it nonfunctional. This is a classic case of a human-factors-initiated recall in that the device was removed from the market only because of its demonstrated potential to be involved in user error.

Other integration issues have been described, including defibrillator-induced, oxygen-enhanced fires as sparks were generated near electrocardiogram (ECG) leads; laser-induced fires as the laser was activated near other devices; electrosurgery-induced fires; and electrosurgery related sensor burns. It has also been reported that the function of an implanted pacemaker can be switched by a magnetic instrument mat (Purday & Towey, 1992). Electromagnetic interference with monitors and other electronic devices has often been reported. There have also been recent concerns about the interaction of MRI and previously implanted intracranial aneurysm clips (FDA, 1993b).

Misfits between devices that are used together but produced by different manufacturers have also occurred. One example of this problem is tracheal tubes manufactured with a constricted diameter that would impede passage of a catheter ("Tracheal tubes," 1993). A second example is glass bottles of an anesthetic agent. The bottles were manufactured with defective threads that would leak when attached to metal filling adaptors ("Safety alert: Anesthetic bottle," 1992). The potential user issues associated with these manufacturing defects are the failure to undertake simple assembly tests of the components prior to use or the failure to detect and correct an evident problem.

In some instances, incompatibility problems led to revised instructions that warned the user not to rely totally on the technological capabilities of the system. For example, the possibility that an infusion pump could misread the size of a syringe led to instructions that the operator should compare the actual size of the syringe with that read and displayed by the pump. In addition, the operator was asked to independently check the actual infusion rate and not to accidentally push the continuous mode button ("Infusion device sensor," 1992).

The simultaneous use of multiple devices from varying manufacturers also presents difficulties in incorporating all of the equipment at the patient bedside. This results in the physical and visual clutter common in hospital settings. Clutter carries the risk of the operator confusing various lines, leads, and alarms. This confusion can adversely affect patient care. In a survey of anesthesiologists, 42% recalled one or more instances in which a patient was put at risk because the source of an alarm could not be promptly identified (Griffith & Raciot, 1992). The array of equipment in use in the anesthesia environment also results in the monitoring equipment being behind the anesthesiologist whereas the controls are in front. This type of clutter and less-than-optimum placement of equipment would not be tolerated in many other industries in which layout of the system in an attempt to enhance operator performance is standard procedure.

Maintenance

Human-factors issues of usability extend to the cleaning and other maintenance of medical devices. User needs here include appropriate instructions for the type and frequency of maintenance. The instructions must include methods of disassembly and reassembly and test methods to assure proper function before the device is again used on a patient. Human-factors issues include making the procedure as simple as possible and designing the parts of the system so that they can be adequately cleaned without damage to the components. The parts of the system must also be designed so that the potential for damage or misassembly is minimized. Test methods must be straightforward procedurally and reasonable in complexity.

These considerations must be made in the context of who will be doing the maintenance. In the hospital setting, some maintenance is done by the clinical user, whereas more in-depth procedures will be done by the maintenance or clinical engineering department. In the home-use environment, some maintenance will be done by the user, whereas other work should require returning the device to the vendor. In each case, the instructions must be appropriate for the likely maintainer. Design and instructions should also be used to discourage inappropriate maintenance efforts.

Problems that have been reported include inadequate maintenance of defibrillators and inappropriate testing of some defibrillators that were not designed to be discharged into the air. The effectiveness and effect of cleaning solutions have also been of recent interest as transmission of viruses from one patient to another has become of particular concern (Bond, Ott, Franke, & McCracken, 1991). The ability of chemical cleaning agents to penetrate the recesses of various devices is a design issue. Removal of the cleaning agent from the device must also be considered so that damage to the device or exposure of the patient to the agent does not occur.

In 1991, two manufacturers of resuscitator valves issued safety alerts warning of possible sticking of components after improper cleaning. One of the manufacturers also made available a plastic insert to prevent the parts from sticking. The use of this insert was required in Canada; however, the insert had to be removed at the time of use of the device. This led to a secondary problem in that failure to remove the insert would render the resuscitator inoperative ("Improper safety components," 1992).

The development of a design fix for one problem resulting in a second problem is one that deserves particular attention. Assuming that considerable safety and human-factors analysis goes into an original design, any modification of that design should be subjected to the same level of review. Moreover, the new feature must be examined with particular emphasis on its interaction with previous safety features. In the resuscitator valve case, the new feature required a change in procedures (placing and then removing the insert) and therefore new user training.

Another example of cleaning problems is an instance in which the depth stop on a surgical brain drill was alleged to have malfunctioned and caused brain injury to a patient as the drill bit penetrated too deeply into the brain ("Six million dollar award," 1992). Part of the explanation for the event was that the cleaning and sterilization procedure did not include adequate removal of debris and that it was the debris that caused the malfunction. In addition, the hospital had apparently transferred cleaning responsibility to less skilled and less costly personnel without adequately establishing procedures and training. This could be a design problem as well as one of appropriate instructions for use.

In another situation, it has been reported in a manufacturer-initiated safety alert that dried residues in an infusion pump mechanism could alter function and result in an overinfusion ("Over infusion hazard," 1992). Improper cleaning can also result in the removal of numerical or alphabetical markings on a device, thereby interfering with proper setting and information transfer to the user. It is possible that such situations could be controlled by designing the device so that the use of reasonably expected cleaning methods would not have an adverse effect on its performance. Specific and appropriate instructions on cleaning procedures, chemicals, and inspection must be provided.

ENVIRONMENT OF USE AND DEMANDS ON THE OPERATOR

The application of human-factors engineering principles to the design of safe medical devices relies in part on consideration of the actual conditions of use of devices in the clinical environment. Also included are the general types of errors that equipment operators can be anticipated to make and how design can prevent them. These errors include failure to follow instructions, inadvertent operation of controls, failure to recognize hazardous or critical conditions, making wrong decisions, and lapses in attention.

The potential for these errors must be considered in the context of the environment of use, which often includes the operation of many different devices and the treatment of different patients more or less simultaneously. Understaffing and long shifts are not uncommon. Problems can also arise from reliance on contract or temporary workers who may be relatively unfamiliar with the specific types of medical devices that they will have to use. For example, a nurse may be fully aware of how to use infusion pumps in general, as well as the associated risks. However, the nurse's familiarity with a particular brand and model of infusion pump may be limited. Although this may not cause a problem under most conditions, it can cause a significant problem during emergencies, when resetting the pump under low light conditions, when quickly interpreting a setting on the pump, or when flushing the line.

More generally, it must be remembered that the ability to understand and use a device on the convention floor or during a training seminar is not the same as being consistently able to use the device correctly under the real conditions of use. Therefore, in many instances of operator error, the operator did know how to do the task correctly and had done it correctly many times in the past. However, this experience did not prevent the incident from occurring because some aspect of the design and local conditions created a situation in which the standard performance was interfered with.

These concerns must be addressed during the development of a device so that procedures are made as consistent as possible with long-standing human-factors and system-safety design guidelines (Hammer, 1972). These guidelines require that no matter how simple a procedure appears, it should be examined critically for error potential and danger. Efforts should be made to compensate for the fact that personnel tend to take shortcuts to avoid arduous, lengthy, uncomfortable, or unintelligible procedures. It should be noted that procedures that involve interruptions will generate circumstances under which steps may be forgotten. Device design and labeling must take into account that personnel often believe they are sufficiently knowledgeable not to make mistakes, even without reading the instructions. Also, requirements for special training should be minimized and assumed to be not consistently available.

In order to understand more fully and prevent user error, general working conditions must be considered along with the usual array of personal characteristics, some of which are more or less constant for an individual and some of which may vary from day to day. These include natural ability, physical size and condition, knowledge and skill, fatigue, state of mind, attitude, and personal or job-related pressure (Peterson, 1984). It should also be noted that many health-care personnel have not received extensive education in physics, electronics, or the underlying base of modern technology. Yet these individuals are often called upon to operate complex equipment under life-saving or life-threatening conditions. In addition, the traditional focus of health-care personnel is on the patient, not on the equipment that surrounds the patient.

Another aspect of the clinical environment that must be considered is that innovation in the use of medical devices is often required. Innovation may result from the equipment not being exactly suited to the individual patient's needs. It may also result from the need to use multiple devices simultaneously. This need to innovate must be balanced against misapplication.

Medical personnel can also be characterized as being able and willing to cope with difficult situations. This desirable characteristic must be balanced against the tendency to cope with equipment that is hard to use to the degree that the coping interferes with the reporting of problems before they become part of an incident.

The problems previously given for equipment use by medical personnel are even more difficult for the home-use environment. In the home, it can be anticipated that the user has had very little training and experience in the use of a particular device. The home user will, in most cases, have very little general medical experience to fall back on. It is also unlikely for them to have immediate access to someone else who can provide knowledgeable assistance. In addition, home users may have functional limitations that affect their ability to operate a medical device. These issues make design and training for the home environment a special challenge.

At the other end of the human experience spectrum from the user, the design engineer has been stereotyped as not understanding the medical environment, the real conditions of use, or the nature of the real people who will be using their devices. Thus, there has been a gap between the technical capability of some medical devices and the actual consistently safe and effective use of those devices.

CONCLUSIONS

The purpose of understanding user error in the use of medical equipment is to prevent injury and improve the effectiveness and efficiency of medical care. The reason for focusing on this aspect of medical care is that user error has been identified as a significant factor in medical device-related incidents, and many instances of user error appear, at least in retrospect, to have been preventable. More importantly, the analysis of past user error and accomplishments in some areas of medicine and in other industries have demonstrated that user error can be substantially prevented if proper attention is paid to dealing realistically with the user, the environment, and the design of the equipment.

User issues are related to underlying intellectual and physical ability, general education, specific task and device training, and work scheduling. These issues also address stereotypical characterizations of human ability and performance and the understanding of human variability as it relates to these and other factors. Environmental issues address general working conditions, including shift schedules, lighting, noise, temperature, distractions, and workload.

The design perspective addresses the need and obligation to incorporate features in a medical device that will facilitate its use and prevent foreseeable errors and misuse under common user and environmental conditions. User and environmental conditions can always be improved, and in many areas they need to be. However, the first principle of designing for safety must be that these conditions are a given that must be the basis for design decisions. Thus, the hierarchy of hazard elimination, provision of protective measures, provision of automated warnings, and lastly, training the user should be the basis of product design.

REFERENCES

Altman, L. K. (1993, December 14). Standard training in laparoscopy found inadequate. *The New York Times*, p. C3.

Bogner, M. S. (1993). Medical devices: A new frontier for human factors. *CSERIAC Gateway, 4*(1), 12–14.

Bond, W. W., Ott, B. J., Franke, K. A., & McCracken, J. E. (1991). Effective use of liquid chemical germicides on medical devices: Instrument design problems. In S. S. Block (Ed.), *Disinfection, sterilization and preservation* (pp. 1097–1106). Philadelphia: Lea & Febiger.

Cooper, J. B., & Newbower, R. S. (1975). The Anesthesia machine—An accident waiting to happen. In R. M. Pickett & T. J. Triggs (Eds.), *Human factors in health care* (pp. 345–358) Lexington, MA: Lexington.

Cooper, J. B., Newbower, R. S., & Kitz, R. J. (1984). An analysis of major errors and equipment failures in anesthesia management: Considerations for prevention and detection. *Anesthesiology, 60,* 34–42.

Cooper, J. B., Newbower, R. S., Long, C. D., & McPeek, B. (1978). Preventable anesthesia mishaps: A study of human factors. *Anesthesiology, 49,* 399–406.

Food and Drug Administration. (1983, September 21). *FDA enforcement report* (Recall T-238-3). Washington, DC: Author.

Food and Drug Administration. (1987, February 25). *Anesthesia apparatus checkout recommendations; Availability* (52 FR 5583-5584). Washington, DC: Author.

Food and Drug Administration. (1990). *Device recalls: A study of quality problems* (HHS Publication FDA 90-4235). Washington, DC: Author.

Food and Drug Administration. (1991). *Medical device reporting test form, in "Medical device reporting for user facilities: Questions and answers based on the tentative final rule"* (HHS Publication FDA 92-4247). Washington, DC: Author.

Food and Drug Administration. (1992a, October 6). *Draft anesthesia apparatus checkout recommendations, 1992; Availability* (57 FR 46033-46035). Washington, DC: Author.

Food and Drug Administration. (1992b). FDA draws attention to concerns about radiation risk from fluoroscopy. *Medical Devices Bulletin, 10*(8), 5–7.

Food and Drug Administration. (1993a, February 17). *FDA Enforcement Report* (Medical Device Safety Alert N-022-3). Washington, DC: Author.

Food and Drug Administration. (1993b). FDA stresses need for caution during MR scanning of patients with aneurysm clips. *Medical Devices Bulletin, 11*(3), 1–2.

Food and Drug Administration. (1993c). *MEDWatch Interim Coding Manual for use with FDA Form 3500A and MEDWatch Form 3500A for mandatory reporting.* Washington, DC: Author.

Food, Drug and Cosmetic Act, as amended, 21 U.S.C. (1992).

Griffith, R. L., & Raciot, B. M. (1992). A survey of practicing anesthesiologists on auditory alarms in the operating room. In J. Hedley-White (Ed.), *Operating room and intensive care alarms* (pp. 10–18). Philadelphia: ASTM.

Hammer, W. (1972). *Handbook of system and product safety.* Englewood Cliffs, NJ: Prentice-Hall.

Improper safety component use interferes with resuscitator operation. (1992). *Biomedical Safety & Standards, 22*(9), 67.

Incorrect infusion pump rate from software error. (1992). *Biomedical Safety & Standards, 22*(19), 148.

Infusion device sensor misreads syringe size. (1992). *Biomedical Safety & Standards, 22*(19), 150.

Joint Commission on Accreditation of Healthcare Organizations. (1993). *Accreditation manual for hospitals.* Chicago: Author.

Kessler, D. A. (1993), Introducing MEDWatch. *Journal of the American Medical Association, 269,* 2765–2768.

Marrero, D. (1993). Proposed strategies for reducing user error in capillary blood monitoring. *Diabetes Care, 16*(2), 493–498.

Morse, H. N. (1984). The law and medical electronics—Legal case 30: When is an apparatus inherently dangerous? *Medical Electronics, 15,* 16.

New York State adopts credentialing guidelines. (1992). *Biomedical Instrumentation and Technology, 26*(6), 446.

Norman, D. A. (1988). *The psychology of everyday things.* New York: Basic Books.

Paulson, E. K., Baker, M. E., Hilleren, D. J., Jones, W. P., Knelson, M. H., Nadel, S. N., Leder, R. A., & Meyers, W. C. (1992). CT arterial portography: Causes of technical failure and variable liver enhancement. *American Journal of Radiology, 159,* 745–749.

Over infusion hazard from dried residues in infusion pump. (1992). *Biomedical Safety & Standards, 22*(2), 9.

Peterson, D. (1984). *Human error reduction and safety management.* Deer Park, NY: Aloray.

Purday, J. P., & Towey, R. M. (1992). Apparent pacemaker failure caused by activation of ventricular threshold test by a magnetic instrument mat during general anesthesia. *British Journal of Anaesthesia, 69,* 645–646.

Reducing needle stick injuries with safe designs. (1993). *Biomedical Safety & Standards, 22*(2), 9, 11.

Rifkin, G. (1993, July 14). Making home dialysis easier. *The New York Times,* p. D5.

Safety alert: Anesthetic bottle & metal filler incompatibility. (1992). *Biomedical Safety & Standards, 22*(5), 148.

Safety alert warns against pulse oximeter misuse. (1992). *Biomedical Safety & Standards, 22*(2), 11.

Sedgwick, J. (1993). The complexity problem. *Atlantic Monthly, 271*(3), 96–100.

Shepherd, M., & Brown, R. (1992). Utilizing a systems approach to categorize device-related failures and define user and operator errors. *Biomedical Instrumentation & Technology, 26*(6), 461–475.

Six-million award for surgical drill brain injury. (1992). *Biomedical Safety & Standards, 22*(12), 94–95.

Tracheal tubes with constricted diameters. (1993). *Biomedical Safety & Standards, 23*(1), 3.

U.S. Small Business Administration. (1993). Office of Innovation, Research and Technology, *SBIR pre-solicitation announcement—December, 1993* (p. 86). Washington, DC: Author.

Vena cava filter insertion instructions revised in safety alert. (1993). *Biomedical Safety & Standards, 23*(5), 25.

16 Diagnosis-Related Groups: Are Patients in Jeopardy?

Margaret H. Applegate
Indiana University School of Nursing

In 1983, a new law was enacted that dramatically changed the way hospitals are reimbursed for health-care services. To implement the law, Yale University researchers developed a model to classify patients into groups according to their diagnosis. It was their premise that clinically similar groups of patients would consume similar resources (Curtin & Zurlage, 1984). The classification system became known as Diagnosis Related Groups (DRGs).

The purpose of this chapter is to consider the effect of DRGs on access and quality of care issues. Specifically, this chapter addresses the question of whether the DRGs lead to increased risk to hospital patients through various errors that can be attributed to the system itself, care under pressure, or inappropriate admission, treatment, or discharge practices. Finally, issues surrounding how the DRGs have influenced nursing homes and community services are addressed.

COST CONTAINMENT

The cost of medical care in the United States has been rising at a rate of 10% per year (Grace, 1990), far outstripping inflation. The cost is now at nearly 13% of the gross national product (GNP) and is expected to reach 16% to 20% of the GNP by the year 2000 (Kissick, 1992). In 1991, America spent $751.8 billion on health care. This is an increase of more than 11% over the 1990 expenditure and represents a per-capita cost of $2,868 (Letsch, 1993). It is estimated by some that costs could reach $1.5 to $2 trillion by the year 2000 (Beauchamp, 1992; Shelton & Janosi, 1992).

Concern about costs in the health-care system, especially for the elderly and disabled covered under Medicare, stimulated major cost containment efforts in the 1980s. The Tax Equity and Fiscal Responsibility Act (TEFRA) and the Omnibus Reconciliation Act of 1982 (OBRA) began sweeping changes in health care. TEFRA provided for specific cost cutting in the areas of Medicare and Medicaid and mandated the development of a cost-containment plan to curtail costs incurred under the retrospective, fee-for-service system in place. The retrospective system paid for services rendered after the fact, that is, services were provided and the cost was assessed and paid (cost reimbursement). The proposed prospective payment system (PPS) establishes a fixed price based on the average cost of care for a given diagnosis modified by geographic variations and other controlled factors. The thesis of this system is that a given diagnosis should require similar resource utilization and thus similar cost. If a hospital is able to care for a patient for less than the PPS payment, it keeps the profit. If, however, the cost exceeds the PPS payment, the hospital takes the loss.

Revisions of OBRA in 1989 and 1990 reduced fees paid based on evaluation of prevailing charges and assessment of overvalued procedures in earlier cost estimates (Ginsburg, 1991). In effect, advance payments were reduced based on the value of the service with geographic adjustments.

The prospective payment system based on DRGs was proposed in 1983 and implemented in 1984. The system organizes patients into groups based on likenesses in resource utilization for a given diagnosis (Kramer & Schmalenberg, 1987). The DRG-based PPS is a cost-containment effort directed toward one delivery site—the acute care hospital. The focus is to obtain the lowest possible cost per product or per DRG (Bull, 1988) by increasing efficiency and decreasing unnecessary tests, services, and hospital days. The goal is cost containment without loss of quality care. Evidence of continued increases in overall health-care costs suggest that the costs were less contained than shifted. This also suggests that hospital savings may be related more to decreased length of stay than increased efficiency.

Califano (1986) suggested that DRGs are a move in the right direction. However, the plan allows hospitals to shift Medicaid and Medicare costs to the states, to private insurers, and to settings outside the hospital. Among such settings are ambulatory diagnosis and treatment centers, outpatient surgery centers, nursing homes, and home health services. Those settings are examples of expanding arenas of health care that have both positive and negative aspects related to efficiency, quality of care, and cost containment.

IMPACT OF DRGS

Clearly, DRGs as the basis for the PPS have resulted in decreased admissions and length of hospital stay. The average length of hospital stay dropped by 2 days in the first year of implementation of the PPS (Dougherty, 1988). From 1983 to 1986, total inpatient days declined 17%, and inpatient days for

Medicare patients declined approximately 21% (Secretary's Commission on Nursing, 1988). These findings suggest that a disproportionate number of elderly people are being discharged sooner and sicker than in the past (Dougherty, 1988). Because the elderly have a tendency to mare chronic illness and more comorbidity in acute illness episodes, earlier discharge suggests a sicker population at discharge. This logic is supported by the documented growth in the demand for heavier home-care services among the elderly after hospital discharge (Wood & Estes, 1990).

Hospitals continued to profit after the initiation of PPS, though not at the rate prior to the plan (Fisher, 1992). However, hospital revenues increased by nearly 12% in 1991 over 1990. This represents 5 consecutive years of accelerated growth (Letsch, 1993).

Several mechanisms emerged to enhance profits and to protect the institution from losses. Cost shifting is an example of a strategy that influences the quality of care and places patients in jeopardy. Cost shifting occurs when care for one group of patients is reimbursed at rates below the provider estimate of the cost of that care. The hospital may attempt to shift the costs of that patient population to a population that is better reimbursed or may shift costs by nonadmission or by transfer of the problem population to a public hospital. There is evidence that some hospitals' efforts at cost containment included nonadmission of patients likely to be the most costly or transfer of indigent patients from private to public hospitals (Grace, 1990).

Cook County Hospital in Chicago had 6,000 emergency transfers in 1984, five times the number in 1981. In Washington, DC, transfers increased similarly over the same time period. This shifting was the result of "dumping" free care patients (Califano, 1986). Many of the patients being transferred from private to public hospitals were in an unstable condition. Some patients were placed at mortal risk, an error of both clinical and moral judgment.

In an attempt to control for "dumping," the Combined Budget Reconciliation Act (COBRA) of 1985 implemented new federal patient transfer provisions. Following implementation of that act, Kellerman and Hackman (1990) monitored all emergency interhospital transfers to a public hospital during three identical time periods in 1986, 1987, and 1988. They considered a transfer to be unauthorized when the patient was transferred without prior notification of the receiving hospital, was en route before notification, or was transferred despite refusal of the receiving hospital to accept the patient at the time of notification. Transfers were considered to be for economic reasons only if "no money" or "indigent" specifically was given as the reason for transfer. Transfers solely for economic reasons remained at about 90% both before and 1 year after implementation of the new COBRA regulations. In 1987, the transfer of unstable patients had declined by only 4%. Based on their data, Kellerman and Hackman concluded that the regulations had little effect on the transfer of indigent patients.

Further study is needed to determine the extent of the problem of patient transferring and to test whether the federal Health Care Financing Administration (HCFA) proposals to impose fines for failure to report dumping will be effective. This example suggests a system error with consequences for both cost and quality of care. Certainly dumping places patients in jeopardy. This practice also raises serious ethical questions about access to health care. Studies similar to that of Kellerman and Hackman (1990) are needed to determine whether the elderly are denied admission or are transferred at an increasing rate, because the cost of hospital care for the elderly is disproportionately high (Dougherty, 1988).

Another form of cost shifting is the refusal to treat patients for whom reimbursement rates are low when estimates of the cost of care exceed reimbursement levels or when the patient is ineligible for a reimbursement program. One in six individuals covered by Medicaid report being turned away by physicians or hospitals (Blendon et al., 1993). This often results in no treatment or shifting these patients to emergency rooms for care. This places the patient at risk of no care or inappropriate care. It also increases the cost of care by shifting the locus of care to the emergency room. Cost-shifting activities apparently have decreased access to private hospitals for the poor, the under- and uninsured, and very likely the elderly in general who may be poor and underserved as well as requiring expensive care.

As attempts to control costs continued, criteria for hospital admissions were tightened. More diagnostic testing and patient preparation for hospitalization occurred outside the institution. There are obvious advantages to this approach, but there are problems as well. Under this approach, patients admitted to the hospital are much sicker, and some may be denied admission for needed care. Admission criteria exclude the less acutely ill. When only the more severely ill are admitted and when they are discharged earlier, the result is an increased density of sicker patients in the hospital—a higher acuity level (severity of illness) among the inpatient population. Decreased length of stay results in the discharge of patients who are still in need of care—thus the common statement that patients are discharged quicker and sicker.

Kosecoff et al. (1990) studied a nationally representative sample of 10,913 patients to determine discharge disability before and after PPS. They measured instability at discharge, sickness at discharge, and abnormal last laboratory values. They found that 15% of patients in their sample were discharged in an unstable state prior to PPS, and 18% were discharged with instabilities post-PPS. This represents a 22% increase ($p < .05$). Further, most of the increase in instability at discharge was found in those discharged to the home rather than to another institution. The population discharged home were 43% more likely to be unstable in the post-PPS population.

Within the hospital there is an increased density and intensity of workload.

Kramer and Schmalenberg (1987) reported that admissions and discharges in some hospitals occur at such a rapid rate that as many as three patients occupy the same bed within 24 hours. The effect is to tie up much of the care provider's time with activities around admission and discharge rather than with the therapeutic goals of the admission. This environment, with the stresses and pressures of increased acuity, data overload, and rapid turnover, is ripe for errors.

MEDICATION ERRORS

Many studies have focused on medication calculation errors. A recent study by Bindler and Boyne (1991) suggested that this problem continues. In a medication calculation test administered to 110 nurses, they found that 81% of the nurses were unable to calculate medication dosages accurately 90% of the time. Further, over 40% of the nurses had scores of less than 70% accuracy. Nurses erred more frequently when a conversion from milligrams to grains was required or when more than one calculation was required to solve the problem.

Jones, Nichols, and Smith (1989) reviewed 1,000 incident reports filed in a 3-month period in a large metropolitan hospital and found 315 (31.5%) of the reported errors to be related to medications. As DRG-related workloads increase, it is easy to imagine that errors might also be increasing.

Although there are common procedures used to decrease errors in the calculation of medications and where unit dose medications (premeasured in a pharmacy) have been a source of error reduction, more is needed. Entering drug orders directly into a computer not only decreases transcription errors but also provides an accurate means of dosage calculation that could eliminate the need for most hand calculations. As the pharmacy receives the order, the needed dosage could be calculated and provided with the medication. Any necessary calculations could be done in advance including necessary conversions (e.g., from milligrams to grains) and the use of decimals could be reduced to a minimum. A common error is the incorrect placement of a decimal point. Precautions related to the ordered drugs could be included to reduce the risk of decision errors that reach beyond calculation. For example, digitalis preparations could include the requirement to check the heart rate prior to administration and the need to withhold the drug if the heart rate is below 60 beats per minute until consultation with the physician occurs.

Jones et al. (1989) found the biggest factor in medication errors to be failure to follow hospital procedures and the largest category of error to be drug omission. An example of an error related to failure to follow hospital policy or procedure is the failure to check the "five rights" of medication administration: right drug, right dosage, right patient, right time, and right route. Other common procedural errors include the failure to place a zero before a decimal point for fractional dosages to prevent an overdose, or to clearly note the end date of limited medication orders.

Hoge (1991) cited a year-long study in which 10 patients in one 35-bed unit died of digitalis intoxication, even though the contraindications of hypokalemia (low calcium level) and renal (kidney) insufficiency were recorded on the patients' charts. Hoge stated that one fifth to one third of all hospitalized patients experience at least one potentially harmful adverse event. One fourth of those events are serious enough to require additional treatment, prolonged hospitalization, or to result in some form of disability. Hoge further suggested that 1% to 2% of hospitalized patients die from iatrogenic causes.

COGNITIVE ERRORS

Whether mistakes have increased as a result of PPS is difficult to determine, but theoretically this clearly is possible. Certainly as protocols for care increase, flexibility to respond to the individual decreases. A health-care provider could be more prone to respond to the protocols than to patient reactions, and important cues could be missed. With heavier workloads and resulting time constraints, care providers must prioritize their activities to ensure that the more urgent measures and those measures most vital to the patient's safety and treatment goals receive first priority. With multiple patients, rapid changes in patients' conditions, and many treatment options to consider, errors in judgment may occur.

As task complexity increases, information processing is more difficult. Error may result because of the number of problems and the amount and quality of information to be processed (data overload). When acuity levels are high, decisions must be made quickly and often with marginal data. MacLean (1989) argued that data overload or poor sorting of relevant data or cues in a time-constricted situation can lead to premature termination of information processing and failure to identify relevant alternatives. She further noted that attitudes, values, and beliefs of the decision maker influence decisions and may be detrimental to the patient. A health-care provider could act based on beliefs about what the patient needs or a limited understanding of the dynamics of the situation rather than objective facts when time is limited and the workload is heavy. Cianfani (1982) found that nurses' diagnostic accuracy increased when data sorts involved the use of a limited amount of highly relevant information for decisions.

These research findings have implications for health-care providers in both educational and practice settings. Faculty in educational programs must focus on improving the critical thinking skills of students. Helping physicians, nurses, and other providers of care to be more cognizant of variables that influence their data-processing ability as well as to practice decision strategies can render their decisions more systematic and objective (MacLean, 1989). Staffing assignments should place experienced nurses and those with special expertise in a position to mentor novices. To decrease the risk of decision errors, nurses should not be rotated outside of their areas of clinical expertise without proper orientation. Working double shifts for several consecutive days on high-intensity units can contribute to errors when the care provider becomes fatigued. Continuing education should be available to assist all care providers to maintain currency and to enhance knowledge. Care for the provider enables care for the patient.

MACHINE INTERFACE ERRORS

The constant interface with machines in the hospital setting is another area for potential error. In interviews with registered nurses, McConnell (1990) found that although nurses appreciate the advantages of machines to both the patient and the caregiver, they also perceive disadvantages. The potential for error or compromised care exists if the caregiver relies too heavily on machine-generated data to the exclusion of actual patient responses. Consider the patient shaking in tetany (spasms due to changes in pH and calcium levels) who was provided with extra blankets on the assumption that she was simply cold because the monitor recorded no other explanation for her shivering (all readings were "normal"). In fact, the patient had no functioning parathyroids, which made her a candidate for tetany without calcium replacement therapy. Only at the insistence of a family member did the nurse obtain additional information and call for medical consultation and calcium replacement. Multiple examples exist of diagnostic errors by various members of the health team when the use of technology is substituted for a careful examination or attention to the history related by the patient. The patient's voice is often lost in a sea of technology and time constraints.

Malfunctioning machines can create false data or safety hazards. Nurses and physicians deal regularly with competing demands of patients and machines as well as ethical issues resulting from the use of sophisticated technology. The entire health-care team is subject to the potential for missing the obvious while involved in the use of technology. There is a need to hear and see the patient as a reliable source of relevant data so that technology is not the sole source of information. There is an equal need to monitor the equipment to ensure safety.

Advanced technology also places patients at risk of being held captive by sophisticated machines and treatment regimes. Availability of technology is not a singular reason for its use. To continue to apply life-extending technology that offers no net benefit to the patient may less prolong life than the act of dying. Errors in moral judgment may sacrifice quality of life and patient dignity at the altar of science and technology.

IDENTIFYING ERRORS

It is difficult to identify errors in hospitals for a number of reasons. Some review processes are in place to help identify major errors; however, many errors occur on a daily basis that are unrecognized or unreported. Many hospitals have departments of quality assurance, infection control, and risk management to identify real or potential problems, trends in patient care, and institutional management issues that require action. The goal of these departments is to minimize costs and maximize the safety and quality of care for patients, families, and employees (Siler, 1989). When they function in a coordinated fashion, these departments can do much to address issues and problems and to prevent error and harm. However, although hospitals have a deep and active interest in quality, they also wish to avoid litigation. The latter influences activities in these three departments.

The identification of medication errors is an example. Errors and accidents are tracked largely through incident reports (Bindler & Boyne, 1991). Errors may not be reported for a variety of reasons. Anecdotal reports suggest that such reasons include decisions about the seriousness of the error and concern about liability and job security. Other concerns are for personal and professional reputation as well as failure to recognize that an error has occurred. In a small-sample study, Barker and McConnell (1962) found through direct observation that 29% of observed errors were not reported, even though it is reasonable to expect that the reporting level would have been higher under direct observation. The errors occurred, and the issue was the number reported. In the absence of costly monitoring, it is increasingly important to provide techniques to minimize errors. Little is gained from simply expecting health-care providers to do better without the tools to do so in a complex environment (Hoge, 1991).

PATIENT OUTCOMES AND PPS

With the advent of PPS, there is pressure to move patients out of the acute-care setting rapidly. Sometimes this occurs when the patient is in an unstable condition. Conflicts between the desire to meet DRG time limitations under PPS and patient needs persist. Studies began to emerge after the implementation of PPS to measure health (sickness) outcomes in the DRG-driven health-care system. Concerns focused on quality of care indicators and the possibility that patients might be denied needed care under PPS. Outcome measures of particular interest in those studies involved quality of care, placement at discharge, readmission rates, and mortality rates.

Kahn et al. (1990) compared outcomes before and after implementation of PPS. They studied a nationally representative sample of Medicare patients admitted in five states in five selected disease categories. In the aggregate, they found a statistically significant decrease in length of hospital stay and a decrease in in-hospital mortality of 3.5%. They found no statistically significant change in severity-adjusted mortality rates at 30 and 180 days post-discharge. There was a decrease in in-hospital mortality rates in the post-PPS group that may reflect decreased length of stay resulting in a change in the location of mortality. Because there was no significant difference in pre- and post-PPS mortality at 180 days, but there was a decrease in the in-hospital mortality rate for the post-PPS group, more patients must have died after discharge in the post-PPS group to neutralize the difference at 180 days. This issue is not clarified in the report.

Although Kahn et al. (1990) reported no significant change in either the rate of readmissions or mortality in the aggregate, they did report a statistically significant increase in death and readmission rates of post-PPS patients in one disease category—myocardial infarction (MI). A significantly higher rate of discharge to nursing homes occurred for all disease categories, with hip fracture patients having the greatest increase. Although an increased number of patients had prolonged nursing home stays, the rate of increase in length of stay was not statistically significant. Their findings related to hip fracture patients are consistent with those of Mayer-Oakes, Oye, Leake, and Brook (1988). They differ, however, from the findings of Fitzgerald, Fagan, Tierney, and Dittus (1987), who reported an increase in prolonged nursing home stays for hip fracture patients post-PPS. They found that three times as many post-PPS hip fracture patients remained in nursing homes 6 months after hospital discharge.

Gerity, Soderholm-Diffate, and Winograd (1989) compared pre- and post-PPS outcomes of elderly patients with hip fractures. They found post-PPS patients to have had a statistically significant decrease in hospital stay, less ability to ambulate at discharge, and increased nursing home placement. After 1 year both the pre- and post-PPS groups had improved clinically and demonstrated no statistically significant differences. Gerity et al. (1989) found differences among patients in both groups to be dependent on the nursing home to which they were discharged. Those who went to nursing homes with rehabilitation services were more likely to return home or to a lower level of nursing home care and to ambulate more independently. This study emphasizes the importance of follow-up care post-discharge to review long-term outcomes and the ultimate cost of care. Prolonged nursing home placement in lieu of a slightly longer hospital stay may not be cost effective. Evaluation of the nursing home to which a patient is to be discharged is also imperative, or patients could be in jeopardy of reduced therapeutic and quality of life benefits.

Leibson, Naessens, Krishan, Campion, and Ballard (1990) examined hospitalizations for all elderly residents in one Minnesota county for 1980, 1985, and 1987 to compare pre- and post-PPS mortality and disposition at discharge. They found post-PPS patients to be sicker at both admission and discharge ($p < .001$). There was also a marked increase in age at admission for the post PPS group. They found a statistically significant increase in mortality at 60 days post-discharge and in nursing home transfers post-PPS. The latter two differences, however, were largely accounted for by the variables of age, gender, severity, and disease complexity. There were no other independent effects of the PPS variable.

Kosecoff et al. (1990), as a part of the overall investigation noted earlier by Kahn et al. (1990), compared discharge impairment levels of pre- and post-PPS patients in five disease categories to address the quicker and sicker discharge issue. The diseases studied were congestive heart failure, acute MI, pneumonia, cerebrovascular accident (stroke), and hip fracture. Kosecoff et al. defined instability at discharge as "important clinical problems usually first occurring prior to discharge" (1990, p. 1980). Prior to implementation of PPS, 15% of patients were discharged in an unstable condition. The post-PPS group had an 18% rate of unstable discharges ($p < .05$). Those discharged with at least one instability in both groups had a higher probability (16%) of dying by 90 days post-discharge than did those discharged with no instability (10%).

Implications of the Kosecoff et al. (1990) study are not entirely clear. Whether the mortality rate for those discharged post-PPS in an unstable condition represents premature discharge is not certain. They compared differences in mortality rates between stable and unstable groups but did not clarify pre- and post-PPS mortality rates within the unstable group. The need for follow-up care in the home is apparent, as is the need for discharge planning to enable a smoother transition from hospital to home or to a nursing home. Twenty-two percent of unstable patients were discharged to nursing homes. This has implications for more skilled care nursing home beds. Long-term follow up of these patients is needed to determine possible effects of such placement on quality of life over time.

In another report of the same national study, Rubenstein et al. (1990) directly addressed quality of care before and after PPS using a sample of 1,366 Medicare patients in the same five disease categories as in the previously discussed studies. Implicit record review was conducted using structured review forms to rate the process of physician and nursing care, appropriate use of hospital services, patient prognosis, treatability of the patient's condition, preventability of patient death when it occurred, quality of outcome, and overall assessment of quality of care. The authors report that deaths within 30 days of admission increased as the quality of care decreased. Those rated to have very poor care had a higher relative risk of death within 30 days when compared to all patients in the sample ($p < .01$). Those conducting the record review found the quality of care to be generally good: 82% of all patients were considered to have received good or very good care. They rated the care given post-PPS more favorably than care given in 1981–1982 ($p < .001$). The difference was statistically significant for three of the five diseases studied: acute MI ($p < .001$), cerebrovascular accident ($p < .001$), and congestive heart failure ($p < .01$). Care in virtually all variables measured improved significantly for the post-PPS period.

Rubinstein et al. (1990) found fewer patients judged to have had an inappropriately long length of stay post-PPS, as might be expected. However, they found significant increases in the number of hospitalizations rated as too short resulting in the discharge of patients in an unstable condition post-PPS ($p < .05$).

A review of these and related studies suggests that quality of care during hospitalization has not been negatively affected by DRGs and in some areas may have improved. However, the fact that 12% of post-PPS patients in the Rubenstein et al. (1990) study were judged to have received poor to very poor care is a cause for concern. Although this indicates a statistically significant improvement over the pre-PPS rating of 25%, it remains a level of poor care that deserves attention. Such care places a substantial number of patients in jeopardy. The value of this finding may be to provide a target for further study and improvement.

The aggregate of studies indicates that patients are admitted to the hospital in a sicker state and are discharged more quickly. Although premature discharge findings were inconsistent, there was sufficient research evidence to target this outcome for future study.

There may be multiple reasons for the increased acuity level of patients at admission post-PPS, which may influence discharge outcomes. Low-risk procedures are done on an outpatient basis, and utilization review (needs assessment to control costs and prevent unnecessary services) makes admission more difficult for individuals who are not considered to be sufficiently ill to require hospitalization. An aging population and technology advances that sustain life may also contribute to a sicker admission population. This is an area for further research including follow-up studies of those who seek and are denied hospital admission. The denied population could be in jeopardy, and the lack of research in this area is a cause for concern.

TRANSITION ERRORS

Transition from the hospital to home or to a nursing home is an area that also requires increased attention. There is insufficient evidence that continuity of care is appropriate to the needs of a population that is being discharged in an increasingly unstable condition. Many acute-care facilities have made changes in discharge planning to provide continuity of care and to ease transition. In a study of the influence of DRGs on discharge planning at eight hospitals, Bull (1988) found that discharge planning post-PPS increased routinization (protocols and procedures), communication, and collaboration. There was a marked increase in discharge planning rounds. All Medicare patients were labeled high risk for purposes of discharge screening and planning for post-discharge needs. Discharge planning forms became a part of the medical record, and nurses were held accountable for their completion. Often the plans were initiated by nurses or social workers with collaboration among the patient, family, and other health team members. A coordinator was assigned to avoid duplication of effort.

Edwards, Reiley, Morris, and Doody (1991) found that increased discharge planning improved efficiency and decreased length of stay. For example, actions were taken to establish early assessment of patient support systems in the home, to enable family participation in the plan, and to improve interdisciplinary collaboration to avoid duplication and delay of services. Laboratory tests and consultation were a primary target of efficiency strategies. Quicker turnaround in those two areas alone contributed to decreasing the length of stay.

Transition problems began to emerge as discharge planners attempted to find services beyond the acute-care setting for a growing number of frail elderly who needed post-hospital assistance. In interviews with service providers and policy makers, Wood and Estes (1990) found that discharge planners were having great difficulty in referring patients for multiple services in the community or to long-term care facilities. In 1987, 34% of 177 discharge planners interviewed reported difficulty in referral based on a lack of community resources and services or long waiting lists for services. Many patients were at risk of being discharged into the community without access to needed care. As a result, many were discharged into a no-care zone. Whereas multiple problems combine to create a no-care zone, the heart of the issue rests with access based on fiscal resources, geography, and availability of support systems and services.

Needed services are often difficult to obtain in rural areas or in the inner city. Cost of service delivery to distant locations is a barrier, as are disincentives for health-care providers to serve in rural and inner-city areas. The patient's own situation may present a barrier when that patient is too frail or ill to access a service center, has no support system or transportation, or is ineligible for homebound care.

The case of an elderly woman serves as an example. She was discharged with a diagnosis of a malignant brain tumor. Radiation therapy was to be provided on an outpatient basis. Her home was 1 mile beyond the border for eligibility for transportation services. She could arrange for private transportation, but the cost was high ($30 per trip), and no one could accompany her without an additional cost. The private agency would not assist the patient to the vehicle. She had mobility problems related to her illness, and her spouse was frail, blind, and unable to assist her. She had been discharged into a difficult situation.

Once the individual is discharged from the acute-care setting, financial barriers and geographic barriers may merge with other issues in the family and community to create problems. Although care outside the hospital is not entirely under PPS, this arena of care has been heavily influenced by DRGs. Of particular concern is the need to provide services to a population that has increasing acute-care needs combined with the complexities of chronic illness.

NURSING HOME CARE

As hospitals discharge more patients into nursing homes for recovery, long-term care, or terminal care, these facilities feel an immediate impact. The patient profile reflects a population needing more skilled, high-technology care. Harron and Schaffer (1986) compared the nursing needs of nursing home patients pre- and post-PPS using a retrospective record review of the first 3 days of care. They found that the pre-PPS group received a total of 195 nursing actions in the first 3 days compared to the post-PPS group, which received 405 nursing actions in the first 3 days after admission.

Trensch, Duthie, Newton, and Badin (1988) conducted a prospective study of 100 consecutive patients admitted to a nursing home unit during the first 23 weeks of 1986. They compared their findings to a group of patients admitted during an identical 23-week period in 1983 (prior to PPS). They found that more than seven times as many patients were admitted in 1986 (N = 100) as in 1983 (N = 14) in a comparable time period. In the post-PPS group they found significant increases in short-term as opposed to long-term admissions ($p < .001$). They also found the post-PPS groups to have increased discharges to the patients' homes ($p < .02$) and increased readmissions to the hospital within 10 days ($p < .05$) and 30 days ($p < .04$) of admission to the nursing home.

Twenty of the 100 post-PPS patients were assessed by the nursing home staff to have been discharged from the hospital prematurely. Thirteen of these 20 required readmission to the hospital. No morbidity data were provided. Lyles (1986) surveyed 51 nursing homes following initiation of PPS and found that over 50% reported increases in severity of illness, prevalence of clinical problems, and use of medical supplies for those admitted post-PPS.

Clearly, patient profiles and demands for skilled care have changed in nursing homes post-PPS. With the elderly being discharged sooner and sicker into nursing homes, those facilities are becoming subacute centers.

This requires a change in staff mix and protocols for care that are difficult to achieve.

Whereas some nursing homes have increased registered nursing staff and some have employed clinical nurse specialists to provide and coordinate the nursing care of elderly patients, many have not increased their staff mix beyond the minimum regulatory requirements. This lack of change in staffing to accommodate shifts in the patient profile could place patients in jeopardy from well-meaning and often caring individuals with little preparation to meet the needs of those under their care. Furthermore, the staff's vision of the kind of care that might be delivered often is blocked by routinization and staff shortages (Collopy, Boyle, & Jennings, 1991).

Long-term care in a nursing home may create psychological problems that can compound the physical problems of the patients. The residents must deal with the fear or reality that they will never go home. Separation from family, friends, and even pets and belongings is difficult. In some cases, nursing care is driven by a view of the elderly as helpless and dependent. There are disincentives to independence and an increased loss of personal control. Many health-care workers involve the patient and family in care planning. They actively seek to provide privacy and some sense of personal space, but this not always the case and may not be the norm.

There is much anecdotal evidence of errors in judgment related to actions driven by routine or regulation rather than individualized assessment. For example, diet trays may be served with large amounts of food to meet caloric regulations. Some elderly people find the amount so overwhelming that they eat little or none of the food with a counterproductive outcome. Other patients have been known to grieve over the waste of food because they were raised in an era that attached guilt to waste.

ERRORS IN ORIENTATION TO CARE

Environmental hazards, accidents, and lack of care and cleanliness as well as incidents of abuse in nursing homes have been recorded. Drug reactions or interactions may cause dizziness and confusion, which are sometimes interpreted as common indicators of age. Vital signs may be checked on schedule rather than need. Consider the elderly woman who became dizzy on ambulation. She and her daughter indicated to the staff that this often occurred when she experienced cumulative effects from her antihypertensive medication. She asked that her blood pressure be checked and that the doctor be called to regulate the dosage if the pressure was low. She and her daughter were informed that blood pressures were checked on Wednesdays and the doctor made rounds on Mondays. Until then, the patient would be physically restrained with a posey restraint for her safety and the drug would be continued as ordered. This represents a lack of understanding of the response of elderly patients to medications, an error of routinization, and an error of moral judgment and lack of sensitivity to the autonomy rights and needs of the patient. The quality of care was diminished and orientation to care was faulty.

A serious quality of care issue in nursing homes is the use and abuse of restraints, both mechanical and chemical, unsupported by diagnosis or treatment considerations. Restraints may be applied as many as 500,000 times per day in the United States (Moss & LaPuma, 1991). In a longitudinal study of residents in 12 skilled nursing home facilities, Tinetti, Liu, Marattoli, and Ginter (1991) found that 66% of the residents had been mechanically restrained. They found restraint use to be common and typically intermittent. The view that nursing homes are the keepers of the frail elderly who are expected to decline may influence care providers to place goals of protection, confinement, and shelter above those of functional rehabilitation, maintenance, and personal achievement (Johnson, 1990). This represents an error in orientation of care and a lack of respect for the patient as a person. In the absence of sufficient studies about the efficacy of the use of restraints and whether benefits outweigh risks, the use of restraints may have emerged as a standard of care by consensus rather than by scientific data (Moss & LaPuma, 1991).

As PPS drives larger numbers of sick, elderly patients into nursing homes for recovery and rehabilitation, it will be crucial that such patterns as the overuse of restraints not delay recovery or, in fact, cause substantial harm. It is equally important that the range of care needed and competent providers of that care be provided.

Funding of nursing home care in particular remains a barrier to access. For those who meet the eligibility requirements, Medicaid provides funding. In 1991, Medicaid carried over a 47% share of costs for nursing home care (Letsch, 1993). Medicare, however, does little for nursing home care. Medicare is oriented to acute care. Over one half of all nursing home expenses in this country are paid through private insurance or private pay. Long-term care insurance is beyond the means of many who need it. For those who have such insurance, reimbursement is often hindered by complicated eligibility criteria and vast and complex documentation requirements often outside the capabilities of the elderly. As a result, many who would profit from nursing home care are in the home-care system or in the no-care zone.

HOME HEALTH CARE

Home health care is one of the most rapidly growing arenas in the health-care delivery system. In the 4 years following implementation of PPS (1983–1986), the number of patients discharged to home care increased by 37% (Shuster & Cloonan, 1989). Between 1980 and 1989, the number of home-health care visits increased from 26 million to 40 million per year (Sherry, 1992). Advantages of short-term home care include a shorter convalescence and more rapid return to independence, whereas long-term home care may enhance overall quality of life (Rogatz, 1985).

The cost effectiveness of home care is controversial. A 1985 audit by the Office of the Inspector General found payments for home health services to be greater than payments for full-time monthly care in skilled nursing facilities (Office of the Inspector General, 1985). Several studies both before and after PPS have failed to demonstrate a cost savings through home health care (Green, 1989; Hedrick & Inui, 1986; Weissert, 1985). However, other data gathered pre- and post-PPS have shown home care to be cost effective as an alternative to acute inpatient care and for reducing both hospital and nursing home lengths of stay (Kent & Hanley, 1990).

Differences in patient populations may contribute to the varied findings. That is, whereas care for short-term, intermittent care is more cost effective, the need for intensive long-term care for complex disabilities may be more costly when delivered in the home. How cost is determined is another issue. If members of the family must take a leave from work or give up employment to provide care, that cost is not included in cost analysis. Rogatz (1985) suggests that studies are needed to compare the overall costs of caring for patients with matched diagnoses at home and in institutions over prolonged periods of time.

There is no question that the PPS has raised additional cost issues and has changed the entire home health-care system. That change has been in terms of clients served, services required and provided, and referral patterns. The average patient in home health care is now sicker, requires more high-technology procedures, and needs more skilled and unskilled care visits of longer duration. As acute-care clients compete with the chronically ill for home health services, access problems increase.

The population served in the community now includes over 4 million disabled elderly, of whom 68% are women. The disabled elderly are about evenly divided among those who live alone, with a spouse, or with others (U.S. Senate Special Committee on Aging, 1991). The overwhelming majority of care given to these individuals is provided by spouses, family, and friends. This care often is provided at great physical, emotional, and financial expense to the caregivers. The average American woman will spend 17 years raising children and 18 years helping disabled and aged parents, grandparents, spouses, and other relatives (Beck et al., 1990).

Many families juggle hectic schedules as they attempt to work, maintain a home and family, and care for aging parents. Many women have had to reduce their work hours, quit work, or give up promising careers to meet the needs of their own parents and those of their spouses. Growth in adult day-care centers and respite care have been useful to this population. Some corporations are beginning to recognize this need. Examples of leave time available to care for family members and mixed-generation, corporate-sponsored day-care centers are emerging. More are needed.

Examples of the aged caring for the ancient and vice-versa are growing in number. A nurse in a Midwest hospital provided an example. A 75-year-old woman was admitted for surgery to remove a brain tumor. She was accompanied by her 81-year-old husband, who had recently undergone bilateral hip replacements. He was ambulating with a walker and was too weak to assume her care upon discharge. Two days after surgery, the patient's 96-year-old mother came to visit and asked the nurse when she would be able to take her daughter home to care for her. Such touching scenarios are not rare.

Even with the extensive involvement of family and friends in providing care, community-based services have become saturated with heavy-care clients. Many of these clients have no family support system and require continuing care or intermittent care to delay or avoid institutionalization. Most are impaired in activities of daily living such as toileting, bathing, transferring from bed or chair, and eating. Others have instrumental impairments that involve such things as using a telephone, shopping, managing finances, or getting around in the community.

Older people who live alone are twice as likely as elderly couples to receive no help when they are disabled. Keeping these people at home is a widely accepted goal, but lack of available services or access to those services often creates barriers to the goal. Medicare generally covers in-home care only for the homebound who have skilled care needs certified by a physician-authorized plan of care. A majority of formal care is provided by unskilled workers who may be supervised as seldom as once every 30 to 60 days. This is an arena ripe for error.

Medicaid coverage for home care varies by state, but is less restrictive than Medicare. However, constant changes in reimbursement guidelines and inconsistent interpretation of those guidelines have increased the national denial rate of Medicaid reimbursement for homebound care from 1.2% in 1983 to 6% in 1986 (Stone & Krebs, 1990). More recent data suggest that one factor in the increase in Medicaid spending is improved reimbursement rates (Letsch, 1993).

Medicare denials for home health-care claims have also increased largely due to stringent interpretation of its regulations. Medicare only reimburses curative or restorative home health care that is short-term, intermittent, and supervised by a physician. Thus, the chronically ill elderly who largely have functional deficits are often ineligible for reimbursement (Helberg, 1990).

HCFA is testing a PPS in home care with demonstration models that test prospective per-visit and episodic models against the current retrospective per-visit model (Kent & Hanley, 1990). OBRA87 was passed to encourage the development and testing of alternative demonstration models of reimbursement to home health agencies. This could provide future models of value. Medicare pays an agency the same amount per visit based on various factors including the lowest of actual charges, costs, or the schedule of limits on agency costs per visit (Helberg, 1990). This approach is problematic in that it errs in the assumption that all visits are equal. Per-visit reimbursement can be an incentive to cut visits short to increase quantity at the cost of quality. Patients can be disadvantaged and even harmed.

Prospective payment systems in home care could be based on the current per-visit plan but specify the total number of visits allowable in advance. It could emulate the DRG system and reimburse based on diagnosis or type of services provided. It could reimburse based on a unit of time criterion in which all of the needed care would be provided for a designated period of time. Finally, a capitation model is possible in which an agency would be paid to be responsible for an agreed-upon set of home services (Helberg, 1990). In each case, reimbursement is predetermined (prospective) and may be based on visits made or on an established unit of time necessary to meet patient care needs (episodic).

If the proposed PPS in home care follows the DRG medical model of classification, errors of eligibility will be perpetuated. Home health-care needs are better determined by functional disability than medical diagnosis. Various classification models have been suggested and tested. Common elements of these classifications are functional level, patient acuity, rehabilitation potential, and outcome (Ballard & McNamara, 1983; Harris, Santaferro, & Silva, 1985; Pasquale, 1987; Sienkiewicz, 1984). These may prove more useful.

Promising comprehensive models of community care exist. Bogdonoff, Huges, Weissert, and Paulsen (1991) initiated a Living-at-Home-Program (LAHP) at 20 sites across the country designed to coordinate and increase access to services for the frail elderly who wish to remain at home. The thesis of Bogdonoff et al. is that increased coordination across providers will result in economy of scale to maximize available resources and thus meet needs more effectively. They seek to reverse the error of fragmented care through case management and outreach designs.

Highlights of the LAHP programs include computer information systems to provide a database and a single source for referral to needed services with follow-up tracking. It also includes a single assessment model to meet eligibility screening requirements across services. Such barriers as ignorance of available services, delayed processing for eligibility, and lack of continuity of care declined substantially with this program. Case management is emerging as a successful means to coordinate care and to assist clients and families in making sense of and accessing the complex array of services needed to meet the broad-ranging needs of the elderly. Case management was a critical element in these projects. Such a plan could reduce errors of fragmentation and orientation by providing more holistic care. It also has cost-containment potential through economy of scale and the ability to reduce duplication of services. To date, such efforts have been marked more by enhanced humanity of care than cost containment.

Block nursing is a program that coordinates resources available in a community with the broad-ranging needs of patients, especially the elderly. The program is organized in neighborhoods to coordinate formal and informal care based on needs rather than eligibility for reimbursement (Jamieson, 1990). Jamieson describes a block nursing program in Minnesota that relies on alternative funding, professional and volunteer community members, and an informal network of community groups. Access to the program is through the Public Health Nursing Service (PHNS), which employs the professional staff. The structure of the PHNS provides quality assurance programs, operational mechanisms, and access to third-party payers for those patients who are eligible. Success of the block program is based on formal and informal caregivers living in the neighborhood and on maximizing the use of existing agencies and resources.

One of the greatest assets to home health care is case management by skilled nurses knowledgeable in gerontology and community health. The advanced nurse practitioner is one of the most underutilized resources in providing comprehensive community-based care. Nurses have demonstrated their ability to provide quality care that is cost effective (Alexander, Younger, Cohen, & Crawford, 1988; Brooten et al., 1986; Burgess et al., 1987; Lipman, 1986; McCorkle et al., 1989; McGrath, 1990)

An important contribution is the nurse's focus on health promotion and preventive health care. Many errors of fragmentation and failure to provide care that enables independence and reduces functional limitations are amenable to the particular skills of professional nurses. Appropriate and expanded use of skilled nurses will be a key to health-care reform now and in the future.

CONCLUSION

There is a great deal of controversy about the influence of DRGs on health-care costs and quality. Many studies have been conducted with conflicting results. Areas for further investigation and additional questions that warrant careful study have emerged. However, there is little compelling evidence that PPS is singularly or directly responsible for a large array of errors in health care. The fundamental error is in the health-care system itself.

So long as health-care agencies and health-care professionals waste valuable time and energy dealing with our complex and fragmented system of care, costs will be less contained than eternally shifted. Huge amounts of resources in time and labor have been expended in dealing with and getting around the regulations and reimbursement mechanisms that seem to change weekly. Efforts have often been aimed more at maintaining the good of the providers than the good of health-care recipients. Cost shifting, creaming, overuse of technology, staff mix inappropriate to client needs, premature discharge, and restricted admissions are areas of potential harm that reflect tension between the needs of providers and recipients of care.

Access to health care must be provided for the 37 million uninsured in this nation. The spiraling cost of health care must be harnessed. Preventive health care must be emphasized. Fragmentation of care must be addressed. States such as Oregon, Minnesota, Florida, and Hawaii must be encouraged to implement their plans of health-care reform and to demonstrate through outcomes the ability of states to meet the needs of their citizens in a cost-effective manner. Social justice and order must be brought to the health-care system. Without concerted and immediate action, everyone is in jeopardy. Our future is now.

REFERENCES

Alexander, J. S., Younger, R. E., Cohen, R. M., & Crawford, L. V. (1988). Effectiveness of a nurse managed program for children with chronic asthma. *Journal of Pediatric Nursing, 3,* 312–317.

Ballard, S., & McNamara, R. (1983). Quantifying nursing needs in home health care. *Nursing Research, 32,* 236–241.

Barker, K. N., & McConnell, W. E. (1962). Detecting errors in hospitals. *American Journal of Hospitals, 19,* 361–369.

Beauchamp, D. E. (1992). Universal health care, American style: A single fund approach to health care reform. *Kennedy Institute of Ethics Journal, 2,* 125–135.

Beck, M., Kantrowitz, B., Beachy, L., Hager, M., Gordon, J., Roberts, E., & Hammill, R. (1990, July 16). The daughter track: Trading places. *Newsweek,* pp. 48–54.

Bindler, R., & Boyne, T. (1991). Medication calculation ability of registered nurses. *Image, 23,* 221–224.

Blendon, R. J., Donelan, K., Hill, C., Scheck, A., Carter, W., Beatrice, D., & Altman, D. (1993). Data watch: Medicaid beneficiaries and health reform. *Health Affairs, 12,* 132–143.

Bogdonoff, M. D., Huges, S. L., Weissert, W. G., & Paulsen, E. (1991). *The living-at-home-program: Innovations in service access and case management.* New York: Springer Verlag.

Brooten, D., Kumar, S., Butts, P., Finkler, S., Bakewell-Sachs, S., Gibbons, A., & Delivoria-Papadopoulos, M. (1986). A randomized clinical trial of early hospital discharge and home follow-up of very low birthweight infants. *New England Journal of Medicine, 315,* 934–939.

Bull, M. J. (1988). Influence of diagnosis-related-groups on discharge planning, professional practice, and patient care. *Journal of Professional Nursing, 4,* 415–421.

Burgess, A. W., Lerner, D. J., D'Agostino, R. B., Vokonas, P. S., Hartman, C. R., & Gaccione, P. (1987). A randomized control trial of cardiac rehabilitation. *Social Science Medicine, 24,* 359–370.

Califano, J. A. (1986). *America's health care revolution: Who lives? Who dies? Who Pays?* New York: Random House.

Cianfani, K. (1982). *The influence of amounts and relevance of data in identifying health problems.* Unpublished doctoral dissertation, University of Illinois, Chicago.

Collopy, B., Boyle, P., & Jennings, B. (1991). New directions in nursing home ethics. *Hastings Center Report, 21-S,* 1–16.

Curtin, L., & Zurlage, C. (1984). *DRGs: The reorganization of health.* Chicago: S-N Publications.

Dougherty, C. J. (1988). *American health care.* New York: Oxford University Press.

Edwards, J., Reiley, P., Morris, A. M., & Doody, J. (1991). An analysis of the quality and effectiveness of the discharge planning process. *Journal of Nursing Quality Assurance, 5,* 17–27.

Fisher, C. R. (1992). Hospital and Medicare financial performance under PPS, 1985–1990. *Health Care Financing Review, 14,* 171–183.

Fitzgerald, J. F., Fagan, L. F., Tierney, W. M., & Dittus, R. S. (1987). Changing patterns of hip fracture care before and after implementation of the prospective payment system, *Journal of the American Medical Association, 258,* 218–221.

Gerity, M. B., Soderholm-Diffate, B. A., & Winograd, C. H. (1989). Impact of prospective payment and discharge location on the outcome of hip fractures. *Journal of General and Internal Medicine, 4,* 388–391.

Ginsburg, P. B. (1991). A bumpy road to Medicare payment reform. *Journal of American Health Policy, 1,* 10–14.

Grace, H. (1990). Can health care costs be contained? *Nursing and Health Care, 11,* 125–130.

Green, J. H. (1989). Long term home care research. *Nursing and Health Care, 10,* 138–144.

Harris, C., Santaferro, C., & Silva, R. (1985). A patient classification system in home health care. *Nursing Economics, 3,* 276–283.

Harron, J., & Schaffer, J. (1986). DRGs and the intensity of skilled nursing. *Geriatric Nursing, 7,* 31–33.

Hedrick, S. C., & Inui, T. S. (1986). The effectiveness and cost of home care: An information synthesis. *Health Services Research, 20,* 851–880.

Helberg, J. L. (1990). Information needs in home care: A review and analysis. *Public Health Nursing, 1,* 65–70.

Hoge, M. H. (1991). New perspectives on our national health dilemma. *Health Care Management Review, 16,* 63–71.

Jamieson, M. K. (1990). Block nursing: Practicing autonomous nursing in the community. *Nursing and Health Care, 11,* 250–253.

Johnson, S. H. (1990). The fear of liability and the use of restraints in nursing homes. *Law, Medicine, and Health Care, 18,* 263–272.

Jones, W. J., Nichols, B., & Smith, A. (1989). Developing patient risk profiles. In K. Darr & J. S. Rakich (Eds.), *Hospital organization & management: Text & readings* (pp. 319–325). Baltimore: Health Professions Press.

Kahn, K. L., Keeler, E. B., Sherwood, M. J., Draper, D., Rogers, W., Rubenstein, M.

D., Kosecoff, J., & Brook, R. H. (1990). Comparing outcomes of care before and after implementation of the DRG-based prospective payment system, *Journal of the American Medical Association, 264,* 1984–1988.

Kellerman, A. L., & Hackman, B. B. (1990). Patient "dumping" post COBRA. *American Journal of Public Health, 80,* 864–867.

Kent, V., & Hanley, B. (1990). Home health care. *Nursing Health Care, 11,* 234–240.

Kissick, W. L. (1992). Rationing or resource allocation in health care. *Health Care Law and Ethics, 7,* 25–28.

Kosecoff, J., Kahn, K. L., Rogers, W. H., Reinisch, E. J., Sherwood, M. J., Rubenstein, M. D., Draper, D., Roth, C. P., Chew, R. R., & Brook, R. H. (1990). Prospective payment system and impairment at discharge. *Journal of the American Medical Association, 264,* 1980–1983.

Kramer, M., & Schmalenberg, C. (1987). Magnet hospitals talk about the impact of DRGs on nursing care: Part I. *Nursing Management, 18,* 38–42.

Leibson, C., Naessens, J. M., Krishan, I., Campion, M. E., & Ballard, D. (1990). Disposition at discharge and 60-day mortality among elderly people following shorter hospital stays: A population-based comparison. *The Gerontologist, 30,* 316–322.

Letsch, S. W. (1993). Data watch: National health care spending in 1991. *Health Affairs, 12,* 95–110.

Lipman, T. (1986). Length of hospitalization of children with diabetes: Effect of a clinical nurse specialist. *The Diabetes Educator, 14,* 41–43.

Lyles, Y. M. (1986). Impact of Medicare diagnoses-related-groups (DRGs) on nursing homes in Portland, OR metropolitan area. *Journal of American Geriatric Society, 34,* 573–578.

MacLean, S. L. (1989). The decision-making process in critical care of the aged. *Critical Care Nursing, 12,* 74–81.

Mayer-Oakes, S. A., Oye, R. K., Leake, B., & Brook, R. H. (1988). The early effects of Medicare's prospective payment system on the use of medical intensive care services in three community hospitals. *Journal of the American Medical Association, 260,* 3146–3149.

McConnell, E. A. (1990). The impact of machines in the work of critical care nurses. *Critical Care Nursing, 2,* 45–52.

McCorkle, R., K., Benoliel, J., Donaldson, G., Georgiadon, F., Moinpour, C., & Godell, B. (1989). A randomized clinical trial of home nursing care for lung cancer patients. *Cancer, 64,* 1375–1382.

McGrath, S. (1990). The cost-effectiveness of nurse practitioners. *Health Care Issues, 15,* 40–42.

Moss, R. J., & LaPuma, J. (1991). The ethics of mechanical restraints. *Hastings Center Report, 21,* 22–24.

Office of the Inspector General. (1985, April 1–September 30). *Semi-annual report to the Congress.* Washington, DC: Author.

Pasquale, D. K. (1987). A basis for prospective payment for home care. *Image, 19,* 186–190.

Rogatz, P. (1985). Home health care: Some social and economic considerations. *Home Healthcare Nurse, 3,* 38–43.

Rubenstein, L. V., Kahn, K. L., Reinisch, E. J., Sherwood, M. J., Rogers, W. H.,

Kamberg, C., Draper, D., & Brook, R. H. (1990). Changes in quality of care for five diseases measured by implicit review, 1981 to 1986. *Journal of the American Medical Association, 264,* 1974–1979.

Secretary's Commission on Nursing (1988). *Support studies and background: Volume II.* Washington, DC: U.S. Department of Health and Human Services.

Shelton, J. K., & Janosi, J. M. (1992). Unhealthy health care costs. *Journal of Medicine and Philosophy, 17,* 7–19.

Sherry, D. (1992). Cost effectiveness and home care. *Home Healthcare Nurse, 10,* 27–29.

Shuster, G. F., & Cloonan, P. (1989). Nursing activities and reimbursement in clinical case management. *Home Healthcare Nurse, 7,* 10–15.

Sienkiewicz, J. (1984). Patient classification in community health nursing. *Nursing Outlook, 32,* 319–321.

Siler, E. (1989). An integrated approach to liability. In K. Darr & J. S. Rakich (Eds.). *Hospital organization and management: Text and readings* (pp. 314–318). Baltimore: Health Professions Press.

Stone, C. L., & Krebs, K. (1990). The use of utilization review nurses to decrease reimbursement denials. *Home Healthcare Nurse, 8,* 13–17.

Tinetti, M. E., Liu, W-L., Marottoli, R. A., & Ginter, S. F. (1991). Mechanical restraint use among residents of skilled nursing facilities. *Journal of the American Medical Association, 265,* 468–471.

Trensch, D. D., Duthie, E. H., Newton, M., & Badin, B. (1988). Coping with diagnosis related groups: The changing role of the nursing home. *Archives of Internal Medicine, 148,* 1393–1396.

U.S. Senate Special Committee on Aging, The American Association of Retired Persons, The Federal Council on Aging, and The U.S. Administration on Aging. (1991). *Aging America: Trends and Projections.* Washington, DC: American Association of Retired Persons.

Weissert, W. G. (1985). Seven reasons why it is so difficult to make community based long-term care cost effective. *Health Services Research, 20,* 423–433.

Wood, J. G., & Estes, C. L. (1990). The impact of DRGs on community-based service providers: Implications for the elderly. *American Journal of Public Health, 80,* 840–843.

17 Human Error in Medicine: A Frontier for Change[1]

Marilyn Sue Bogner
U.S. Food and Drug Administration

It is clear from the preceding chapters that human error in medicine does exist and is a problem—a profound problem. As is necessary when seeking to solve a problem, the problem must be described and the factors contributing to it identified. The chapters in this book describe many aspects of the problem of human error in medicine. This chapter discusses another aspect, the estimated dollar cost of preventable iatrogenic error. Describing the problem, however, will not make it go away—only activity focused on the problem will do that. Such activity in the form of an action plan for reducing the incidence of error in medicine is proposed in this chapter.

COST OF PREVENTABLE ERROR IN MEDICINE

The data that provided the basis for the following estimate of the cost of preventable error in medical care were reported by Johnson et al. (1992) in their follow-on to the Harvard Medical Practice Study (Brennan et al., 1991; Leape et al., 1991).

The Harvard Medical Practice Study (HMPS), which is briefly described by Leape, Perper, Van Cott, and Moray (this volume, chap. 2, 3, 4, and 5, respectively), identified and analyzed factors associated with iatrogenic AEs in hospitalized patients in New York State in 1984.

Adverse events (AEs) identified by the HMPS were used by Johnson et

---

[1]Opinions expressed are the author's and do not necessarily represent those of the Food and Drug Administration or the Federal Government.

al. (1992) to estimate costs of the consequences of AEs for a sample of the affected population. These findings, together with the determination that 70% of the AEs in the HMPS were preventable (Leape, this volume, chap. 2), provide an estimate of the cost to the United States of preventable human error in one aspect of medicine for 1 year.

A sample of 794 individuals with AEs from the HMPS was interviewed by Johnson et al. (1992). Information was obtained about medical treatment related to their 1984 hospitalization that they received during the period from 1984 to July 1988. (Expenditures for medical problems other than those associated with the AEs were not included in the calculation.) In addition, earnings lost due to AEs were determined as well as projected future financial losses for those who were disabled or deceased by AEs. Similarly, the monetary value of lost household production (i.e., cost of household services) was determined at full market value for those in the sample who took care of their homes.

Of the $20.3 billion medical cost in New York State for 1984, Johnson et al. (1992) estimated that $3.8 billion was attributable to iatrogenic AEs in the HMPS population. (All cost amounts are in 1989 nondiscounted dollars.) Of that $3.8 billion, $1.81 billion was the cost of medical care related to AEs, approximately $.5 billion was lost earnings, and nearly $1.5 billion was lost household production cost. Because a relatively large number of homemakers died as a result of iatrogenic AEs, there was a substantial reduction in personal consumption costs (household services that would not be used). The authors reduced the $1.5 billion to $602 million, which reduced the estimated cost of iatrogenic AEs to $2.9 billion.

Because the purpose of their research was to explore the monetary robustness of no-fault compensation plans, Johnson et al. (1992) stated that their cost estimates probably were high. To have a conservative estimate of the cost of error for this discussion, the estimated cost of AEs was reduced by 15% to $2.5 billion.

Given the finding that 70% of the AEs were preventable, then of the estimated $2.5 billion cost of the consequences of AEs, $1.75 billion was for preventable AEs. Projecting that cost to the entire country provides a staggering $24.5 billion estimated cost of the consequences of preventable AEs in hospitalized patients in 1984. Although this amount includes lost household production cost, it does not include another cost necessitated, albeit indirectly, by incapacity or death due to AEs. That is the cost of providing care to dependent individuals.

Estimated Cost of Dependent Care

The population of this country is aging, and as Vroman et al. and Applegate note (this volume, chap. 6 & 16, respectively), providing for this population to live independently is important to everyone concerned, but it is not easy.

Many elderly persons live interdependently with another elderly person (spouse, sibling, friend) to meet their economic and social needs.

Older people experience a greater incidents of AEs than younger: The rate of AEs for patients 65 and older was 5.9; 45–64 years old, 4.7; and 16–44 years old, 2.6 (Brennan et al., 1991). It is not known how many of those who experienced iatrogenic AEs in the HMPS were caregivers. Because of the higher rate of AEs for the middle-age and older groups, the ages of people most likely to provide dependent care, it follows that there are numerous cases of affected caregivers.

When one of the partners in an interdependent living arrangement ceases to be functional, as in the case of an AE, the other may not be able to assume that person's duties. Under these circumstances, both people could find living independently increasingly difficult and be forced to enter a nursing home. A similar result from an AE could occur for a middle-aged person providing dependent care.

For the purpose of this discussion, it is assumed that 15% of the HMPS population provided dependent care for at least one person (of any age or gender). The extent that individuals can continue providing dependent care after experiencing AEs depends on the degree of impairment resulting from the AEs.

The population distribution of AEs by category (extent) of disability reported by Brennan et al. (1991) projected for the United States and subjected to the 70% preventability calculation provided the basis for the estimated number of affected caregivers presented in Table 17.1.

As seen in Table 17.1, most of the total of 771,085 preventable AEs resulted in impairments that last 1 month or less (549,212). An estimated

TABLE 17.1
Estimated U.S. Population Distribution of Preventable Adverse Events and
Affected Caregivers According to Disability

	Preventable Adverse Event	
Category of Disability	Number	(Estimated Number of Affected Caregivers)
Minimal impairment (recovery within 1 month)	549,212	(83,382)
Short-term impairment (recovery from 1 to 6 months)	132,506	(19,876)
Moderate impairment (recovery more than 6 months)	27,068	(4,060)
Permanent impairment (disability of 50% or less)	37,309	(5,596)
Permanent impairment (disability more than 50%)	24,990	(3,748)

19,876 caregivers had AE-related impairments that took from 1 to 6 months for recovery. In addition, an estimated 13,404 caregivers experienced impairment that took more than 6 months to recover or had permanent disability. Many of those impaired persons who provided dependent care probably would be unable to continue. This would necessitate finding alternative care, which in many instances would be long-term institutional (i.e., nursing home) care.

For the purpose of this discussion, assume that as a result of lost care due to preventable AEs, 3,000 people entered a long-term care facility with a presumed average 1984 cost of $15,000 per year per person, and those people lived in that facility an average of 5 years (and the cost per year remained constant). The cost of lost dependent care due to preventable AEs in 1984 would have been $225 million. Adding this to the previously estimated costs gives a total cost of preventable AEs of nearly $25 billion. It is important to remember that this cost reflects the number of people experiencing preventable AEs as hospitalized patients in 1984 in 1989 dollars.

Because of the increase in the elderly population, which is hospitalized more often and for longer lengths of stay than the younger population, and the proliferation of high technology in medical diagnostics and treatment, both the number and the cost per hospitalized patient would be greater in 1994 than in 1984. The mere thought of the current cost of preventable AEs in hospitalized patients is enough to make one shudder.

It also is important to remember that the $25 billion estimated cost is for consequences of human error in only one type of medical care setting—that of inpatient hospital care. Although the data are far from exhaustive even for this one setting, it is a formidable tip of the iceberg of medical error. Other medical care settings can provide more dimensions to the iceberg.

Errors resulting in AEs occur throughout the medical care delivery system as evidenced by the chapters in this book. Errors occur in hospitals' outpatient surgical units; medical laboratories; radiology, nuclear medicine, and radiation therapy facilities; physicians' offices; clinics; nursing homes; home health-care situations—in short, wherever medical care is provided. Not all preventable errors can be prevented, however. Accidents and mistakes do occur. Nonetheless, based on the estimates from the HMPS, the current cost of the consequences of preventable AEs for the entire medical system would be mind boggling indeed.

Human Costs of Error

There are additional costs that add to the cost of human error in medicine, the ephemeral human costs of pain and suffering. Unfortunately, as Johnson et al. (1992) noted, those considerations are most often included and abused in litigation and because of that have a negative connotation. It is a concept to be acknowledged, however, in considering the impact of preventable AEs.

An example of the domino impact of a preventable AE is provided by the case of a very fit, very successful man who entered a hospital for elective surgery to correct a relatively minor running injury to his heel. Upon being anesthetized, the man was intubated to provide assisted breathing (see Gaba's chapter in this volume for a discussion of the process); however, the tube was placed in the esophagus instead of the trachea. Surgery was delayed for 30 minutes, during which time the man suffocated due to the inappropriate intubation. Not only was his life lost as a consequence of a preventable AE, but his wife unexpectedly died within the month, apparently as a consequence of the tragic loss of her husband. In the duration of 2 months, their children were orphans due to a preventable occurrence. Situations such as this are not considered in the dollar cost of preventable AEs, but they provide another dimension to the cost and an additional incentive to reduce the incidence of human error over and above the monetary concerns.

ACTION AGENDA

This rudimentary agenda presents several interrelated approaches to reducing preventable error in the assortment of medical settings and providers that have come to be known in the United States as the medical care system. This agenda first describes a systems approach for the consideration of preventable medical errors. Next, the pivotal role of information is discussed. Then the agenda becomes two agendas that can be implemented concurrently. One agenda describes activities that can get started without much delay and through which activities can be accomplished in the short term; the other agenda describes activities for the longer term.

Systems Approach

One of the conclusions to be drawn from this book is that a systems approach is necessary to effectively address human error in medicine. This is underscored by the statement that "preventing medical injury will require attention to the systemic causes and consequences of errors, an effort that goes well beyond identifying culpable persons" (Leape et al., 1991, p. 383).

What does a systems approach to human error mean? From a systems perspective, the delivery of medical care occurs in interrelated subsystems. This is not to propose that the current medical care "system" in the United States is comprised of interrelated, coupled, communicating subsystems. This would be the case if it truly were a system. Rather, this systems approach considers the medical care delivery settings as discrete systems with constituent subsystems.

When the subsystems have been well defined, their interrelationships can be explored and actions taken to integrate them into a suprasystem of medical care. To do otherwise could lead to "white tape" activities—activities analogous to the efforts of nurses to reduce the likelihood of error in hospitals by improvising using white adhesive tape—in other words, stopgap, temporary approaches.

The subsystems themselves are quite complex and will require further decomposition into smaller subsystems for productive analysis and action. For example, a hospital is a system with each of its functioning components, that is, wards, services such as radiology and pharmacy, and units such as the intensive care unit, operating room (OR), and emergency room, as subsystems. In their discussions of the activity in the OR and particularly anesthesia delivery, Gaba, Helmreich and Shaefer, Cook and Woods, and Krueger (this volume, chap. 11, 12, 13, and 14, respectively) underscore the value of analyzing smaller subsystems within the context of the superordinate system.

Reducing Human Error

As for the immediate task of reducing human error in medicine, the medical care system for the purpose of analysis should be expressed as a system with the most elemental subsystem being the medical care provider, the person receiving the care, and the means by which the care is delivered. Error-related issues and examples pertaining to this elemental subsystem are discussed by Bogner, Vroman et al., Klatzky et al., Sheridan and Thompson, Senders, and Serig (this volume, chap. 1, 6, 7, 8, 9, and 10, respectively). The effect of policy on error-related behavior in a range of medical care settings is discussed throughout this book.

The contribution of the systems approach to the analysis of human error in medicine is that error is considered the consequence of systemic, interrelated factors impacting the individual. This provides a broader spectrum of contributing factors than considering error as the result of one action committed by one individual. To address the problem solely in terms of the apparent perpetrator is to address only a symptom of the underlying problem.

The systems approach is to analyze the situation, decompose it to the level at which the function associated with the error occurred, identify those factors that precipitated the error, bring those factors to the attention of the appropriate responsible party for action to remove or alter those factors, and evaluate the impact of the resultant action on the future incidence of error. The key to the viability of this approach is information.

Information

The effort to identify and reduce the incidence of error without analyzing the context within which the error occurred is a condition in which there is "one hand clapping." To identify and define the other hand to participate in the clapping, to explore the operational context for error, it is necessary to have information about the elemental unit of the provider, the patient, and whatever medical treatment device(s) or medication were used at the time of the error. It is also necessary to have information about other factors that impinged on that elemental unit and contributed to the error. Thus, it is necessary to design and implement an error-reporting system to collect information for analysis. Such a system cannot be developed without input from those who understand the situation and can identify the factors that induce errors: the medical care providers.

Those who are associated with errors are the most likely people to be able to provide information about what contributes to the errors. There is an impediment to their providing information, however, which is fear of malpractice litigation. Most medical care providers in the United States will not provide error-related information because to do so might be construed as admitting responsibility for any error under consideration. This could lead to litigation.

It is imperative that an atmosphere be created in which medical care personnel can freely provide data about errors and near errors they experience. To effectively address the issues surrounding error and especially to promote the reporting of error information, it is necessary to develop action agendas for both the short and long term.

SHORT-TERM AGENDA

Pending the implementation of no-fault compensation plans, which is discussed under the long-term agenda, several short-term efforts can begin to address the problem. Central to these efforts is communication. Problems cannot be addressed if they are not known by those in position to correct them.

One activity is to establish on a routine basis *face-to-face meetings* among those involved in the delivery of medical care—providers, administrators, and possibly industry (when appropriate)—to identify and discuss problems and ways of addressing them. These meetings would occur in all medical care settings. Among topics that could be discussed are institutional policies including government regulations and policies, equipment purchase, and work-shift designation (the importance of which is discussed by Kreuger in chap. 14 of this volume, as well as in research by Laine, Goldman, Soukup, &

Hayes, 1993, and Holdnak, Harsh, & Bushardt, 1993). Resource people outside the medical care setting (e.g., systems analysts, human factors specialists) would be included to bring expertise to address issues.

Another effort is for medical care facilities and industry to *use existing information* to avoid and correct problems. Many of the problems that occur in the delivery of medical care are the result of problems that have been addressed in other domains, such as military and aviation. For example, there is extensive, readily available information from those sources on basic design and written material that will reduce the likelihood of error in the operation and maintenance of complex medical devices. This is valuable for acute care settings with trained personnel, and especially so in home health care.

The value of using information from sources outside the immediate field of medical care (e.g., aviation, nuclear power) is noted throughout this book. Moray (this volume, chap. 5) cites several sources of information that have been used successfully in correcting problems that induce error. Often problems can be readily and inexpensively corrected by retrofitting a corrective component.

A case in point is the problem of apparently misinterpreting the meaning of an amber light on a dialysis machine that possibly contributed to 3 people dying and several others being hospitalized (James, 1993). Replacing the amber lens with a red one provides the universal signal of danger and avoids the likelihood of the error of continued operation of the machine in a toxic condition.

Another way to use existing information is to link the existing medical information systems and by so doing enhance access to them. Personnel at the Naval Health Research Center have devised a way to integrate their discrepant databases. Their experience could be tapped to vitalize the current medical databases.

LONGER TERM AGENDA

An agenda of longer term activities directed to reduce the incidence of human error in medicine is centered around two components that interact in an iterative fashion: research and action.

Information is needed to define the problem so that research might be conducted to identify the factors contributing it; action is needed to redress the situation, to redirect or eliminate the contributing factors. The process is iterative because information gathered concerning the outcome of the remediated action, the evaluation, is integrated with other information to provide a focus for research. The findings from that research become the basis for subsequent action. The process is repeated and applied systematically, for example, to the nested components of the subsystem, then the system.

Given the current medical error-reporting systems, there is no way to obtain anything close to an estimate of the actual incidence of adverse events resulting from human error. There exists a variety of information systems and data collection programs related to error in medicine as described by Perper and Hyman (this volume, chap. 3 and 15, respectively). This is a case of having everything and having nothing because the information systems are piecemeal. Nowhere is it possible to obtain information pertaining to error in medical care that affords sufficient information to focus the issue so that the contributing systemic factors can be identified.

The sparse information that currently is available is so underreported as to be hardly representative of any problem. It is difficult to understand how this can continue to occur in the largest industry in this country outside government until one remembers the power of the fear of malpractice litigation. The fear of reprisal for error is not unique to medicine; however, the lack of ingenuity in coming to grips with that fear is.

There are ways to lay the specter of malpractice litigation to rest. One approach is *no-fault compensation*. This not only would do away with the fear of litigation, but it would provide compensation for all those who have impairment from adverse events rather than the very few who have successful malpractice suits (Johnson et al., 1992).

Another approach is to develop a *no-fault reporting system* that encourages reporting error by addressing the fears of reprisal. A successful model is provided by the Aviation Safety Reporting System (Reynard, Billings, Cheaney, & Hardy, 1986). Pilots are concerned about their licenses in the same way that medical care providers are concerned about malpractice, yet they report error to that system. How can this be?

First, although the licensing of pilots and the federal regulation of the aviation industry is under the aegis of the Federal Aviation Administration, the data collection for errors in that industry is conducted by the National Aeronautics and Space Administration. This way errors are not reported directly to the licensing agency.

Second, after those data are captured from the Aviation Safety Reporting System (ASRS) reporting form, the identity of the reporter is destroyed, rendering the report anonymous. In addition, efforts are made to generalize the information to obscure the identity of the reported incident.

Third, reporting pays. Although a reported incident may be found to be a violation of regulations, neither a civil penalty nor certificate suspension will be imposed if the violation was inadvertent, did not involve an accident or a criminal offense, the person was not guilty of a violation in the previous 5 years, and they reported it to the ASRS within 10 days of the incident.

ASRS data are consolidated and distributed to appropriate persons, agencies, and industry for information and action. They are also integrated into a widely disseminated newsletter. In addition, the ASRS data also are available for research and have proven valuable in that capacity.

A systems approach to research provides an organizing perspective for a *program of research*. The findings from such a program can provide a focus for remedial action to reduce the likelihood of error. Such a research program involves communication from those identifying problems through a viable reporting system to those funding research, whether it be government, industry, or foundations.

In reviewing the medical care literature and that on error and adverse events, one is struck by the fragmentation and overlap. A lot is known about a little, but little is known about a lot. The HMPS is an exception—it is a program of research and attests to the value of that approach. All too often, programs of research have only a topic as a focus, and the research addresses disparate aspects of the topic. The findings from such research, because of their disparate foci, provide such a fragmented picture that no activity may be formulated. This is particularly the case with research in human error in medicine. Correcting fragmentation is a formidable but not impossible task using a systems approach.

Not only is there fragmentation of research on a topic funded by a single organization, but there is also a lack of knowledge let alone coordination of research activities within and across agencies in the same government department. That is not surprising, given the size and compartmentalization, not to mention the bureaucracy, of government agencies. This is not because of lack of interest in sharing information and knowledge of activities.

An example of interest in sharing information was provided by recent workshops. Twenty-eight Federal agencies were represented at two Intra-Governmental Workshops on Human Error held to communicate activities and interest on the topic (Bogner, 1993). Ongoing activities stemming from the workshops attest to the productivity of information exchange. *Intragovernment cooperation and collaboration* is an activity in which every participant can benefit.

CONCLUSIONS

Human error in medicine is not fraud, waste, and abuse in the delivery of medical care. Rather, it is the indication of circumstances that have gone awry. The consequences of those circumstance are various degrees of impairment, illness, injury, and the ultimate impairment, death. The cost of preventable human error in medicine is a substantial contributor to the crippling cost of medical care in the United States. Also, there is the cost in human anguish from the disruption in peoples' lives and the lives of those around them resulting from preventable error in medical care.

Past activities to reduce the incidence of error in medical care, such as continuing medical education, quality assurance programs, and malpractice litigation, have addressed the individual provider of care and have not done the job. Therefore, it is time to institute a different approach—the systems approach. To describe and analyze the delivery of medical care as a system with interrelated subsystems is a viable way to identify and address the otherwise unconsidered systemic factors that lead to error.

Many actions must be taken to energize the systems approach. Information must be available to nourish the approach, to give it the substance with which to function. This is not an easy undertaking, but it can be done. A number of ingrained attitudes and habits must be changed for the approach and the action agenda to be effective, but that can be done. The cost is too great not to do so. Human error in medicine indeed is a frontier for change.

REFERENCES

Bogner, M. S. (1993). *Workshops on human error*. Unpublished report.

Brennan, T. A., Leape, L. L., Laird, N. M., Hebert, L., Localio, A. R., Lawthers, A. G., Newhouse, J. P., Weiler, P. C., & Hiatt, H. H. (1991). Incidence of adverse events and negligence in hospitalized patients. *New England Journal of Medicine, 324*(6), 370–376.

Holdnak, B. J., Harsh, J., & Bushardt, S. C. (1993). An examination of leadership style and its relevance to shift work in an organizational setting. *Health Care Management Review, 18*(3), 21–30.

James, F. (1993, July 20). Answers not easy in dialysis deaths. *Chicago Tribune*, Section 2, p. 3.

Johnson, W. G., Brennan, T. A., Newhouse, J. P., Leape, L. L., Lawthers, A. G., Hiatt, H. H., & Weiler, P. C. (1992). The economic consequences of medical injuries. *Journal of the American Medical Association, 267*(18), 2487–2492.

Laine, C., Goldman, L., Soukup, J. R., & Hayes, J. G. (1993). The impact of a regulation restricting medical house staff working hours on the quality of patient care. *Journal of the American Medical Association, 269*(3), 374–378.

Leape, L. L., Brennan, T. A., Laird, N., Lawthers, A. G., Localio, A. R., Barnes, B. A., Hebert, L., Newhouse, J. P., Weiler, P. C., & Hiatt, H. (1991). The nature of adverse events in hospitalized patients. *The New England Journal of Medicine, 324*(6), 377–384.

Reynard, W. D., Billings, C. E., Cheaney, E. S., & Hardy, R. (1986). *The development of the NASA Aviation Safety Reporting System* (NASA Publication 1114). Washington, DC: NASA.

18 Afterword

Jens Rasmussen
HURECON

There has been much discussion in this book about the systems nature of human error in medicine, that is, factors in the entire system in which clinicians function that impinge on them and induce them to commit errors. There also has been discussion about the need for viable information systems that provide data pertaining to the system, the environment in which clinicians function and the activities of clinicians themselves. The following discussion presents a taxonomy for describing malfunctions associated with human behavior that was developed for use in an industry other than medicine, but has potential for the medical arena.

Because the conditions that precipitate human error tend to be analogous across industries, efforts in domains other than medicine are applicable to medicine as several authors in this book have noted. The taxonomy described subsequently is not presented as a definitive way of organizing and analyzing data for the study of human error in medicine. Rather, the taxonomy is presented as an initial, well thought-out framework on which to build databases related to human error in medicine. The results of analyses of those data could be used to guide efforts to effectively reduce the incidence of human error in medicine.

The need for information systems based on a systems perspective has been identified for operation of high-hazard systems such as nuclear power and chemical process plants. Great efforts have been spent on the establishment of human error databases for design and operation of such systems. In spite of decades of effort, no reliable human error database exists today for predictive analysis in the operation of plants for such process industries. This does not, however, imply that the analysis of human error incidents is without benefit. It suggests that the basic nature of human error should be very carefully considered when schemes for data collection and analyses are planned.

The following discussion briefly reviews conditions to consider for transfer to the medical arena from lessons learned in the efforts to develop a taxonomy for the consideration of human error in process industries. The assumptions and preconditions for a human error taxonomy developed by a group of experts on the Committee on the Safety of Nuclear Installations under the international Organization for Economic Cooperation and Development for data collection in nuclear power plants is taken as an example (for details see Rasmussen et al, 1981; Rasmussen, 1982).

THE FOUNDATION OF A TAXONOMY

Expert performance depends on very effective adaptations by a human actor to the characteristics of the task situation and the overall work environment. This adaptive match becomes more complex and selective the higher the level of expertise. Errors reflect the loss of match in a situation when variation of either the behavior of the environment or the actor exceeds the flexibility of the match. Errors result if the actor does not perceive and correct the mismatch. Error analysis, therefore, must be directed toward the understanding of the human–system interaction in the particular mismatch situations. The findings from error analyses can lead the design of new systems to reduce the likelihood of error. The following aspects of human–system interaction have been important for the structure of the taxonomy:

• Human error reports basically do not reflect human errors committed, but only errors that were not immediately corrected by the actor. That is, reports are heavily biased by those features of the work environment that determine whether the actor can immediately detect the unacceptable effect of an act (error observability) and can correct the effect (error reversibility). To reduce such bias, reports must represent all relevant features of the complex system/task/actor situation.

• A taxonomy implies a decomposition or parsing of a continuous flow of behavior into elements and a categorization of recurrent elements. Normally, a taxonomy is a conceptual framework structuring an exclusive, hierarchical classification. (Frequently, analogies are drawn to biological taxonomies.) Such a hierarchical approach is not practical for characterization of the complex system/task/actor situation. For error analysis, a useful taxonomy is more appropriately characterized as a multifaceted framework for describing and understanding than a taxonomy for classification and quantification.

- The repetition of any error incident can be prevented in many different ways: by better selection, by training and motivating people, by better task design and instruction, and by improved system (that is, the context in which the work occurs as well the tool or device). The facets of error description should represent all such aspects of system design and they should all be considered during analysis of an incident. Only analysis across incidents will make it possible to exclude ad hoc solutions. Therefore, development of a knowledge base including incidents from analogous work situations is worthwhile even if reliable quantitative data are not found.

- It cannot be assumed that a human error is the root cause of an accidental stream of events. All incidents should be carefully analyzed for external precursors of a human error such as, for example, conflicting task requirements, and distractions from phone calls and interruptions from colleagues—conditions that rarely appear in routine incident reports.

- Finally, it is very difficult to define errors and describe their preconditions except for activities that are reasonably well structured by a stable environment. That is, identification and analysis of errors with reference to a taxonomy are relevant, for example, to the operation of medical devices and equipment, but very difficult for medical diagnosis (see Rasmussen, 1993).

THE FACETS OF THE TAXONOMY

The facets of the taxonomy are listed in Table 18.1 together with a brief description of the categories considered for industrial process plants, which because of the analogous conditions for the delivery of health care, are relevant to the study of human error in medicine.

The taxonomy reflects a causal chain in which the human error is a link that must be described from several perspectives.

Only by describing the error from several perspectives is it possible to link an observable input to an actor in terms of events in the external work situation to the output in terms of observable human acts. That is, a couple of mental events may be included in the description of the error chain, as follows: (a) an *external event* activates (b) a particular *psychological mechanism* that (c) results in a faulty *decision-making process* that is reflected in (d) an *erroneous act* that, ultimately, leads to the detection of (e) an *inappropriate state in the work system*. This representation includes mental events that can only be inferred by use of a cognitive model. The categories to

TABLE 18.1
The Facets of the Taxonomy (for details see Rasmussen, 1982).

System Aspects

Behavior shaping factors of work environment: Stress factors, social factors, workload, level of training, shift work, etc.

Functional system properties: Description of the degree of structure imposed on task by functional relations of work system.

Consequences of errors for system operation: Goals or targets missed *or* effects of extraneous events: human injuries, fatalities, loss of equipment, environmental impact, etc.

Situation Aspects

Task requirements: Category of task in work domain terms: Operation, management, maintenance, staff-planning, etc.; normal (not normative) work procedure for particular work situation and task.

External cause, antecedent events: Changes in goals and work requirements, faults in devices and equipment, distraction by co-worker, conflicting or competing requirements, excessive task demands, etc.

Human Aspects

Failed, overt acts: Description of particular failed acts in categories relevant for the particular task and work domain. For example, acts omitted, performed incorrectly; acts interchanged; acts in the wrong place, at the wrong time; extraneous acts described in the terms of the procedure.

Decision function inadequately performed or bypassed by heuristic:

  Detection

  Situation analysis: Observation, identification

  Decision: Selection of goal and task

  Planning: Of task and procedure

  Action: Execution, coordination

  Monitoring: Of actions

  Communication

Psychological mechanism involved:

  Input information processing: Information not sought or received, misinterpretation of observation and communication, unsupported assumptions

  Discrimination: Familiar pattern not recognized, familiar shortcut, stereotype fixation, stereotype take-over (capture error)

  Inference: Preconditions or side effects not adequately considered

  Recall: Isolated act forgotten, mistake among alternatives, other slips of memory

  Physical coordination: Motor variability, spatial misorientation

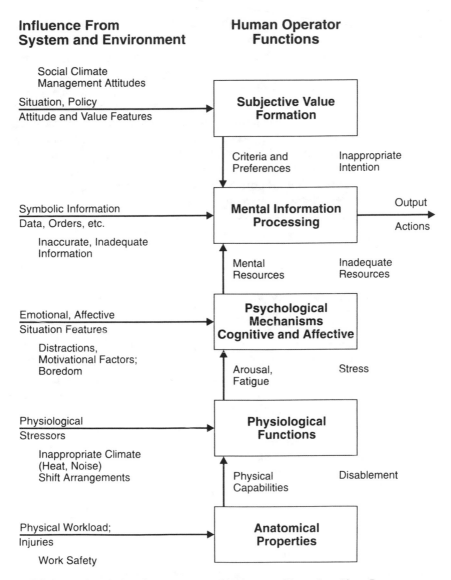

**Influence From
System and Environment**

**Human Operator
Functions**

Social Climate
Management Attitudes

Situation, Policy
Attitude and Value Features

Subjective Value
Formation

Criteria and
Preferences

Inappropriate
Intention

Symbolic Information
Data, Orders, etc.

Mental Information
Processing

Output

Actions

Inaccurate, Inadequate
Information

Mental
Resources

Inadequate
Resources

Emotional, Affective
Situation Features

Psychological
Mechanisms
Cognitive and Affective

Distractions,
Motivational Factors;
Boredom

Arousal,
Fatigue

Stress

Physiological
Stressors

Physiological
Functions

Inappropriate Climate
(Heat, Noise)
Shift Arrangements

Physical
Capabilities

Disablement

Physical Workload;
Injuries

Anatomical
Properties

Work Safety

FIG. 18.1. Interactions in a person–machine system. (Reproduced from Rasmussen, 1982, by permission of Elsevier Scientific Publishing).

apply, therefore, depend on the model adopted (Rasmussen, 1980, 1982). In the present context, the message is not the particular categories of the taxonomy, but its structure.

This causal chain only considers the information process aspects of the human–machine interaction; that is, the chain of events related to changes in the conditions for human decisions or the process of decision making itself.

However, the work environment influences human actors in a much more complex way than through the information domain alone, as illustrated by Fig. 18.1. Therefore, conditioning factors related to affective, motivating aspects of the work situation as well as physiological factors are included in the taxonomy, as shown in Table 18.1.

The conditioning factors will not directly appear in the causal chain of events, but may influence it by changing limits of human capability, choice of goals, subjective preferences in choice of mental strategies, and so on. Some of these factors can only be identified by careful analysis of the actual work climate or situation and they are rarely included in routine event reports. Similarly, the identification of possible external causes of erroneous behavior depends on very detailed, hard-to-get information about details of the actual situation and the interactions in a team.

USE OF THE TAXONOMY

As mentioned, the taxonomy was developed for analysis of human operation of well-structured technical systems for which the acceptable task procedures are stable and well known. In the medical arena, the taxonomy will be most applicable for analysis of errors during the use of technical medical devices and equipment. For such activities, the taxonomy will be useful in focusing incident analysis, planning data collection, and designing knowledge bases. The categories to be used for the various facets of the taxonomy should be elaborated along with data collection for each application domain.

The cognitive categories of the taxonomy will be inadequate for unstructured work situations such as medical diagnosis. The category of inference, for instance, only includes "preconditions and side effects not adequately considered" (sec Table 18.1). This partly reflects the origin of the taxonomy but also the fact that errors during knowledge-based problem solving are hard to define. If medical diagnosis is considered to be hypothesis generation and the subsequent therapy to be hypothesis testing, the concept of *error* becomes dubious. But even then, the taxonomy has been found to be very useful in guiding analyses of incident reports for understanding a complex work situation.

Ultimately, analysis of error incidents should lead to the design of safer work systems, that is, work environments. I have repeatedly argued that human errors neither can nor should be removed by design of safe work systems (Rasmussen, 1990). Humans are basically boundary seeking and errors often reflect efforts to adapt effectively to work requirements. Attempts to remove observed causes of error tend to be met by compensatory adaptation (Rasmussen, 1994). The aim of a system designer should be to design error-tolerant systems; that is, systems in which errors do not cause irreversible effects. Instead, system interfaces should support immediate detection of errors and allow the actor to recover by corrective acts.

The taxonomy can serve to design error-tolerant equipment in two ways. The first way is the classical *error mode and effect analysis*. In this, all relevant error modes (types) are postulated during an act and their effects predicted by analysis of the causal relations of the work environment. This analysis involves the formulation of the procedural sequence of acts necessary to operate the device or equipment successfully. For each step of this procedure, the activation of each of the error-producing psychological mechanisms is postulated and its effects are traced downstream through the decision and planning functions to be overt acts, until the ultimate effect on system performance can be judged.

A simple example demonstrates the interaction among the facets during an error scenario. The act of closing a valve, such as valves on tanks of oxygen and anesthesia gases, can be unsuccessful due to different error mechanisms. The wrong valve may be closed due to a mistake among alternatives in which a name or a label is mistaken for another; for example, an oxygen label being mistaken for an anesthesia gas label. The valve may be opened fully instead of closed by a mistake of alternatives (directions) or, if the valve operates in reverse to the usual direction, because of a stereotype fixation. Closing the valve may be omitted due to a simple slip of memory, which can very likely happen if the act is functionally isolated from the other aspects of the task. (It has been found that the isolated task of switching a system back to operation after a successful test and calibration was an overwhelming source of failure.) With the slip of memory example, two coupled errors occurred: A wrong valve was closed and the intended valve was left open. It is important to predict the mistaken valve. Depending on the error mechanism involved, the valve to be predicted will be topographically close (topographic misorientation) or be a part of a very familiar routine that is similar to the present task (psychologically close, stereotype take-over). The message of this example is that the causal relationships among mechanisms, mental functioning, and task elements must be maintained during error analysis and they must be represented in the knowledge base to be able to regenerate the relevant error scenarios during the evaluation of new work designs.

During an analysis to test a new design for error tolerance, the focus should be on the judgment of observability and reversibility of the effect of the various error types. The design should be revised until the potential for error recovery is found acceptable. In this way, the likely errors during normal task sequences can be considered during design.

Another approach is required to detect a potential for high-hazard, low-probability, extraneous acts; that is, acts that are not reflections of simple errors in a known task sequence. Examples are acts on a work system that reflect interference from other activities that may or may not be related to the task considered (interference from maintenance of nearby equipment, cleaning activities, transport of patients and equipment, stumbling over cables, etc.). To identify such hazardous acts, the direction of an analysis is reversed in a *causal tree analysis*. In that analysis, the possible accidents are designed and then prevented. This is accomplished by identifying the major source of hazard of the work situation. Then the possible accidental course of events that will release the hazard is identified from the analysis of the functionality of the system. All activities around the system in the context within which the work is performed are then reviewed to determine whether the hazardous act sequence can be assembled from a combination of acts and errors in these activities. If this is the case, a redesign for error tolerance is considered. The taxonomy is a very effective guide for this analysis.

CONCLUSION

Although the taxonomy was originally developed to design databases for predictive risk analyses for industrial installations including data on human errors, it has been found to be useful for analysis of incident reports from other work systems as well. The reason for this is that the taxonomy is not a usual taxonomy of error aimed at classification and quantification. It is a multifaceted conceptual framework serving to unravel the many factors determining adaptive human behavior in a complex work environment. This line of development has later been extended toward a framework for work analysis and cognitive engineering (Rasmussen, Pejtersen, & Goodstein, 1994) that also considers coordination in teams and organizations in more detail.

Because of similarities of error-inducing conditions and factors across domains of work and because of the multifaceted nature of this taxonomy and its development, the taxonomy has potential as a guide for the consideration and ultimate reduction of the likelihood of human error in medicine.

REFERENCES

Rasmussen, J. (1980). What can be learned from human error reports. In K. Duncan, N. Gruneberg, & D. Wallis (Eds.), *Changes in working life*. New York: Wiley.

Rasmussen, J. (1982). Human errors. A taxonomy for describing malfunction in industrial installations. *Journal of Occupational Accidents, 4,* 311–333.

Rasmussen, J. (1990). The role of error in organizing behaviour. *Ergonomics, 33,* 1185–1190.

Rasmussen, J. (1993). Diagnostic reasoning in action. *Transactions on Systems, Man, and Cybernetics, 23,* 981–992.

Rasmussen, J. (1994). Risk management, adaptation, and design for safety. In N. E. Sahlin & B. Brehmer (Eds.), *Future risks and risk management.* Dordrecht: Kluwer.

Rasmussen, J., Pedersen, O. M., Mancini, G., Carnino, A., Griffon, M., & Cagnolet, P. (1981). *Classification system for reporting events involving human malfunction* (Tech. Rep. Risø-M-2240. Roskilde, Denmark: Risø National Laboratory.

Rasmussen, J., Pejtersen, A. M., & Goodstein, L. P. (1994). *Cognitive systems engineering.* New York: Wiley.

Author Index

Subject Index

405